THE SCIENTIST'S GUIDE
TO CARDIAC METABOLISM

THE SCIENTIST'S GUIDE TO CARDIAC METABOLISM

Edited by

MICHAEL SCHWARZER AND TORSTEN DOENST

Department of Cardiothoracic Surgery
Friedrich-Schiller-University of Jena
Jena, Germany

AMSTERDAM • BOSTON • HEIDELBERG • LONDON
NEW YORK • OXFORD • PARIS • SAN DIEGO
SAN FRANCISCO • SINGAPORE • SYDNEY • TOKYO

Academic Press is an Imprint of Elsevier

Academic Press is an imprint of Elsevier
125, London Wall, EC2Y 5AS, UK
525 B Street, Suite 1800, San Diego, CA 92101-4495, USA
225 Wyman Street, Waltham, MA 02451, USA
The Boulevard, Langford Lane, Kidlington, Oxford OX5 1GB, UK

Notices
Knowledge and best practice in this field are constantly changing. As new research and experience broaden
our understanding, changes in research methods, professional practices, or medical treatment may become
necessary.

Practitioners and researchers must always rely on their own experience and knowledge in evaluating and
using any information, methods, compounds, or experiments described herein. In using such information or
methods they should be mindful of their own safety and the safety of others, including parties for whom they
have a professional responsibility.

To the fullest extent of the law, neither the Publisher nor the authors, contributors, or editors, assume any
liability for any injury and/or damage to persons or property as a matter of products liability, negligence
or otherwise, or from any use or operation of any methods, products, instructions, or ideas contained in the
material herein.

British Library Cataloguing-in-Publication Data
A catalogue record for this book is available from the British Library

Library of Congress Cataloging-in-Publication Data
A catalog record for this book is available from the Library of Congress

ISBN: 978-0-12-802394-5

For information on all Academic Press publications
visit our website at http://store.elsevier.com/

Publisher: Mica Haley
Acquisition Editor: Stacy Masucci
Editorial Project Manager: Shannon Stanton
Production Project Manager: Julia Haynes
Designer: Matt Limbert

Typeset by Thomson Digital

  Working together
to grow libraries in
developing countries

www.elsevier.com • www.bookaid.org

Contents

13. Energetics in the Hypertrophied and Failing Heart

CRAIG A. LYGATE

14. Cardiac Metabolism – The Link to Clinical Practice

PAUL CHRISTIAN SCHULZE, PETER J. KENNEL, TORSTEN DOENST, LINDA R. PETERSON

15. Historical Perspectives

TERJE S. LARSEN

List of Contributors

Christophe Beauloye Université catholique de Louvain, Institut de Recherche Expérimentale et Clinique, Pole of Cardiovascular Research, Brussels, Belgium; Université catholique de Louvain, Cliniques Universitaires Saint Luc, Division of Cardiology, Cardiovascular Intensive Care, Brussels, Belgium

Jessica M. Berthiaume Department of Physiology & Biophysics, School of Medicine, Case Western Reserve University, Cleveland, OH, USA

Luc Bertrand Université catholique de Louvain, Institut de Recherche Expérimentale et Clinique, Pole of Cardiovascular Research, Brussels, Belgium

David I. Brown McAllister Heart Institute, University of North Carolina at Chapel Hill, Chapel Hill, NC, USA

Torsten Doenst Department of Cardiothoracic Surgery, Jena University Hospital, Friedrich Schiller University of Jena, Jena, Germany

Jan F.C. Glatz Department of Genetics and Cell Biology, Cardiovascular Research Institute Maastricht (CARIM), Maastricht University, Maastricht, The Netherlands

Louis Hue Université catholique de Louvain, de Duve Institute, Protein Phosphorylation Unit, Brussels, Belgium

Peter J. Kennel Department of Medicine, Division of Cardiology, Columbia University Medical Center, New York, New York

Terje S. Larsen Cardiovascular Research Group, Department of Medical Biology, UiT the Arctic University of Norway, Tromsø, Norway

Craig A. Lygate Radcliffe Department of Medicine, Division of Cardiovascular Medicine, University of Oxford, Oxford, UK

Miranda Nabben Department of Genetics and Cell Biology, Cardiovascular Research Institute Maastricht (CARIM), Maastricht University, Maastricht, The Netherlands

Tien Dung Nguyen Department of Cardiothoracic Surgery, Jena University Hospital, Friedrich Schiller University of Jena, Jena, Germany

Bernd Niemann Department for Adult and Pediatric Cardiac Surgery and Vascular Surgery, University Hospital Giessen and Marburg, Justus Liebig University Giessen, Rudolf Buchheim Strasse, Giessen

Moritz Osterholt Department of Internal Medicine, Helios Spital Überlingen, Überlingen, Germany

Linda R. Peterson Department of Medicine, Cardiovascular Division, Washington University School of Medicine, St. Louis, Missouri, USA

Susanne Rohrbach Institute for Physiology, Justus Liebig University Giessen, Aulweg, Giessen

Andrea Schrepper Department of Cardiothoracic Surgery, Jena University Hospital, Friedrich Schiller University of Jena, Jena, Germany

Paul Christian Schulze Department of Medicine, Division of Cardiology, Columbia University Medical Center, New York, New York

Michael Schwarzer Department of Cardiothoracic Surgery, Jena University Hospital, Friedrich Schiller University of Jena, Jena, Germany

Marc van Bilsen Departments of Physiology and Cardiology, Cardiovascular Research Institute Maastricht, Maastricht University, Maastricht, The Netherlands

Christina Werner Department of Cardiothoracic Surgery, Jena University Hospital, Friedrich Schiller University of Jena, Jena, Germany

Monte S. Willis McAllister Heart Institute, University of North Carolina at Chapel Hill; Department of Pathology & Laboratory Medicine, University of North Carolina Medicine, Chapel Hill, NC, USA

Martin E. Young Division of Cardiovascular Diseases, Department of Medicine, University of Alabama at Birmingham, Birmingham, AL, USA

Foreword

If you consider yourself a scientist already or want to become one and you have found interest in investigating cardiac metabolism but are lacking the fundamentals, you need *The Scientist's Guide to Cardiac Metabolism*. Reading this book will provide you with the basic and, therefore, often timeless information required to get a flying start in any good cardiac metabolism lab.

You get the chance to refresh your basics on biochemistry, cell biology, physiology as well as the required methodology to investigate new areas. You will be familiarized with fundamental principles relevant to cardiac metabolism, learn regulatory mechanisms and pathways and also find out which investigative methods have been used in the past and which are currently applied to further develop the field. Having read this book you will know "what the experts in the field are talking about" and develop a solid base for quick understanding of the sometimes dry appearing but indeed highly interesting publications in this field. We are certain it is worth your while.

Michael Schwarzer, Torsten Doenst
Department of Cardiothoracic Surgery,
Jena University Hospital, Friedrich Schiller
University of Jena, Jena, Germany

Introduction to Cardiac Metabolism

Michael Schwarzer, Torsten Doenst

Department of Cardiothoracic Surgery, Jena University Hospital,
Friedrich Schiller University of Jena, Jena, Germany

In order for the heart to sustain its regular heartbeat, it needs a constant supply of energy for contraction [1]. This energy comes primarily from the hydrolysis of ATP, which is generated within the cardiomyocyte by utilizing various competing substrates and oxygen, which again are supplied by coronary flow [2,3]. Cardiac metabolism therefore comprises all processes involved in the biochemical conversion of molecules within the cell utilizing energy substrates. In addition, cardiac metabolism comprises all biochemical processes of the cell aimed at the generation of building blocks for cell maintenance, biosynthesis, and cellular growth.

There is an intimate connection between cardiac metabolism and contractile function, which is illustrated schematically in Fig. 1.1. As simple as this illustration, which stems originally from Heinrich Taegtmeyer, appears as complex is its meaning [4]. It is clear that changes in contractile function require changes in cardiac metabolism as more power needs more fuel, that is, ATP, and less power needs less fuel. The schematic also illustrates that contractile function

is directly influenced by metabolism. Again, if ATP is limited (e.g., during ischemia), it is easily envisioned that contractile function seizes. However, the scheme finally encompasses myocardial metabolism as potential target for treating contractile dysfunction [5]. Considering that metabolic processes also influence biosynthesis, it becomes clear that metabolism is a prime target of investigations for nearly all physiologic and pathologic states of the heart, may it be ischemia/reperfusion, diabetes, hypertrophy, and acute and chronic heart failure [6].

In order to develop an understanding for these interrelations and to obtain basic knowledge about the methods and tools used for the investigation of (cardiac) metabolism, we have compiled this book. It reflects a selection of chapters geared toward the transfer of principles in cardiometabolic research. The book does not claim to be complete, but its content should make the reader quickly understand most of the specific topics he or she intends to specialize in and to be better able to put the personal investigations into perspective.

The Scientist's Guide to Cardiac Metabolism
http://dx.doi.org/10.1016/B978-0-12-802394-5.00001-7

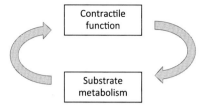

FIGURE 1.1 **Schematic illustration of the interrelation of cardiac contractile function and substrate metabolism.** *Adapted from Ref. [4].*

In Chapter 2, Jan Glatz and Miranda Nabben begin with illustrating basics in metabolically relevant biochemistry. They show that metabolism is tightly coupled to all major types of biomolecules as virtually every biomolecule can be used as a substrate or pathway component in metabolism. Carbohydrates and fatty acids are the main substrates used to produce ATP. Amino acids and nucleotides are mainly used to build proteins and nucleic acids. However, all biomolecules come with specific characteristics and even when they are "exclusively" used as substrate for ATP generation, their biochemical influence on other cellular processes needs to be taken into account as well. Furthermore, the properties of biomolecules influence their transport as well as their import into the cell or into cellular substructures, such as mitochondria. Fatty acids as lipophilic compounds are not readily soluble in the aqueous blood and cytoplasm. Carbohydrates, nucleic acids, and amino acids are more hydrophilic and may not cross membranes without help. Thus, it is important to be aware of the properties of biomolecules and their biochemistry. This chapter introduces the reader to the biochemical properties of the major classes of molecules and illustrates their behavior.

In Chapter 3, Bernd Niemann and Susanne Rohrbach address metabolically relevant cell biology and illustrate the roles of intracellular organelles for cardiac metabolism. In this chapter, the roles of all major cellular organelles with respect to cardiac metabolism are described. The reader may find that both fatty acid oxidation and phospholipid ether biosynthesis may be peroxisomal processes and that the endoplasmatic/sarcoplasmatic reticulum has a major role in calcium homeostasis which influences cardiac contractility as well as metabolic enzyme activities. While the role of ribosomes seems to be better known, the importance of transport systems and vesicle pools may have been less recognized and their role in glucose and fatty acid uptake, fission and fusion of mitochondria is highlighted. Finally, the authors elegantly explain the different modes of cell death known as apoptosis, autophagy, necrosis, and necroptosis. They describe their causes, regulations, and their differences.

In Chapter 4, together with Christina Werner, we address principle metabolic pathways and metabolic cycles as they relate to energy production and building-block generation in the heart. This chapter covers the important biochemical parts of substrate use in cardiac metabolism. The contents of this chapter represent another fundamental component of cardiac metabolism, as it demonstrates how glucose and fatty acids as the main substrates are metabolized. Here, the connection between different pathways is illustrated and the importance of the citric acid cycle for the generation of reducing equivalents as well as for building blocks for biosynthetic processes becomes readily visible. The role of the respiratory chain as acceptor of reducing equivalents, as consumer of oxygen and most importantly as the main site of ATP production is made apparent. Furthermore, anaplerosis as mechanism to "refill" exploited moieties within metabolic cycles is introduced and the interrelation of hexosamine biosynthetic pathway, pentose phosphate pathway, and glycolysis is presented as well as the influence of fatty acid oxidation on glucose use and vice versa. Understanding of the principles explained in this chapter is essential to follow the metabolic path of substrates in an organism.

Louis Hue, Luc Bertrand, and Christophe Beauloye then address the principles of how the previously described cycles and pathways are regulated and how metabolism is controlled. Cardiac metabolism must never stop and needs to be adjusted to substrate availability, hormonal regulation, and workload. The authors elegantly describe how metabolic pathways are organized and controlled. Furthermore, they discuss how short- and long-term control of enzyme and pathways activity is achieved and how flux may be controlled. With flux control, they distinguish between two general mechanisms: control by supply as a "push mechanism" or control by demand as a "pull mechanism." Another way to control substrate metabolism is achieved by substrate competition and interaction, which seems to be the most sensitive regulation seen in metabolism. Chapter 5 offers the reader a thorough understanding of the regulations and interdependencies of cardiac metabolic pathways and cycles.

The previously mentioned information is strictly focused on processes ongoing in the mature, adult heart. However, metabolism undergoes massive changes during development. These changes are described by Andrea Schrepper in Chapter 6. The adult heart consumes preferentially fatty acids followed by lower amounts of glucose, lactate, and ketone bodies. In contrast, embryonic, fetal, and neonatal hearts; considerably deviate from the adult situation. Oxygen availability is frequently limited and substrate provision differs significantly from the adult situation. Glucose is the major substrate in these hearts with glycolysis as the main process for ATP generation. With birth, the heart has to adapt quickly to the abundance of fatty acids and increased oxygen availability. The change from glucose as the preferred substrate in the fetus to the adult situation is described in this chapter. Furthermore in the aging organism, cardiac metabolism changes again and the heart has to cope with increasing limitations in metabolism and function. The findings in cardiac metabolism in the aging heart are also discussed.

With Chapters 7 and 8, we enter the realm of methods and models. Together with Moritz Osterholt, we first present a general overview of methods used to investigate cardiac metabolism. From basic biochemical determinations of individual metabolite concentrations and enzyme activities using spectrophotometry, through powerful new tools for broad analyses of RNA and protein expression or metabolite concentration (the "-omics") up to nuclear and magnetic resonance tracing of metabolic rates, the principles are illustrated. We have tried to illustrate the strengths and the weaknesses of the individual methods. As mitochondria have moved more and more into the focus of metabolic research, we have addressed those biochemical analyses frequently used in the context of mitochondrial investigations as an example for the integration of methods.

We then move to address commonly used models to investigate cardiac metabolism. Metabolic measurements are frequently impossible in humans, thus animal models are required. Modeling of disease in animal models brings along advantages and shortcomings. The chapter is intended to introduce the reader to surgical, interventional, environmental, and genetic animal models and should enable the reader to choose an appropriate model for cardiac metabolic research. The chapter includes models of cardiac hypertrophy from different causes, ischemic as well as volume or pressure overload heart failure models as well as models of diabetes and nutritional intervention. Exercise may influence cardiac metabolism as well as infection. Furthermore, there are *in vitro* models as the isolated Langendorff or the working heart preparation, which are well suited for the investigation of metabolic fluxes in relation to contractile function or for the metabolic investigation of ischema/reperfusion. Cell culture models are used more and more to assess signaling mechanisms in cardiovascular disease, although the loss of workload-dependent contractile function makes the interpretation difficult at times. Thus, understanding the limits of these models may prove helpful.

Another physiologic principle, which in itself is highly interesting and even clinically relevant, also affects the proper conduct of metabolic research and the planning of metabolic experiments. Martin Young describes elegantly the impact of diurnal variations in cardiac metabolism and how genetically determined cardiac and biologic rhythms affect cardiac function and the methods used to investigate them. Cardiac metabolism not only changes in response to changes in environmental conditions or disease, it also changes regularly throughout the day. Diurnal variations are mainly caused by variations in behavior such as sleep–wake cycle and feeding at different times. They significantly affect both gene and protein expression. These variations lead to changes in glucose and fatty acid metabolism. Disturbance of diurnal variations may even lead to heart failure, underscoring their relevance. Frequently, there is little attention paid to diurnal variations in the experimental design, yet a different time point of investigation within 1 day may significantly alter the amount of protein or RNA to be investigated. Reading this chapter not only provides interesting and important information, but also it helps to clarify the relevance of diurnal variation for planning of experiments.

We then enter a series of chapters addressing states of disease. Marc van Bilsen starts with the description of the influence of nutrition and environmental factors on cardiac metabolism. As should be clear by now, the heart is able to utilize all possible substrates and has therefore, been termed a metabolic omnivore. Cardiac metabolism is therefore relatively robust. However, chronic changes in substrate supply lead to chronic adaptations of cardiac metabolism, which may not always be associated with the preservation of normal function. Nutritional changes, such as fasting or high-caloric or high-fat feeding, profoundly affect cardiac metabolism. The heart and its metabolism is even more severely affected in conditions such as obesity, metabolic syndrome, and diabetes, which all result from a nutritional "dysbalance," that is, the over-reliance on one substrate (mainly fatty acids). Exercise in turn may not only lead to cardiac hypertrophy, but affects cardiac substrate metabolism as well as mitochondrial function in a way that may provide protection against such metabolic insults. This excellently written chapter clearly addresses the influence of nutritional and exercise-induced changes on cardiac metabolism with respect to acute and chronic consequences.

Chapter 11 touches on the vast field of ischemia, hypoxia, and reperfusion. David Brown, Monte Willis, and Jessica Berthiaume describe how cardiomyocytes as well as the complete organ depend on a continuous coronary flow for proper function. Thus, hypoxia and ischemia present potentially deadly challenges for the entire organism. Hypoxia is defined as reduced oxygen availability, which may be, up to a certain degree, tolerated by the heart. In contrast, ischemia (myocardial infarction) interrupts the provision of oxygen and nutrients to the heart and the removal of carbon dioxide and disposal of "waste products" together; and depending on the degree of ischemia even completely (low flow- or total ischemia). This has a profound effect on cardiac metabolism. Importantly, the necessary reperfusion to terminate ischemia provokes more changes to cardiac metabolism and causes damage to the cell by itself, a phenomenon termed reperfusion injury. In the long run, ischemia is the most common cause for the development of heart failure. In this chapter, the effects of hypoxia, ischemia, and reperfusion on cardiac metabolism and metabolic therapies for ischemia-induced heart failure are discussed.

Chapter 12 then addresses heart failure but this time with pressure overload as the cause. T. Dung Nguyen illustrates that cardiac hypertrophy and heart failure can be induced by several different mechanisms but pressure overload is a major cause. The relation of metabolic remodeling and morphologic remodeling in the heart during the development of heart failure is discussed and their possible interrelation presented. While a causal role for impaired cardiac metabolism in the development of heart failure seems not always clear; the observed metabolic changes frequently indicate the state of heart failure progression (e.g., mitochondrial function). Furthermore, concepts to target cardiac metabolism for the treatment of hypertrophy and heart failure are presented and their results analyzed.

A similar target is investigated by Craig Lygate from a both conceptually and methodologically different perspective. Energetics address the role of high-energy phosphate generation and turnover as assessed by nuclear magnetic resonance spectroscopy. This perspective also assumes a tight link between ATP production and contractile function, but adds the creatine kinase system to the picture. Creatine kinase deficiency has been observed in cardiac hypertrophy and heart failure, but the regulation of creatine kinase is very complex. In Chapter 13, the creatine kinase system is described including various findings in hearts with elevated or reduced levels of creatine. Furthermore, energy transfer and energy status of the heart in hypertrophy and heart failure are discussed and the effect of treatments to improve energy status is presented.

In the end, we attempt together with Christian Schulze, Peter Kennel and Linda Peterson to illuminate the clinical relevance of metabolism and the current efforts and achievements of metabolism in the treatment of cardiac disease. In this chapter, the advantages and disadvantages of noninvasive metabolic assessment of the heart by nuclear and magnetic resonance techniques is addressed, illustrating how powerful but also how complex metabolic research can be. In addition, a detailed update on metabolic therapy in clinical practice is provided in the second part of the chapter again illustrating the important role of metabolism in cardiac disease.

Finally, Terje Larsen provides a historic overview over the field. Metabolic investigations have a long tradition and many early discoveries were necessary to build the foundation for today's investigations of cardiac metabolism. Historically, cardiac metabolism started with the ancient Greeks when Aristotle observed that cardiac function is associated with heat and that nutrition and heat are connected. Several historic findings strongly influenced the development of the field of metabolism and cardiac metabolism and allowed more and better understanding of cardiac function and its coupling to cardiac metabolism. Furthermore, several methods to perform cardiac metabolic research have their base on such "historic" work and the historic findings have been the base for several Nobel prizes in medicine.

We hope you will find useful information for your endeavor into cardiac metabolism and we wish you lots of curiosity and success in your investigations.

References

[1] Kolwicz SC Jr, Purohit S, Tian R. Cardiac metabolism and its interactions with contraction, growth, and survival of cardiomyocytes. Circ Res 2013;113:603–16.

[2] Neely JR, Liedtke AJ, Whitner JT, Rovetto MJ. Relationship between coronary flow and adenosine triphosphate production from glycolysis and oxidative metabolism. Recent Adv Stud Cardiac Struct Metab 1975;8:301–21.

[3] Neely JR, Morgan HE. Relationship between carbohydrate and lipid metabolism and the energy balance of heart muscle. Ann Rev Physiol 1974;36:413–39.

[4] Taegtmeyer H. Fueling the heart: multiple roles for cardiac metabolism. In: Willerson J, Wellens HJ, Cohn J, Holmes D Jr, editors. Cardiovascular medicine. London: Springer; 2007. p. 1157–75.

[5] Taegtmeyer H. Cardiac metabolism as a target for the treatment of heart failure. Circulation 2004;110:894–6.

[6] Taegtmeyer H, King LM, Jones BE. Energy substrate metabolism, myocardial ischemia, and targets for pharmacotherapy. Am J Cardiol 1998;82:54K–60K.

2

Basics in Metabolically Relevant Biochemistry

Miranda Nabben, Jan F.C. Glatz

Department of Genetics and Cell Biology, Cardiovascular Research Institute Maastricht (CARIM), Maastricht University, Maastricht, The Netherlands

The four main types of biological molecules in the body are carbohydrates, lipids, proteins, and nucleic acids. The building blocks of the first three are monosaccharides (in particular glucose), fatty acids, and amino acids, respectively. All these molecules serve as fuel for adenosine triphosphate (ATP) production. As mentioned in Chapter 1, the heart is a metabolic omnivore and will use each of these latter compounds as well as some of their conversion products, in particular lactate and ketone bodies, for metabolic energy production. In this chapter, we will describe the basic biochemical features of each of these fuels, how they are taken up by cells, and subsequently temporarily stored (as glycogen and intracellular fat depots). Finally, we will also briefly outline the biochemistry of enzyme activities and their regulation. For a more detailed overview, the reader is referred to a biochemistry textbook [1].

CARBOHYDRATES

Carbohydrates are a class of chemical compounds composed of carbon, hydrogen, and oxygen in 1:2:1 ratio, respectively. Carbohydrates are ingested via the diet (for instance, bread and pasta) or can be synthesized in the body. The simplest carbohydrates are the monosaccharides. The main dietary monosaccharides are glucose (dextrose or grape sugar), fructose (fruit sugar), and galactose (milk sugar). After absorption from the intestinal tract, virtually all of the fructose and galactose saccharides are rapidly converted in the liver into glucose or intermediates of glucose metabolism. Therefore, glucose represents the main carbohydrate source for the heart. Glucose is a hexose and contains 6 carbon, 12 hydrogen, and 6 oxygen atoms. It exists in D- and L-isomers, which designate the absolute confirmation. In contrast to L-glucose, D-glucose occurs widely in nature. The hydroxyl groups make the carbohydrates readily dissolvable in water. Although glucose can exist in a straight-chain form, it predominantly cyclizes into a ring-like structure (Fig. 2.1). The position of the hydroxyl- (OH-) group relative to the ring's midplane determines the denotation α or β. α-Carbohydrates are those in which the OH-group on the first carbon points in opposite

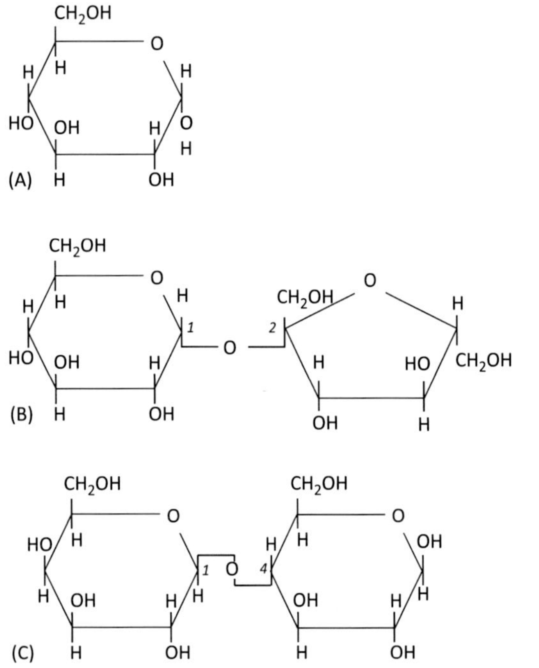

FIGURE 2.1 **Chemical structures of the monosaccharide α-glucose (A), and disaccharides sucrose (B) and lactose (C).** Sucrose consists of the monosaccharides glucose and fructose that are joined by an α-1,2-glycosidic linkage. Lactose consists of the monosaccharides galactose and glucose that are joined by a β-1,4-glycosidic linkage.

direction of carbon number 6. In β-carbohydrates, the OH-group points in the same direction as number 6. This little chemical difference makes a significant change in metabolism. For example, whereas starch and glycogen consist of α-glucose bonds and can be easily digested in humans, cellulose consists of β-glucose bonds and is very difficult to digest.

Disaccharides are formed out of two monosaccharides that are chemically linked, for example, glucose + fructose will form sucrose and glucose + galactose will form lactose (Fig. 2.1). Oligosaccharides consist of 3–20 monosaccharides, whereas polysaccharides consist of more than 20, often thousands, of monosaccharides linked together. Polysaccharides are often used for energy storage or structural support.

Examples of polysaccharides are starch and glycogen. Starch is the glucose energy storage form in plants. Starch saccharides can be unbranched, like amylose, or branched, like amylopectin. In animals, glucose is stored as glycogen. The structures of starch and glycogen are very similar with the only exception that glycogen has branch points every 8–12 residues and starch every 24–30 glucose residues.

In the intestinal tract, monosaccharides are readily taken up to enter into the blood circulation. However, in order for di-, oligo-, and polysaccharides to be taken up, they first need to be degraded by specific enzymes present in the dietary tract into monosaccharides. Interestingly, the uptake of glucose by intestinal epithelial cells is an active process and occurs by the mechanism of sodium–glucose cotransport. The active transport of sodium provides the energy for absorbing glucose against a concentration gradient. Note that this sodium cotransport mechanism functions only in certain special epithelial cells (intestine and kidney), while at all other cell membranes (including cardiomyocytes) glucose is transported only from higher concentration toward lower concentration by facilitated diffusion (to be discussed later in this chapter).

Pentoses are monosaccharides containing 5 carbon atoms. Although pentoses are of little or no importance as a source of energy for the body, they are present in small amounts in all cells, since D-ribose and D-2-deoxyribose are components of nucleic acids and are therefore a leading component of the cell's genetic information (DNA).

GLYCOGEN AS ENDOGENOUS GLUCOSE STORAGE

For long-term energy storage, glucose can be stored as glycogen. Glycogen is a polysaccharide structure that is present in large amounts in the liver, where it can be converted back into

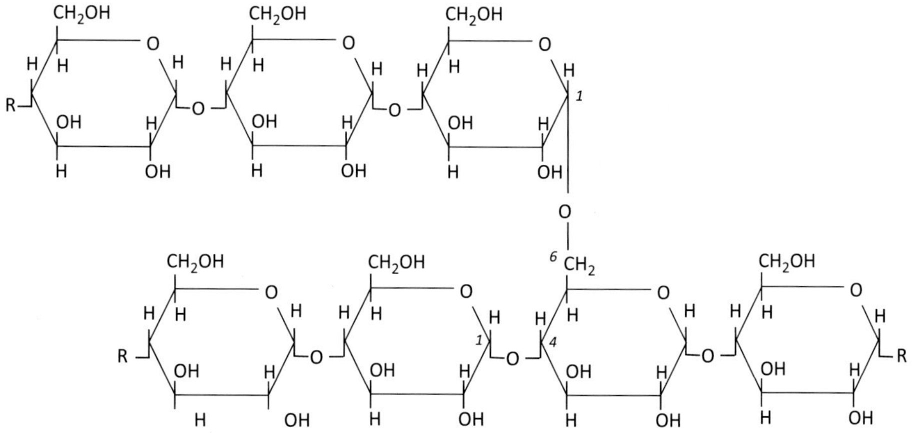

FIGURE 2.2 **Chemical structure of glycogen, the storage form of glucose.** Most of the glucose residues in glycogen are linked via α-1,4-glycosidic bonds. Branches are created via α-1,6-glycosidic linkage.

glucose and distributed to other organs, such as brain, and also to heart and skeletal muscle. Additionally, relatively large amounts of glycogen can be stored in heart and skeletal muscle. Importantly, muscle glycogen can be used as muscular energy source but cannot be converted into glucose to be excreted into the circulation. Glycogen is mainly composed of α-D-glucose residues that are linearly linked via α-1,4-glycosidic bonds with branches that are created via α-1,6-glycosidic bonds (Fig. 2.2).

The synthesis of glycogen is referred to as glycogenesis. First, glucose is phosphorylated into glucose-6-phosphate by hexokinase or glucokinase. This glucose-6-phosphate either can enter the glycolysis pathway where it is converted into fructose-6-phosphate and eventually into pyruvate, or it can enter the glycogenesis pathway where it is converted into glucose-1-phosphate by the enzyme phosphoglucomutase. Together with uridine triphosphate (UTP), this glucose-1-phosphate will then form a uridine diphosphate (UDP)-glucose molecule, which is the basic building block for glycogen. The glucose-1-phosphate-uridyltransferase enzyme catalyzes this process. The transfer of glucose molecules from UDP-glucose to glycogen is catalyzed by glycogen synthase. UDP will be dropped off and the newly derived glucose molecule will be transferred onto the existing elongating glucose chain via linear α-1,4 bonds, via dehydration synthesis. A branching enzyme is required to form α-1,6 linkages and transform glycogen into a branched polymer.

The breakdown or hydrolysis of glycogen to glucose (glycogenolysis) starts with glycogen phosphorylase cleaving of the α-1,4 bonds, and the debranching enzyme cleaving of the α-1,6 bonds. This will form glucose-1-phosphate that is transformed into glucose-6-phosphate by phosphoglucomutase. As the hexokinase/glucokinase step is unidirectional, a separate enzyme, glucose-6-phosphatase is necessary for removal of phosphate and formation of glucose. Since this enzyme is only present in liver, in other tissues (in particular heart, skeletal muscle, and brain) glucose-6-phosphate from glycogen enters the glycolytic pathway.

The control of glycogen synthesis versus breakdown is under hormonal influence. For example, insulin initiates glycogen synthesis, whereas epinephrine and glucagon stimulate glycogen breakdown and glucose release (from liver) while inhibiting glycogen synthesis.

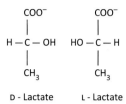

FIGURE 2.3   **Chemical structure of the D- and L-isomeric forms of lactate.**

LACTATE

Under conditions of insufficient tissue oxygen availability (which may occur, e.g., in selected skeletal muscles during exercise) the complete oxidation of carbohydrates is not possible; however, a small amount of energy can still be produced by conversion of carbohydrates (particularly glucose) into lactate. Lactic acid is an α-hydroxyl carboxylic acid that contains 3 carbon, 6 hydrogen, and 3 oxygen atoms.

Under physiological conditions, lactic acid is ionized and thus present in the lactate form. Both lactic acid and lactate exist in D- and L-isomeric forms (Fig. 2.3). After formation, lactate can be released from one cell into the interstitial space and blood compartment to serve as a precursor carbon source for oxidative phosphorylation or as a gluconeogenic substrate for glycogen synthesis in other cells throughout the body. Of note, cardiac muscle is a main consumer of lactate produced by erythrocytes and (anaerobic) skeletal muscle.

FATTY ACIDS

Fatty acids, particularly long-chain fatty acids, form a main constituent of various lipid species and are a major substrate for metabolic energy production while specific fatty acids and fatty acid metabolites also function as signaling compounds. Lipids are vital components of many biologic processes and serve as building blocks of biologic membranes (e.g., phospholipids, sphingolipids) or of specific proteins (e.g., myristoylation, palmitoylation). Due to their hydrophobic or amphiphilic nature, all lipid species and long-chain fatty acids are characterized by their virtual insolubility in aqueous solutions.

Fatty acids are simply long-chain hydrocarbon organic acids. These lipids consist of a long, nonpolar hydrocarbon tail and a more polar carboxylic head group ($-COO^-$), and therefore, are amphipathic compounds (i.e., both polar and nonpolar regions within one molecule). A typical fatty acid is palmitic acid, which has the chemical formula $CH_3(CH_2)_{14}COOH$. The carbon chain of a fatty acid may be saturated or may have one (monounsaturated fatty acid) or more double bonds (polyunsaturated fatty acid). In most naturally occurring fatty acids, the double bond is in the *cis* geometrical configuration. The *trans* formation is often generated during food processing and occurs when fatty acids with at least one double bound are heated in the presence of water (i.e., hydrogenated fats, as often used for deep frying). *Trans* fatty acids have been found to be associated with increased cardiovascular risk [2]. Furthermore, nearly all fatty acids have an even number of carbon atoms and have chains that are between 14 carbon atoms and 22 carbon atoms long, with those having 16 or 18 carbons being the most abundant. In fatty acids containing two or more double bonds, the fatty acids are always separated by one methylene group, that is, $-CH=CH-CH_2-CH=CH-$.

Thus, fatty acids differ primarily in (1) chain length, (2) number, and (3) position of their unsaturated bonds. The most widely used nomenclature designates these three characteristics as follows: C16 (palmitic acid) denotes a saturated chain of 16 carbons, C18:1 *n*–9 (oleic acid) denotes a chain of 18 carbons with one double bond at position 9 from the methyl terminal end of the chain, C20:4 *n*–6 (arachidonic acid) denotes a chain of 20 carbons with 4 double bonds starting at position 6 from the methyl terminal end of the

TABLE 2.1 Most Abundant Saturated and Unsaturated Long-Chain Fatty Acids

Carbon atoms	Common name	Systematic name
Saturated fatty acids		
12	Lauric acid	C12
14	Myristic acid	C14
16	Palmitic acid	C16
18	Stearic acid	C18
20	Arachidic acid	C20
24	Lignoceric acid	C24
Unsaturated fatty acids		
16	Palmitoleic acid	C16:1 *n*–9
18	Oleic acid	C18:1 *n*–9
18	Linoleic acid	C18:2 *n*–9
18	Linolenic acid	C18:3 *n*–9
20	Arachidonic acid	C20:4 *n*–6
20	Eicosapentaenoic acid	C20:5 *n*–3
22	Docosahexaenoic acid	C22:6 *n*–3

chain (with the other double bonds at positions 9, 12, and 15 from the methyl terminal end). The main naturally occurring long-chain fatty acids are listed in Table 2.1. Of particular interest are the polyunsaturated fatty acids of marine origin, that is, eicosapentaenoic acid and docosahexaenoic acid, because their multiple double bonds provide these fatty acid species with unique properties especially when incorporated in phospholipids forming biological membranes.

Although long-chain fatty acids are essentially insoluble in water, their Na^+ and K^+ salts are soaps and form micelles in water that are stabilized by hydrophobic interactions. However, the vast majority of long-chain fatty acids is esterified in phospholipids, as part of biological membranes, or in triacylglycerols, being the predominant storage form of lipid metabolic energy.

Triacylglycerols (triglycerides) are composed of glycerol (a trihydric alcohol) in which each of the hydroxyl groups forms an ester link with long-chain fatty acids. The resultant triacylglycerol has almost no polar qualities. Phospholipids are derived from diacylglycerol phosphate (phosphatidic acid) with an additional polar group, usually a nitrogen-containing base such as choline or a polyalcohol derivative such as phosphoinositol. Phospholipids commonly have long-chain unsaturated fatty acids on the 2-position. Common examples of a triacylglycerol and a phospholipid are shown in Fig. 2.4.

Cell membranes are composed of a double layer of phospholipids, interspersed with specific peripherally located or transmembrane proteins such as hormone receptors, transporter molecules, and ion channels. Cell membranes may also contain particular lipid species such as sphingomyelin, which stiffens the membrane, and cholesterol, which is involved in the regulation of membrane fluidity. In the phospholipid bilayer, the polar "heads" of the phospholipid molecules are presented to the aqueous external environment while the nonpolar "tails" of the two bilayers face each other and form a hydrophobic region within the membrane interior. The physicochemical nature of such biological membrane dictates that, in general, molecules cannot diffuse freely across it because polar molecules would not be able to cross the inner, hydrophobic region whereas nonpolar molecules would not be able to cross the outer, polar (hydrophilic) face of the bilayer. As a result, specific membrane-associated proteins act to facilitate transmembrane transport of compounds (to be discussed later).

KETONE BODIES

Under specific conditions, such as long-term starvation, the liver will produce three compounds that together are referred to as ketone bodies. These compounds are acetoacetic acid, β-hydroxybutyric acid, and acetone (Fig. 2.5). The primary compound formed in the liver is acetoacetic acid, which in part is converted into β-hydroxybutyric acid while only minute

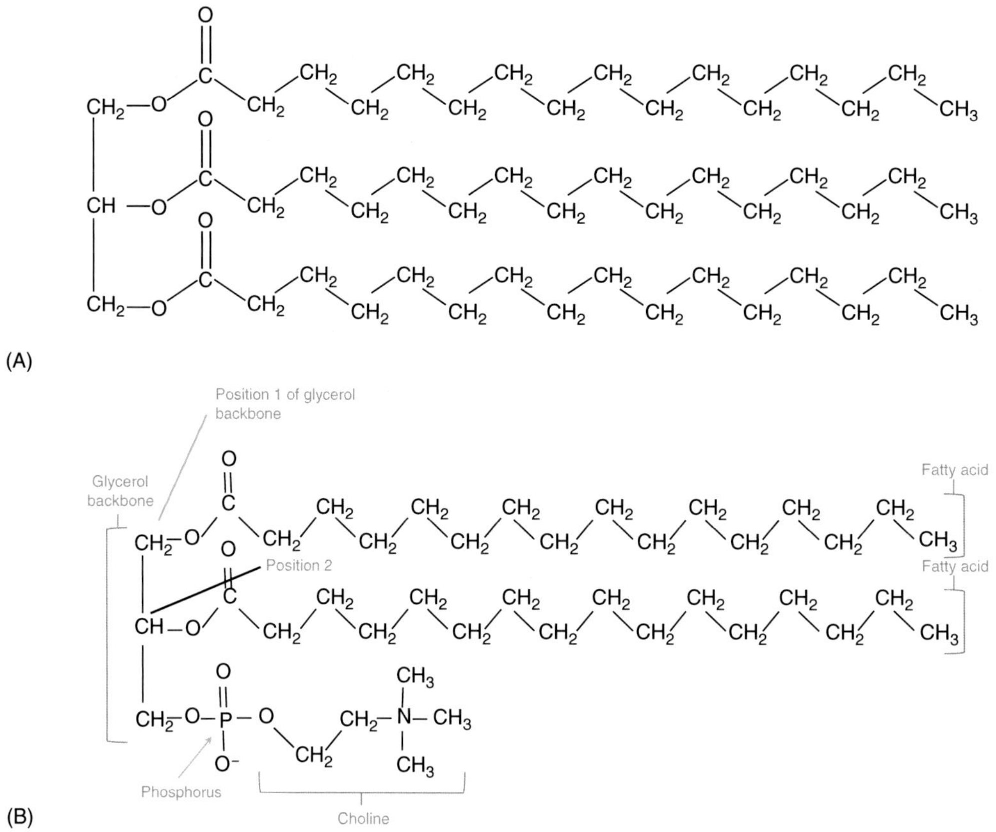

(A)

(B)

FIGURE 2.4 **Chemical structure of the triacylglycerol tripalmitoylglycerol (A) and of the abundantly occurring phospholipid, phosphatidylcholine (also known as lecithin) (B).** Triacylglycerol is an ester derived from a glycerol backbone and three fatty acids. Phospholipids also contain fatty acids, however, in contrast to triacylglycerol these usually contain a diacylglycerol, a phosphate group, and a simple organic molecule such as choline.

quantities are converted into acetone. These compounds are excreted into the blood and may serve as metabolic substrate for energy production in other organs, particularly brain, skeletal muscle, and cardiac muscle.

AMINO ACIDS – BUILDING BLOCKS FOR PROTEINS

Proteins play crucial roles in virtually all biologic processes. They are involved in catalysis of chemical reactions through enzymes, transport of molecules and ions, storage as complexes, coordinated motion via muscle contraction and mechanical support. Furthermore, proteins are involved in immune protection through globulines and antibodies, generation and transmission of nerve impulses, and control of growth and differentiation via hormones.

Amino acids are the building blocks for proteins. They contain an acidic carboxyl (COOH) and a basic amine (NH_2) group, a hydrogen atom, and a distinctive "R" group bound to a central carbon atom (α-carbon). There are 20 different kinds of "R" groups that are commonly

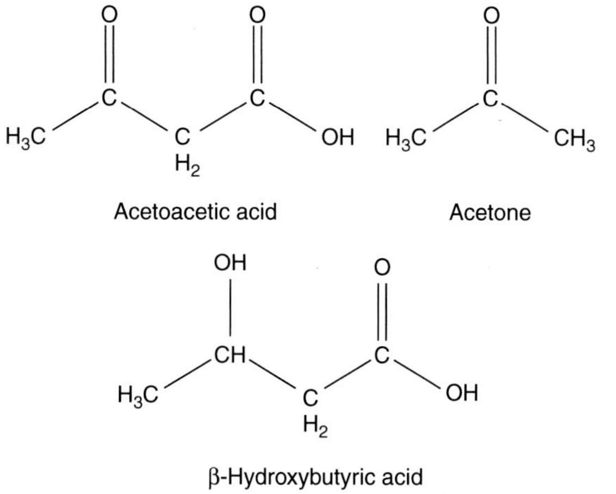

FIGURE 2.5 Chemical structure of the three ketone bodies acetoacetic acid, β-hydroxybutyric acid, and acetone.

found in proteins, varying in size, shape, charge, hydrogen bonding capacity, and chemical reactivity. These side chains can be (1) aliphatic without (glycine, alanine, valine, leucine, isoleucine) or with (proline) a secondary amino group; (2) aromatic (phenylalanine, tyrosine, tryptophan); (3) sulfur-containing (cysteine, methionine); (4) hydroxyl aliphatic (serine, threonine); (5) basic (lysine, arginine, histidine); (6) acidic (aspartate and glutamate); or with a (7) amide-containing (asparagine and glutamine) group.

The ionization state of the amino acids varies with pH (Fig. 2.6). Amino acids exist in D- and L-isomers of which mainly the L-amino acids are constituents of proteins. Proteins are on average 200 amino acids long (the number varying considerably among various proteins) that are bound together via peptide (or amide) bonds. These bonds link the carboxyl end of one amino acid together with the amine group of the other, thereby removing water via dehydration synthesis. A combination of two amino acids is called a dipeptide; three amino acids linked together is a tripeptide; while, multiple amino acids form a polypeptide.

The structure of a protein is determined at several levels. The primary level (protein primary structure) is the sequence of the amino acids. Subsequently, the repertoire of 20 different side chains enables the proteins to fold into distinct two- and three-dimensional structures. Thus, the secondary level refers to coils and folds formed as a result of hydrogen bonds in the polypeptide backbone. The most common forms are the α-helix (favored by glutamate, methionine, leucine), β-sheet (favored by valine, isoleucine, phenylalanine) or a collagen helix (favored by proline, glycine, aspartate, asparagine, serine). The tertiary level is formed due to irregular interactions between the "R" groups and basically

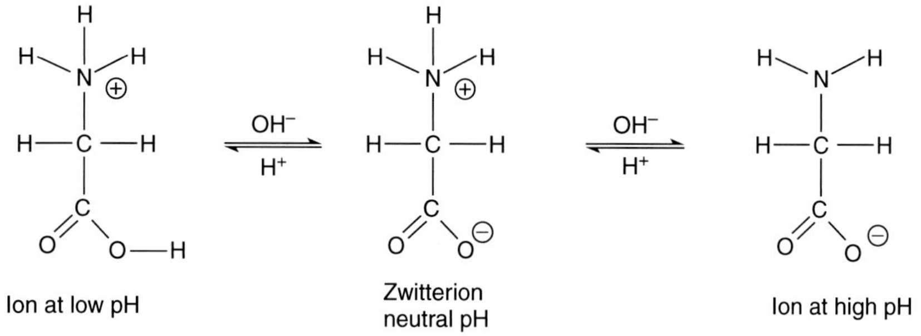

FIGURE 2.6 **The ionization state of the amino acids is pH dependent.** In solution, at neutral pH, the amino acids are predominantly present as dipolar ions (or zwitterion) rather than unionized molecules. In acid-solution, the predominant form consists of an unionized carboxyl group and an ionized amino group. In alkaline solution, the carboxyl group is ionized and the amino group is unionized.

forms the three-dimensional arrangement of the polypeptide chain. Finally, the quaternary level refers to the presence of more than one individual polypeptide chain, and is determined by their number and specific arrangement in the protein molecule. Unfolding or denaturation of proteins can be caused by treatment with solvents or due to extreme pH and temperature effects.

BRANCHED CHAIN AMINO ACIDS

Amino acids can be classified as nutritionally essential or nonessential amino acids on the basis of their dietary needs (essential) or the body's ability to adequately synthesize the amino acids (nonessential) for normal growth and nitrogen balance. Histidine, isoleucine, leucine, lysine, methionine, phenylalanine, threonine, tryptophan, and valine are essential amino acids, whereas alanine, asparagine, aspartic acid, glutamic acid, and serine belong to the nonessential amino acids. Arginine, cysteine, glycine, glutamine, proline, and tyrosine are considered conditionally essential in the diet, as their synthesis can be limited under certain conditions, such as prematurity, during growth, or severe catabolic distress.

Whereas most metabolic and catabolic activities of amino acids occur in the liver, a subgroup of essential amino acids, the branched chain amino acids (BCAAs), leucine, isoleucine, and valine, are catabolized primarily in nonhepatic tissues, like (cardiac) muscle and the periphery. BCAAs share an aliphatic side-chain structure with a branch. Their side-chains differ in shape, size, and hydrophobicity. After largely escaping the first-pass hepatic catabolism, BCAAs seem to be taken up by the non-hepatic tissue. Remarkably, the first part of the BCAA breakdown is common to all three BCAAs, involving the BCAA aminotransferase and branched-chain α-keto acid dehydrogenase enzymes. Thereafter, the BCAAs follow different catabolic pathways to different products (sterol, ketone bodies, and/or glucose). They eventually are degraded into acetyl-CoA or succinyl-CoA, which are consumed in mitochondria through the tricarboxylic acid (TCA) cycle for the production of reduced nicotinamide adenine dinucleotide (NADH) for respiration. Together, these three BCAAs commonly account for ~20–25% of most dietary proteins.

CELLULAR UPTAKE OF METABOLIC SUBSTRATES

As discussed earlier, the cellular uptake of each of the metabolic substrates is facilitated by specific transporter proteins embedded in the cell membrane. For glucose, there are two families of transporters: (1) a more widespread family of passive glucose transporters (GLUT) (uniporters), allowing the movement of glucose across cell membranes only down a concentration gradient (facilitated diffusion), and referred to as GLUTn and (2) a family of active glucose transporters enabling glucose to move up a concentration gradient by cotransport with Na^+ ions which are moving down a concentration gradient, and referred to as sodium–glucose cotransporters (symporters), SGLTn [3]. The expression of all of these transporter family members is tissue specific, and their properties are an integral part of the regulation of glucose metabolism in the particular tissue. The SGLTn are present in intestine and renal tubules and will not be discussed here. In contrast, the GLUT's occur in virtually all tissues. The GLUT's are related 45 kDa proteins, each having 12 membrane spanning regions. In cardiac myocytes, the primary glucose transporters are GLUT1, which constitutively resides in the sarcolemma, and GLUT4, which is present in endosomal membranes from where it can be recruited to the sarcolemma to increase the cellular glucose uptake rate in order to meet the cellular energy requirement. Likewise, internalization of GLUT4 from the sarcolemma to the

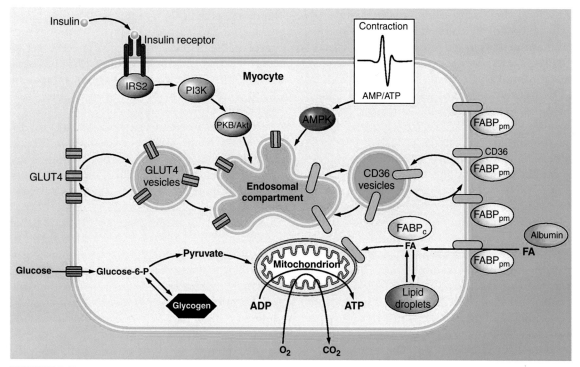

FIGURE 2.7 **Similarity between the regulation of cellular uptake of fatty acids and glucose.** The uptake of both fatty acids and glucose by cardiac and skeletal muscle is increased after translocation of specific transporter proteins (shown for CD36 and GLUT4, respectively) to the sarcolemma in response to stimulation with insulin or during increased contractile activity. CD36 and GLUT4 may be mobilized from different stores within the endosomal compartment. At the sarcolemma, CD36 is in interaction with FABPpm. Note that the involvement of GLUT1 in glucose uptake and that of the FATPs in fatty acid uptake are not shown. *Adapted from Ref. [5], with permission.*

endosomal stores will lower the cellular rate of glucose uptake. The two main triggers that recruit GLUT4 to the sarcolemma are insulin and (increased) muscle contraction (Fig. 2.7) [4,5]. This intracellular GLUT4 recycling is a primary mechanism regulating the overall utilization of glucose by cardiac muscle cells.

Long-chain fatty acid transport across a biological membrane is also facilitated and regulated by specific membrane-associated proteins. The proteins involved are the peripheral membrane fatty acid binding protein FABPpm (43 kDa), a family of six so-called fatty acid transport proteins (FATP1–6; 63 kDa), and CD36 (also referred to as fatty acid translocase; 88 kDa). Most likely,

these proteins act at the extracellular side by facilitating the capture of fatty acids and their subsequent entry into the membrane, followed by the desorption of fatty acids at the intracellular side of the membrane and subsequent binding to cytoplasmic fatty acid binding protein (FABPc). The transmembrane transport of fatty acids, from the outer to the inner leaflet of the phospholipid bilayer, may occur by a spontaneous process referred to as "flip-flop" for which facilitation by proteins is not needed. In cardiac muscle, CD36 is the primary protein involved in cellular fatty acid uptake, assisted by FABPpm with which it shows molecular interaction. FATP1 and FATP6 appear to be involved mostly in the uptake of very long-chain fatty acids. Interestingly, CD36 was found to regulate fatty acid uptake by a mechanism that closely resembles that of GLUT4-mediated glucose uptake. Thus, following an acute stimulus (insulin, muscle contraction), CD36 translocates from an intracellular store (endosomes) to the sarcolemma to increase fatty acid uptake (Fig. 2.7) [5]. Similar to glucose uptake, the protein-assisted cellular uptake of fatty acids serves a major regulatory role in the overall rate of cardiac fatty acid utilization.

The other substrates, that is, lactate, ketone bodies, and amino acids, also enter cells by facilitated diffusion. The monocarboxylic acids, lactate, and ketone bodies are transported by monocarboxylate transporters (MCTs), a family of well-characterized 45 kDa membrane proteins [6]. The heart (and skeletal muscle and some other tissues) expresses MCT1, which facilitates the proton-linked trans-sarcolemmal (bidirectional) movement of lactate and ketone bodies. Given its major role in metabolism, L-lactate is quantitatively by far the most important substrate for MCT1. This transporter is stereoselective for L-lactate over D-lactate. MCTs require the ancillary glycoproteins embigin or basigin for correct membrane expression. Amino acids enter myocardial cells by specific amino acid transporters; however, the exact transport mechanism of amino acids into the heart remains largely underexplored. It seems that there are (at least) three types of L-type amino acid transporters present in the heart which all belong to the solute carrier (SLC) 7 family [7]. With respect to the catabolism of BCAAs, these seem to be taken up by nonhepatic tissues and downstream activated through involvement of L-type amino acid transporters and the bidirectional transporters for L-glutamine and L-leucine [8].

ENZYME ACTIVITIES AND THEIR REGULATION

Enzymes are the catalysts in biological systems. They lower the amount of activation energy needed for a chemical reaction and therefore, accelerate its rate, without undergoing a change in structure. Nearly all enzymes are proteins. They consist of a specific active site consisting of amino acid residues that have several important properties such as specific charges, pK_a, hydrophobicity, flexibility, and reactivity.

There are six classes of enzymes: (1) oxidoreductases that catalyze oxidation–reduction reactions in which oxygen or hydrogen are added or removed; (2) transferases that catalyze the transfer of functional groups between donor and acceptor; (3) hydrolases that break single bonds by adding water; (4) lyases that remove or form a double bond with transfer of atom groups; (5) isomerases which carry out many kinds of isomerization processes like the L- to D isomerizations; and (6) ligases that link two chemical groups together by removing the elements of water, using energy that is usually derived from ATP.

The enzymes' catalytic power stems from the specific shape of the active site which complements and binds to a specific substrate only, similar to a key fitting into a lock. Upon binding, an enzyme–substrate complex is formed which results in the formation of bonds that can eventually proceed to the formation of a product. Alternatively, the complex can dissociate back into an enzyme and a substrate. The rate of the enzymatic reaction mechanism follows Michaelis–Menten kinetics. This means that an increase in the amount of enzyme increases the rate of reaction and while the product is being formed rapidly at first, the rate of reaction eventually levels off as the concentration of the substrate decreases and the concentration of product increases (Fig. 2.8). At the end of the reaction, an equilibrium is reached.

Next to enzyme and substrate concentration, the rate of the enzyme reaction can also be affected by temperature, pH, K_m, and allosteric regulation. Furthermore, the action of enzymes can be affected by several other factors. Some enzymes require cofactors (small inorganic chemicals not containing carbon; e.g.,

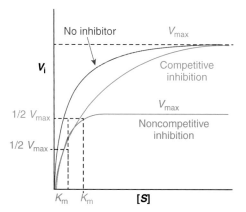

FIGURE 2.8 **Graph showing kinetics of enzymatic reactions.** V_i, initial velocity (moles/time); $[S]$, substrate concentration (molar); V_{max}, maximum velocity; K_m, substrate concentration when V_i is one-half of V_{max} (Michaelis–Menten constant). In the presence of a competitive inhibitor, the reaction velocity is decreased at a given substrate concentration, but V_{max} is unchanged. In the presence of a noncompetitive inhibitor, V_{max} is decreased. *Reproduced with permission from Kimball's Biology pages (www.biology-pages.info).*

ATP GENERATION THROUGH SUBSTRATE-LEVEL PHOSPHORYLATION AND AT THE PROTON PRODUCTION LEVEL

In almost all biological processes, ATP functions as the carrier of free energy. In order to keep up with the body's energy needs, ATP has a very high turnover rate and is continuously being generated from the breakdown and oxidation of substrates.

ATP is a nucleotide consisting of an adenine, a 5-carbon sugar (ribose), and three phosphate groups. Adenine is a purine, with a nitrogenous base that together with ribose forms adenosine. ATP is energy rich because its triphosphate unit contains two phosphoanhydride bonds. The high-energy bond between the second and third phosphate group in particular is most often hydrolyzed to release energy. In animals, ATP is generated through substrate-level phosphorylation and through oxidative phosphorylation. Free energy is liberated when ATP is hydrolyzed into adenosine diphosphate (ADP) and inorganic phosphate (P_i), or into adenosine monophosphate (AMP) and pyrophosphate (PP_i).

During glycolysis, a small amount of ATP is being formed, together with the three-carbon compound pyruvate and NADH. Glycolysis does not involve molecular oxygen. Under aerobic conditions, this pyruvate and NADH enter the mitochondria for cellular respiration. Here, pyruvate is oxidized into acetyl-CoA by pyruvate decarboxylation thereby producing more NADH. The acetyl-CoA will enter into the TCA cycle yielding more NADH, as well as flavin adenine dinucleotide–reduced form (FADH$_2$) and guanosine triphosphate (GTP). The amount of energy built into GTP is equivalent to the amount built into ATP.

The oxidation of fatty acids also generates ATP, again through production of reducing equivalents (NADH and FADH$_2$) during β-oxidation. The amount of ATP generated through fatty acid oxidation depends on the fatty acid chain length. Fatty acids are first transformed into acyl-CoA

ions, DNA polymerase, minerals) or coenzymes (organic molecules; e.g., NADH that acts as a carrier molecule) to help catalyze reactions. On the other hand, the action of the enzymes can be prevented or inhibited via competitive inhibition (competition for space with substrate) or allosteric inhibition (by binding to another side on the enzyme itself, thereby covering up the active side or changing the shape of the active side, so the substrate does not fit).

The slowest step in a metabolic pathway, which determines the overall rate of the reactions in the pathway is considered the rate-limiting step. Identification of these rate-limiting steps will therefore offer important therapeutic strategies for targeting metabolic diseases.

In heart and skeletal muscle, glucose uptake mediated by GLUT4 is considered the rate-limiting step in cellular glucose utilization. In cardiac and muscular fatty acid utilization, the rate-limiting steps are the uptake of fatty acids into the cell and the entry of activated fatty acids (fatty acyl-CoA esters) into mitochondria [5].

esters (at the expense of ATP), which then will enter into the β-oxidation pathway. During each round of β-oxidation, two carbons are cleaved off, generating acetyl-CoA, NADH, and $FADH_2$. Similar to the acetyl-CoA formed by pyruvate oxidation, this fatty acid-derived acetyl-CoA will enter into the TCA cycle yielding more NADH, $FADH_2$, and GTP.

After these substrate oxidation steps, the production of cellular energy from all the major catabolic pathways including glycolysis, fatty acid oxidation and amino acid oxidation, and TCA cycle are integrated into the oxidative phosphorylation (OxPhos) system. The OxPhos system uses O_2 to produce H_2O and is responsible for the generation of the majority of cellular ATP. Here, all the formed NADH and $FADH_2$ will donate electrons to complex I and complex II, respectively, of the electron transport chain. This causes protons to be pumped out of the mitochondrial matrix into the outer compartment of the mitochondria, yielding a proton gradient. The enzyme ATP synthase uses this gradient to facilitate a proton-flux back into the matrix, thereby releasing a lot of free energy that is used to drive ATP synthesis. Each NADH molecule is valued to result in 2.5 molecules of ATP, each $FADH_2$ in 1.5 molecule of ATP, and each GTP in 1 molecule of ATP. In total, this means that the complete oxidation of glucose is coupled to the synthesis of 36 ATP molecules and the complete oxidation of the 18 carbon-fatty acid stearic acid to 120 ATP molecules. In general, fatty acids require more oxygen to produce the same amount of ATP than glucose since the carbohydrates contain more oxygen per molecule. During anaerobic conditions, only 2 molecules of ATP are generated for each glucose molecule that is converted into lactate.

Amino acid metabolism also generates ATP. Depending on the type of amino acid, they can use similar catabolic pathways as for glucose or fatty acids. Deamination of certain amino acids results in pyruvate that can be used for energy production and also for glucose synthesis. Deamination of other amino acids results in acetyl-CoA that enters the TCA cycle by binding to oxaloacetate to form citric acid, while the breakdown of the BCAAs valine and isoleucine and that of methionine yield succinyl-CoA that can enter the TCA cycle directly (so-called anaplerotic substrates). Upon excess calories consumed, some of the acetyl-CoA from amino acid breakdown can be used to synthesize fatty acids, instead of going through the ATP generating pathway.

References

[1] Berg JM, Tymoczko JL, Stryer L. Biochemistry. 7th ed. New York: W.H. Freeman Publishers; 2012.
[2] Lichtenstein AH. Dietary trans fatty acids and cardiovascular risk: past and present. Curr Atheroscler Rep 2014;16(8):433.
[3] Chen LQ, Cheung LS, Feng L, Tanner W, Frommer WB. Transport of sugars. Annu Rev Biochem 2015;2:865–94.
[4] Thong FSL, Dugani CB, Klip A. Turning signals on and off: GLUT4 traffic in the insulin-signaling highway. Physiology 2005;20:271–84.
[5] Glatz JFC, Luiken JJFP, Bonen A. Membrane fatty acid transporters as regulators of lipid metabolism: implications for metabolic disease. Physiol Rev 2010;90:367–417.
[6] Halestrap AP. Monocarboxylic acid transport. Compr Physiol 2013;3:1611–43.
[7] Fotiadis D, Kanai Y, Palacin M. The SLC3 and SLC7 families of amino acid transporters. Mol Aspects Med 2013;34:139–58.
[8] Huang Y, Zhou M, Sun H, Wang Y. Branched-chain amino acid metabolism in heart disease: an epiphenomenon or a real culprit? Cardiovasc Res 2011;90:220–3.

Metabolically Relevant Cell Biology – Role of Intracellular Organelles for Cardiac Metabolism

Bernd Niemann*, Susanne Rohrbach†

*Department for Adult and Pediatric Cardiac Surgery and Vascular Surgery, University Hospital Giessen and Marburg, Justus Liebig University Giessen, Rudolf Buchheim Strasse, Giessen; †Institute for Physiology, Justus Liebig University Giessen, Aulweg, Giessen

CELLULAR COMPARTMENTS

Eukaryotic cells exhibit different compartments, each of those processing functional specialization. Coated by the plasmatic membrane cytosol, cytoplasm and cellular organelles are separated to compartmentalize the environment of specified biochemical reaction. Thus, construction, maturation, modification, and degradation of proteins are spatially separated by biomembranes. Within eukaryotes, organelles exhibit a characteristic pattern meeting the cellular needs of specialization. Thus, skeletal myocytes and in particular cardiac myocytes exhibit enormous amounts of mitochondria (mt), which can represent up to a third of the cellular volume. Organelles, virtually identifiable in all cells are nucleus, endoplasmic reticulum (ER), mt, peroxisomes, endosomes, lysosomes, Golgi apparatus, and – to a certain extent – physiologic or pathologic lipid droplets and vesicles. The cellular cytoskeleton stabilizes cellular geometry and enables certain cells for directed movement and mechanical activity on the one hand and for directed transport of substrates and derivates within the cell on the other hand.

CYTOSOL

The cytosol, containing molecules in aqueous solution, is the major reactive environment building up to 50% of the cellular volume. The cytoplasma on the other hand is defined as the total inner-cellular volume with the exception of the nucleus, that is, the cytosol and all associated organelles. Metabolic key-mechanisms are located within the cytosol – for instance glycolysis and major parts of gluconeogenesis, fatty acid biosynthesis, protein biosynthesis, and the pentose phosphate pathway.

MITOCHONDRIA: MPTP OPENING, FUSION, FISSION, MITOPHAGY, AND MITOBIOGENESIS

ATP is the major energy intermediate for all functions of organelles and organisms. A human produces nearly the same amount of ATP per day as its own bodyweight [1]. This impressive relation objectifies the central importance of the main source of ATP, the mt that produce about 90% of the cell's ATP. Overall, the cytosolic concentration of ATP remains stable at 3–4 mM, representing an amount of ~50 g ATP/body, a hydrolysis of 50 g ATP/min and thus the need for repetitive molecular ATP-hydrolysis and -synthesis up to 1000 times/day. The physiologic energy content of ATP is approximately 50 kJ/mol mainly accumulated within the anhydride junctions of the triphosphate group.

Mitochondria are 1–2 μm measuring organelles, which are subject of maternal heredity. The number of mt per cell differs depending on the cellular energy demands, the host's age, training status, metabolic deterioration, or genetic background. In general 1000–4500 mt can be found in a single cell. Unlike other cellular organelles, they possess two distinct membranes and a unique genome. During oxidative phosphorylation at the inner mitochondrial membrane, electrons are transferred from electron donors to electron acceptors until electrons are passed to oxygen, the terminal electron acceptor in the respiratory chain. The energy released by electrons flowing through the respiratory chain is utilized to transport protons across the inner mitochondrial membrane. In addition to supplying energy, mt are involved in reactive oxygen species (ROS) production, signal transduction, cell death, calcium handling, and cell growth. While the outer membrane has a smooth surface the inner membrane is folded and forms cristae and tubules. By this microanatomical structure four reaction spaces are formed: inner and outer mitochondrial membrane, intermembrane space, and the mitochondrial matrix. A characteristic of the inner mitochondrial membrane is the unique prevalence of cardiolipin, which is otherwise only to be found in bacteria. The mitochondrial DNA-pool (mtDNA), which is located within the mt matrix, organized as a unique ring from which up to 10 copies are present per mitochondrion. The human mitochondrial genome consists of 16.569 bp and encodes for 13 proteins (mainly as part of complexes of the respiratory chain), 22 tRNAs, and 2 rRNAs. Mitochondrial DNA is free of introns and the genome is encoded on the (+) as well as on the (−) strand as shown in Fig. 3.1. The close proximity of the mtDNA to the oxidative complexes of the respiratory chain result in high susceptibility for oxidative mtDNA-damage mainly by $OH^{\bullet-}$ radicals (see Fig. 3.4). Moreover missing DNA-repair-mechanisms and histones exhibit reduced protection against DNA-mutating irritation. Thus different mitochondrial genome mutations and damages can be found within a single cell or even mitochondrion, which is called heteroplasmy. Mitochondria encode for small mitochondrial ribosomes (28S- and 39S-subunits). However, the major part of mitochondrial proteins (~1500 proteins) is encoded within the nuclear genome. These proteins are synthesized within the cytosol and are subsequently imported into the mitochondrial matrix (Fig. 3.4). The mitochondrial protein import is aided by mitochondrial transport systems. TOM (translocase of the outer membrane) and TIM (translocase of the inner membrane) capture cytosolic proteins, which are inhibited to fold themselves by HSP70, which acts as a chaperon to an N-terminal signaling sequence. The transmembrane transport is partly driven by the negative charge of the mt matrix and positive charge of the proteins but mainly enabled by ATP hydrolysis. While the intermembrane space via the outer membrane is connected to the

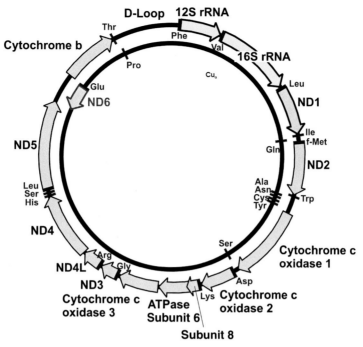

FIGURE 3.1 The mitochondrial human genome consist of entire rings each of those encoding for 37 genes on the *cis* and *trans* strand.

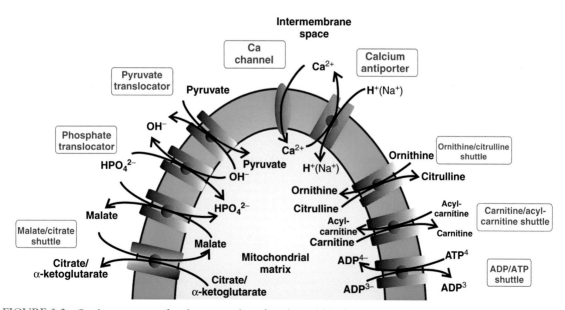

FIGURE 3.2 Carrier systems and series connection of carriers within the inner mitochondrial membrane.

cytosol by numerous porins that allow diffusion of smaller molecules (up to 10 kDa) and ions but cytochrome c, the inner mitochondrial membrane resembles a well-isolated barrier against the mitochondrial matrix (with the exception of water, oxygen, and carbon dioxide). For maintenance of a proton gradient between mitochondrial matrix and intermembrane space, an absolute impermeability of the inner mitochondrial membrane on the one hand and the transport of substrates and products through the membrane on the other hand is indispensable. This contradictory demand is solved by carrier systems (see Table 3.1 and Fig. 3.2). From those, mitochondrial carrier for anions, redox-equivalent transporters, and transporters for cations are distinguishable. Mitochondrial anion carriers catalyze transport via symport or antiport of anions, which can be paired by proton transport. Thus four classes of transport proteins are definable:

TABLE 3.1   Carrier Systems and Mitochondrial Transporters

Carrier	Substrate	Mechanism	Metabolic function	Distribution
ELECTROGENIC				
Adeninnucleatide carrier	ADP^{3-}/ATP^{4-}	Antiport	Energy transfer	
Aspartate/glutamate carrier	Asp/Glu	Antiport	Malate/aspartate cyclus, gluconeogenesis, urea synthesis	
Thermogenine	H^+	Uniport	Thermogenesis	Brown fat tissue
ELECTRONEUTRAL CARRIER, COMPENSATED BY PROTONS				
Phosphate carrier	Phosphate/H^+	Symport	Phosphate transfer	Ubiquitious
Pyruvate carrier	Pyruvate/H^+; ketone bodies/H^+	Symport	Krebs cycle, gluconeogenesis	Ubiquitious
Ornithine carrier	Ornithine, citrulline	Antiport	Urea synthesis	Liver
Branched-chain-a mino acid carrier	Amino acids/H^+	Symport	Degradation of amino acids	Skeletal muscle, myocardium
ELECTRONEUTRAL CARRIER				
Ketoglutarate/malate carrier	Ketoglutarate/malate, succinate	Antiport	Malate aspartate cyclus; gluconeogenesis	Ubiquitious
Dicarboxy late/phosphate carrier	Malate, succinate/phosphate	Antiport	Gluconeogenesis, urea synthesis	Liver
Citrate/malate carrier	Citrate/isocitrate, malate, succi nate, phosho enolpyruvate	Antiport	Lipogenesis, gluconeogenesis	Liver
Glycerin-3-phosphate/dihydroxyacetonphosphat	Glycerine-3-phosphate/dihydroxyacetonphosp	Antiport	Glycerine-3-phosphate cycle	Ubiquitious
Ornithine carrier	Ornithine citrulline	Antiport	Urea synthesis	Liver
NEUTRAL CARRIER				
Carnitine carrier	Carnitine/acyl-carnitine	Antiport	Fatty acid oxidation	Ubiquitious
Glutamine carrier	Glutamine	Uniport	Glutamine de gradation	Liver, kidney

Modified from Ref. [2].

1. *Electrogenic carrier*: substrates and electrical charge are transported, which is a secondary active transport driven by the proton gradient. For example, adeninnucleotid carrier, which is inhibitable by atractyloside.
2. *Nonelectrogenic, proton-compensated carrier*: these carrier symport anions and protons. For example, the phosphate-carrier and the pyruvate carrier, which is highly demanded during aerobic glycolysis.
3. *Nonelectrogenic exchange carrier*: these allow for the exchange of di- and tricarboxylates across the mitochondrial membranes thus connecting metabolic pathways in cytosol and mitochondrial matrix. For example, acetyl-derivates are transported by the aspartate malate carrier (see Fig. 3.3).
4. *Neutral carriers transport acids*: for instance carnitin is carrying fatty acids and glutamine is carried by the glutamine carrier into the mitochondrial matrix (see Fig. 3.3).

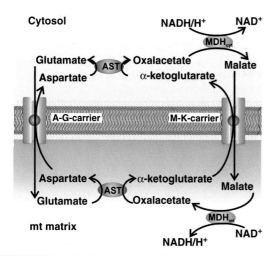

FIGURE 3.3 **Malate/aspartate shuttle system.** Transport of reducing equivalents from the cytosol into mitochondria. M-K-carrier, malate ketoglutarate carrier; A-G-carrier, aspartate glutamate carrier; AST, aspartate amino transferase; MDH, malate dehydrogenase; mt, mitochondrial; cyt, cytosol. *Modified from Ref. [2].*

Ca^{2+} transport is driven by the proton gradient through a highly selective Calcium channel. Very short and localized opening of the channel is induced by elevated cytosolic Ca^{2+} concentrations. The resulting elevation of mitochondrial Ca concentration is inducing activity of dehydrogenases and stimulating the metabolism. Fatty acids and pyruvate have to be transported into the mitochondrial matrix as well as ADP and phosphate while ATP has to be shuttled from the mitochondrial matrix into the cytosol. Several reductases (ETF-ubichinone-oxidoreductase) and dehydrogenases (glycerin-3-phosphate dehydrogenase) feed electrons via $FADH_2$. Multiple coacting shuttle-systems are located within the inner mt membrane (see Fig. 3.2). Electrons are transported by the malate-aspartate shuttle, which is carrying NADH-bound electrons into the mitochondrial matrix for processing within the respiratory chain. Elementary pathways are beta oxidation and Krebs cycle. Substrates and derivates need transport between mt matrix and cytosol. Furthermore, the mitochondrial matrix and the ER represent the main cellular calcium reservoirs. Calcium ions pass the membrane through Ca channels driven by an electrochemical gradient. Ca^{2+}/Na^+ or Ca^{2+}/H^+ antiporter exchange these ions at a constant rate. Energetic demands for these transport mechanisms are covered by electrochemical gradients generated by the respiratory chain. Under physiologic conditions, mt consume large amounts of oxygen to produce ATP at Complex V of the respiratory chain. The healthy, well-perfused heart thereby utilizes mainly fatty acid oxidation to meet its energy requirements. When the heart becomes hypoperfused and oxygen is lacking, electron flow along the respiratory complexes is inhibited and mitochondrial oxygen consumption as well as ATP production decrease [3–5]. During ischemia, glycolysis becomes the major source of ATP production. The mitochondrial matrix homes the pyruvate dehydrogenase complex, enzymes of the Krebs cycle, beta oxidation, and most enzymes of the urea cycle and heme-biosynthesis. These metabolic pathways directly feed electrons via NADH and FADH into the respiratory chain. Cytochrome c and adenylate cyclase (AC) are located within the intermembrane space. AC is recycling ATP by generating AMP. Oxidation of nutrients releases electrons, which are stepwise transferred to less-energetic acceptors thus enabling controlled energy release for ATP synthesis by generating a highly energetic phosphoric-acid anhydride binding. Terminal redox acceptor is oxygen thus defining the name of the process as oxidative phosphorylation (Oxphos). Hydrogen is transferred to the respiratory chain by NADH/H$^+$ and FADH2. While NADH is a reversibly binding cosubstrate, FADH2 serves as a cofactor to a group of enzymes. Central metabolic processors are the four respiratory chain complexes associated together by F1/F0 ATP synthase, which is often called Complex V as well (see Fig. 3.4). Traditional understanding defines separate complexes, but more recent knowledge assumes the existence of respi-

ratory super-complexes, so-called respirasomes. Transfer of electrons between complexes is provided by two mobile substrates, cytochrome c and ubiquinone. Cytochrome c carries one electron from Complexes III to IV by changing redox status of iron within the FE-Hemcore from FE^{3+} to Fe^{2+} and is associated to the outer surface of the inner mitochondrial membrane. Ubiquinone is a potential carrier of two electrons by reduction to hydroquinone. Direct transfer of protons is only to be found at the beginning of the respiratory chain, while further transfer corresponds to changes of electric charge but not to mechanistic hydrogen transfer. However, ubichinon is reduced by several specific dehydrogenases (NADH ubiquinone oxidoreductase (Complex I), Succinate ubiquinone oxidoreductase (Complex II), electron-transferring flavoprotein (ETF): ubiquinone oxidoreductase and glycerophosphate ubiquinone oxidoreductase).

NADH ubiquinone oxidoreductase (Complex I) complex (1000 kDa) consists of 45 subunits, 7 from those encoded by the mitochondrial

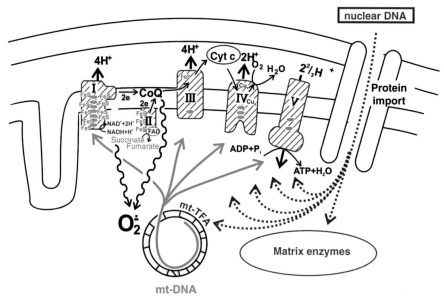

FIGURE 3.4 **Mitochondrial respiratory chain.** Assembly of complexes by nuclear and mitochondrial encoded proteins. Electron flux and generation of a proton gradient. Oxidative damage by ROS derived from the respiratory chain.

genome. The coenzyme FMN transfers electrons from NADH to intermediate FeS complexes and finally to ubiquinone, thus transferring $2H^+$ into the intermembrane space. Further four protons/ NADH are transferred by a yet unknown mechanism. Rotenone, which is a chinone analog derived from Leguminoses and also high concentrations of barbiturates inhibit CI. Cardiac dysfunction originating from ischemia–reperfusion injury, afterload-induced cardiac dysfunction, obesity, and aging per se are associated with Complex I dysfunction. Therapeutic intervention by training or mechanisms inducible by caloric restriction might be hopeful targets for metabolic re-remodeling.

Succinate ubiquinone oxidoreductase (Complex II) is composed from four subunits totally encoded within the nuclear genome. Two hydrophilic subunits assemble succinate dehydrogenase. Thus Complex II is directly reduced within the Krebs cycle and reduces Ubiquinone without binding protons. ETF – ubiquinone oxidoreductase and glycerin-3-phosphate: ubiquinone oxidoreductase – are further valuable bypasses for delivering electrons to the respiratory chain. ETF is reduced by acetyl-CoA-dehydrogenase from the beta oxidation and the reduced flavin is oxidized by flavin ubiquinone oxidoreductase thus reducing ubiquinone. G3P-UR channels cytoplasmic reduction equivalents into the respiratory chain.

Cytochrome bc1 complex (Complex III) consists of 11 subunits, from which solely cytochrome b is encoded by the mitochondrial genome. The electron transport is realized by the ubiquinone cycle (Q-cycle, QC). During QC, neutralization of electric charge enables translocation of protons without binding or "pumping." Shortly, this mechanism relies on the generation of an "energetic seesaw" by the Rieske complex and intermittent reduction of Q forming a highly reduced ubisemiquinone, which is able to reduce cytochrome b. In mammals reoxidation of Q is the only feasible Complex III activity. Each oxidation of QH_2 releases four protons, two from

those are transferred into the intermembrane space. The excess of two protons are counter balanced by electron-uptake in cytochrome c.

Cytochrome c oxidase (Complex IV) oxidizes cytochrome c by transferring electrons to oxygen, thus forming oxygen radicals and releasing four protons per O_2. In mammals, Complex IV consists of 13 subunits, 3 from those encoded by the mitochondrial genome. The cytochrome c binding site is formed by a "double-cored" copper centrum (CuA). The proton transport mechanism so far is less well understood. However, two proton channels have been identified and the double-cored copper centrum seems to drive transport. Cytochrome c oxidase is inhibitable by several compounds displaying high similarity with oxygen. For example, cyanide, carbon monoxide, azide, and azotic monoxide are known potent inhibitors.

F1/F0-ATP synthase (also called Complex V) is catalyzing ATP-genesis from ADP and P_i. ATP synthase consists of 16 subunits, 2 from those encoded by the mitochondrial genome. A transmembranous FO-part is connected to an F1 part, which is protruding in the mitochondrial matrix. Protein A is binding the F1/F0 inhibitor oligomycin ("O") thus being eponymic to the complex. The F1 part is a heterohexamer. A pedicle formed by a homodimer of protein B and a central pedicle formed by protein γ connect FO and F1 and control crossrotation of the subunits. The rotation is driven by mitochondrial proton gradient and drives ATP-generation by conformational variation of three F1 catalytic centers. Each of those binds ADP and P_i when in state "L" (loose) conformation and exhibits low affinity when in state "O" (open) conformation. The intermediate state "T" (tight) catalyzes phosphorylation and the formation of the highly energetic anhydride binding. The energy for ATP release and change from state T to state O is generated during rotation by protein γ. One complete rotation of the F1 subunit provides production of three ATP. Because F1 is a homomultimer of eight protein c, and eight protons are needed for a 360° shift, production of a single ATP requires a reflux of 2,7 protons into the mt matrix. Energetic output is a gradient of 10 protons per $NADH/H^+$ (Complexes I–IV) and 6 protons per succinate (Complexes II–IV). P/O ratio describes how many ATP can be produced per oxygen. For $NADH_2$-dependent respiration P/O is 2,7 ATP/O_2, while succinate-dependent respiration exhibits a P/O of 1,6 ATP/O_2 when assuming energy-consuming transmembrane transport mechanisms. Due to membrane-leaks and proton-fluxes these values might be a theoretical maximum capacity, which is not achievable *in vivo*.

As the functional status of the respiratory chain depends on the availability of substrates and integrity of complexes, lack of ADP or Pi reduces respiratory flow. This phenomenon is called "respiratory control" or "regulation of oxidative phosphorylation." Dynamic equilibria are known as steady states and have been defined in five classes by Britton Chance in 1956. Flux in respiration is controlled by availability of different substrates as shown in Table 3.2. During state 3 the respiratory chain works at maximum activity, depending on the transmembrane potential. State 4 also termed as "controlled state" exhibits limited function of respiratory chain due to ADP deprivation. Each state of dysfunction may lead to ineffective control and redox injury. Uncoupling the respiratory chain

by protonophores disturbs respiratory control. Since passive proton reflux into the matrix abolishes mt-membrane-potential the respiratory chain is working unimpeded as long as substrates are available. Lipophilic acids, atractyloside, dinitrophenol, carbonylcyanid-*m*-chlorophenyl-hydrazone (CCCP) and carbonylcyanid-*p*-trifluoromethoxyphenylhydrazon (FCCP) act as protonophores and are used for diagnostic procedures. While unregulated uncoupling leads to a complete breakdown of cellular energy supply and thereafter, cellular death, controlled uncoupling is needed to survive. Uncoupling proteins, for example UCP2, directly reduce transmembrane potential in states of reduced electron flux to avoid redox-associated injury to proteome, genome, or lipidome of mt and the entire cell.

Superoxide radicals mainly result from reactions and/or disturbances of function at Complex I and Complex III. The transfer of electrons itself results in an unstable intermediate semiquinone within the Q cycle. $^\bullet Q^-$ transfers electrons to molecular oxygen, forms superoxide radicals in a nonenzymatic rate and results in a ROS-production proportional to metabolic rate. Mitochondrial superoxide dismutase (MnSOD) converts the superoxide anion into hydrogen peroxide, which can be detoxified by catalase (CAT) and glutathione peroxidase (GPx) within mitochondria. Disturbances or knockouts of CAT abolish live-extending potency of caloric restriction, accumulate oxidative impairment of mitochondrial proteome, genome, and lipidome, facilitate mitochondrial deterioration and dysfunction, and lead to premature aging and death [6–12].

Regulation of the metabolic chain and oxidative phosphorylation in front of ROS-endangerment is vital to cells. Acetylation normally inhibits activity of metabolic enzymes by neutralizing positive charges in ε-aminogroups of lysine residues. Enzymes of the Krebs cycle, beta oxidation, and other nuclear-encoded subunits of the electron transport chain complexes, especially Complexes I and II are acetylated. While no acetyltransferases have been

TABLE 3.2   Dynamic Equilibria Known as Steady States Defined in Five Classes by Britton Chance in 1956

States	Excess	Limiting factor
1	O_2	ADP and substrate
2	O_2, ADP	Substrate
3	O_2, ADP, substrate	$\Delta\mu H$
4	O_2, substrate	ADP
5	ADP, substrate	O_2
Uncoupled	O_2, substrate	Maximum flux capacity

identified in mammalian mitochondria, several class III histone deacetylases have been found. From the so-called sirtuins (homologs of the yeast Sir2 gene) isoforms SIRT3, SIRT4, and SIRT5 are located in mt and SIRT3 has been shown to account for major mitochondrial regulation. Mitochondrial SIRTs are essential regulators of cell survival. Activity of sirtuins depends on NAD^+ as a cosubstrate in deacetylation and ADP-ribosylation. For example, Complex I subunit NDUFB8, an ATP-regulator protein, is acetylated by SIRT3 [13–21].

Apart from ATP, mt also generate ROS [22–24]. ROS within mt originate from different sources, one of them is the respiratory chain, mainly Complexes I and III. Here, electrons from the electron transport chain can be transferred to oxygen, which results in the formation of superoxide anions. High concentrations of mitochondrial ROS – together with other factors – facilitate opening of the MPTP, a large conductance pore in the inner mitochondrial membrane most likely build up by dimerization of the mitochondrial ATP synthase [25]. MPTP opening enhances the inner mitochondrial membrane permeability to solutes with molecular weights up to 1.5 kDa and therefore leads to mitochondrial depolarization and subsequently to ATP depletion. Mitochondrial matrix volume increases and induces rupture of the outer mitochondrial membrane causing loss of pyridine nucleotides and contributes to the inhibition of electron flow along the protein complexes of the electron transport chain. Rupture of the outer membrane also releases cytochrome c thereby initiating apoptotic cell death [26–28].

Many cardiovascular diseases impact mitochondrial morphology and the inter mitochondrial network, that is, modify fusion and fission of mitochondria. Mitochondrial fusion proteins include the outer membrane proteins Mfn 1 and 2 and the inner membrane protein optic atrophy protein (Opa) 1. Mitochondrial fission in mammalian cells is mediated by a large GTPase, the dynamin-related protein 1 (Drp1), along with other proteins such as the outer mitochondrial fission protein Fis1 and the mainly cytoplasmic endophilin B1/Bif-1. During division of mt dynamin-related protein 1 translocates to the outer mitochondrial membrane and interacts with Fis1 to promote mitochondrial fission (review in Refs [29–34]). Mitochondrial fusion and fission are constantly ongoing processes, which are essential for the maintenance of normal mitochondrial function. Mitochondrial fusion serves as a prosurvival mechanism and content/protein exchange might help to overcome local functional deficiencies such as mtDNA mutations within the mitochondrial network [35,36]. Accordingly, inhibition of fusion results in an accumulation of mtDNA mutations triggering mitochondrial dysfunction, the loss of the mitochondrial genome, and finally organ dysfunction [36,37].

Interestingly, cells have developed a defense mechanism against aberrant mt that can cause harm, in that, selective sequestration and subsequent degradation of the dysfunctional mt are initiated before they can cause activation of the cell death machinery [38]. Components required for this organelle-specific type of autophagy, which is also called mitophagy, include the mitochondrial kinase PINK1, the E3 ubiquitin ligase Parkin, the atypical Bcl-2 homology domain 3-only (BH3-only) proteins Bnip3 and Nix, and several autophagy-related genes (Atg) [39]. Mitophagy was reported to be increased in cells harboring dysfunctional mt [40–42], while cells lacking the essential autophagy component Atg5 show accumulation of damaged mt, altered mitochondrial morphology [43] and interrupt autophagy [44]. The elimination of damaged mt by mitophagy is cardioprotective [45–47].

If depletion of mt is not followed by an appropriate induction of mitochondrial biogenesis, this will have detrimental consequences for the heart as well. Mitochondrial biogenesis requires the coordination of the nuclear and the mitochondrial genome and involves changes in the expression of more than 1000 genes (reviewed in Ref. [48]). Transcription factors involved in this process include the nuclear respiratory factors (NRF-1/NRF-2), the mitochondrial transcription factor A (Tfam), the estrogen-related receptors alpha and gamma, ubiquitous transcription factors, and the transcriptional coactivators PGC-1α or PRC [49]. Many of the pathways regulating mitochondrial biogenesis seem to converge at the transcriptional coactivator PGC-1α, which has been shown to directly dock on some of these transcription factors and modulate their activity [50–53]. The hearts of PGC-1α knockout mice [54,55] demonstrate reduced oxidative capacity and mitochondrial gene expression but normal mitochondrial volume density, suggesting additional mechanisms controlling cardiac mitochondrial biogenesis. Although cardiac dysfunction under basal conditions is moderate in PGC-1α knockout mice [54,55], aortic constriction leads to accelerated heart failure [56]. PGC-1α overexpression on the other hand causes uncontrolled mitochondrial proliferation and loss of sarcomeric structure, finally leading to dilated cardiomyopathy and premature death [57]. Thus, a well-balanced and tightly controlled change in the expression of PGC-1α appears to be necessary to maintain optimal mitochondrial and cardiac function.

PEROXISOMES

Fatty acid beta oxidation, fatty acid alpha oxidation, glyoxylate detoxification, and ether phospholipid biosynthesis are performed in peroxisomes. Metabolic diseases, like the Zellweger syndrome, are relying on peroxisomal metabolic impairment and exhibit severe clinical phenotypes. Peroxisomes are energy consuming and do not possess a respiratory chain or Krebs cycle, thus interorganelle cross-talk mechanisms are essential. However products of metabolism generated in peroxisomes are transported to different organelles by vesicular transport and diffusion as well. Peroxisomes, identified in 1954 by Rhodin, are single-membrane-bounded organelles. However, now, a broad heterogeneity of these has been identified between species, individuals, and organs. Within peroxisomes, high concentrations of catalase and hydrogen peroxide producing enzymes may be found – as nomenclature suggests, but otherwise some species express peroxisomes lacking catalase, therefore expressing predominant components of glycolytic pathways. Currently peroxisomes are understood to be organelles derived from the ER. Under physiologic conditions, beta oxidation of short, medium, and long chain fatty acids is predominantly handled by the mitochondrial beta oxidation. Nevertheless, peroxisomes exhibit beta oxidation as well, which is specialized for a number of metabolites. These comprise very long chain fatty acids (VLCFA i.e., C22:0, C24:0, C26:0) that cannot be oxidized by mitochondria. Peroxisomal beta oxidation products have to be shuttled to mt for further oxidation and ATP generation. Moreover, participation of so-called redox shuttles is urgently needed to enable oxidation of NADH for regeneration of the peroxysomal NAD^+ pool. However, pristanic acid, di- and trihydroxycholestanoic acid (DHCA and THCA; precursors of cholic acids), tetracosanoic acid (C24:6n-3; DHA-formation precursor) and long chain dicarboxylic acids are metabolized by peroxisomal beta oxidation. In general, peroxisomal beta oxidation exhibits widely identical structures compared to mt but differs regarding the first metabolic step, which is processed by isoforms of the flavoprotein acyl-CoA oxidase (ACOX1 and 2). 3-Methyl branched fatty acids cannot undergo direct beta oxidation because of the methyl group at position 3 (phytanic acid, for example). Removal of this carbon atom is conducted by alpha oxidation in peroxisomes generating a 2-methyl fatty acid, which is prone to undergo beta oxidation subsequently. Production of estherphospholipids in peroxisomes is catalyzed by alkyl-DHAP synthase, which is exclusively localized in peroxisomes. Last but not the least, peroxisomes catalyze glyoxylate detoxification by the enzyme alanine glyoxylate aminotransferase (AGT). Transfer of electrons resulting from various metabolic pathways in peroxisomes results in hydrogen peroxide generation. Different peroxisomal enzymes such as acyl-CoA oxidase, xanthin oxidase, urate oxidase, D-aspartate oxidase are a source of ROS (H_2O_2, $O_2^{\bullet -}$, $OH^\bullet$, $NO^\bullet$) within the cell.

ENDOPLASMIC/SARCOPLASMIC RETICULUM

The ER represents a network organelle, which is found in all eukaryotic cells erythrocytes. Flattened sac-like vesicular tubes called cisternae form stacks, which are stabilized by the cytoskeleton. The monolayer membranes of the ER are continuously connected with the nuclear membrane. Microscopically, a rough (RER) and a smooth ER (SER) can be distinguished. In general, function of the ER is synthesis of membrane proteins and secretory proteins as well as membrane lipid synthesis. Protein folding is performed with assistance of ER chaperones including BiP/GRP78 and calnexin. Correctly folded proteins are transported to different districts of the endomembrane system, preferably to the Golgi, instead defective proteins are degraded by ER-associated degradation (ERAD). Especially, proteins transformed without availability of chaperones often underly malfolding and/or defective translation and thus have to be subjected to degradation. Otherwise accumulation of these defective proteins would ultimately cause ER stress and induction of apoptosis (ATF6, IRE1, and PERK pathways) [58]. Nevertheless, the ER protein synthesis and export system handles a wide variation of heterogeneous proteins differing in size, topology, solubility, state of aggregation, and final destination. The RER carries ribosomes bound on its surface, which are responsible for protein biosynthesis depending on the synthetic demands of the cell, further maturation, posttranslational modification, and storage are located within the ER. Mainly cells with highly active protein-output, that is, for example hepatocytes, express predominant RER. However, SER lacks ribosomes and is predominantly active in lipid metabolism, storage carbohydrate metabolism and detoxification via the cytochrome P450 system.

Interaction of the ER with the Golgi is achieved by bidirectional vesicular transport, regulated by targeting proteins within these vesicles toward the Golgi (directed by COPII-protein) or from the Golgi to the RER (directed by COPI protein). Furthermore, direct membrane contact sites enable exchange with organelles, e.g., with mitochondria.

Key functions of the RER may be understood as production and storage of signal peptides with or without targeting tag, glycosylation of proteins, producing and embedding membrane proteins, and production of lysosomal proteins.

SER, on the other hand, has predominantly metabolic function and is the production site of steroids, lipids, and phospholipids. Thus SER has a central role in receptor attachment, steroid-metabolism, and carbohydrate metabolism. Gluconeogenesis is dependent on SER function as glucose-6-phosphatase, the key enzyme of gluconeogenesis is located here.

Disturbances and stressors can lead to an ER stress response resulting in slow biogenesis of proteins and increased amount of unfolded proteins. Potential stressors are hypoxia/ischemia, aging, obesity and insulin resistance, viral infections, disturbed redox regulation, and glucose deprivation.

A special "isoform" of the SER, exclusively found in (cardio-)myocytes is the sarcoplasmic reticulum (SPR). SPR exhibits a specialized subset of marker proteins in its membrane, which enable the organelle for Ca^{2+} storage and regulation of Ca^{2+} homeostasis; thus, giving special importance to functional aspects of contraction and relaxation of these cells as Ca^{2+} is released from the SPR by Ca^{2+}-dependent mechanisms when myocytes are excited.

Two types of Ca channels, inositol triphosphate (IP3)-receptors and ryanodine receptors are expressed on the SPR. IP3 is a second messenger derived by cleavage from phosphatidylinositol 4,5-bisphosphate by G-protein-induced peptidase C activity. Ryanodine-receptors on the other hand are "calcium-activated Ca channels", which are responsible for calcium-induced calcium release (CICR), which has predominant importance in cardiac myocytes. Contraction is initiated by Ca entry through L-type Ca channels in the cytosol triggering CICR. During diastole Ca^{2+} is transported into the SPR by a Ca^{2+}-ATPase (SERCA) from which isoforms 1–3 have been described while cardiac tissue predominantly expresses SERCA 2a. SERCA activity is regulated by phosphorylation through its inhibitory protein phospholamban (PLB). Phosphorylation of PLB disrupts the inhibitory interaction with SERCA resulting in increased Ca^{2+} transport into the SR. Furthermore, metabolic deterioration and cardiometabolic disease leading to hypocontractility display reduced SERCA activity. Thus deterioration of Ca signaling might be – among others – "one" underlying mechanism of diabetic cardiomyopathy.

RIBOSOMES AND METABOLISM-REGULATED PROTEIN SYNTHESIS

Protein and rRNA synthesis are fundamental for cellular integrity, growth, and survival. Tight control of protein synthesis but also of RNA synthesis is essential. The processes are coupled to energetic cellular state through the availability of nutrients, growth factors, and ATP. Critical energetic sensors, mediating activation of growth factors, transcription, and translation are the PI3K/AKT/mTORC1/RAS/RAF/ERK pathways, and MYC transcription factor. Glucose, which is imported by glucose transporters GLUT 1–5, is the main energy source in a majority of proliferating

cells [59,60]. Inhibition of glucose metabolism increases the ADP/ATP and AMP/ATP ratio. AMP-dependent protein kinase (AMPK), which is a heterotrimeric complex consisting of one catalytic (α) and two regulatory (β and γ) subunits, responds to AMP/ATP elevation and is activated by upstream kinases, the tumor suppressor liver kinase, the calcium/calmodulin-dependent protein kinase kinase β, or transforming growth factor β-activated kinase-1. Activated AMPK inhibits mTORC1 by direct phosphorylation of TSC2 or Raptor. AMPK-mediated inhibition of mTORC1 silences rDNA transcription and ribosome biogenesis thus conserving energy expenditure [61–69].

Ribosomes are small ribonucleoprotein particles that are found within the cytosol as free structures or attached to the RSR, which are sites of translation of mRNA into peptides and proteins. Even though ribosomes are found in all cells, differences in structure and size exist between eukaryotes and bacteria. Mitochondrial ribosomes exhibit structural similarity compared to bacteria. Ribosomal subunits are characterized regarding their aggregation in units of Svedberg. Eukaryotic ribosomes (25–30 nm) with 80 S size are complexes from a bigger 60 S (50 proteins; 2800 kDa) and a smaller 40 S (33 proteins; 1400 kDa) subunit. These proteins are heterodimerized together with rRNAs of 28 S or 18 S, respectively. Bacterial ribosomes (70 S) are slightly smaller (20 nm) and exhibit a bigger 50 S (34 proteins; 1600 kDa) and a smaller 30 S (21 proteins; 900 kDa) subunit. Mitochondrial ribosomes show wide similarities compared to bacteria and exhibit 55 S ribosomes, composed by a 39 S and 28 S subunit [70].

Mammalian mitochondrial ribosomes are encoded by the nuclear genome and imported into the mitochondria. Here they are assembled with mitochondrial encoded 12 S and 16 S rRNAs [71,72]. Translation is subdivided into initiation, elongation, and termination. Reaction sites of ribosomes are solely formed by rRNAs and 60% of total ribosomal mass is derived from rRNAs.

The two rRNAs form characteristic tRNA binding sites, which are – according to their functional activity within the ribosome called A-site (from *A*-minoacyl-tRNA; "uptake-site"), P-site (from *P*-eptidyle tRNA; "synthetic site"), and E-site (from *E*-xit-site; "output-site"). All of those three sites are formed by the 16/18 S and 23/28 S ribosomal subdomain. The catalytic domain, the so-called peptidyl transferase centrum is formed around the A- and P-site, which is an accelerating reaction of two amino acids by optimization of steric alignment. rRNAs interacting with this site are called ribozymes because of their "enzyme-like" catalytic function.

A single mRNA can be translated by repetitive Ribosomes at the same time. These poly-ribosomal mRNA complexes are called polysomes. Elongation of the resulting peptide or protein is promoted by elongation factors (EFs). The binding of aminoacyl-tRNAs at the A-site is promoted through interaction with eEF1A, which induces conformational changes that result from induced fit and minor-motif interaction. These interactions allow a binding only if correct aminoacyl-tRNAs complementary to the translated mRNA are recruited within the A-site. Elongation factor eEF2 promotes translocation of the mRNA–tRNA complex from A- and P-sites into the P- and E-sites. eEF2-GTP is stabilizing this hybrid states and GTP hydrolysis leads to mRNA advancement within the ribosome. eEF2 exhibits a modified histidin-residual, which is called diphtamide. This can be ADP-ribolysed by diphtheria toxin, thus inactivating translation. The ascending peptide chain is leaving the ribosome through an outlet passage formed by proteins, which is a structural target for many antibiotics for selection between bacteria and eukaryotes.

Golgi Apparatus

The Golgi apparatus is named after its first descriptor Camillo Golgi. It is found in almost all cells as endomembranous organelle with close relation to the ER. However, compared to the ER the system of microtubule and cisternae is highly organized and builds a stack-like structure, which is subdivided into a *cis*-, medial-, and *trans*-compartment thus defining a *cis*-Golgi network (CGN) and a *trans*-Golgi network (TGN). *cis*- and *trans*-face of each cisternae have a well-defined morphology and biochemical characterization. Neighboring cisternae are individually separated and transport in-between those driven by vesicles. The clear structure of the Golgi enables enzymatic compartmentalization, thus enabling stepwise modification (by phosphorylation, sulfatation, glycosylation, acetylation, and proteolytic cleavage) thus maintaining consecutive and selective processing steps of proteins. Mammalian cells contain Golgis with 35–100 stacks, which have central importance as processors of proteins being modified for secretion, membranous localization, or lysosomal function. Mature proteins are packed into membrane-bound vesicles sprouting from the *trans* surface. Thus, Golgi represents a point of intersection of mechanisms, which involve endocytotic and lysosomal processes but also secretion of proteins. Lysosomal vesicles exhibit specific importance as their content reaches a pH of 4.5. These vesicles are coated by proteoglycans. Lysosomal proteins are tagged by mannose-6-phosphate, which is bound by m6p receptors targeting these restricting enzymes to exclusively lysosomes [58]. Different Golgi stress responses have been identified so far. The TFE3 pathway, resulting in transcriptional activation of Golgi structural proteins (GCP60, GM130, and giantin), glycosylation enzymes (sialyltransferase 4A and fucosyltransferase 1), and a vesicular transport component (syntaxin 3A) upon Golgi stress [58]. The HSP47 pathway protects cells from Golgi stress-induced apoptosis [73]. ER-stress responsive CREB3/Luman, as a part of the CREB3-ARF4 pathway of the mammalian Golgi stress response, resides in the ER membrane as a transmembrane protein. ARF4 is located in the Golgi-membrane and regulates vesicular ER-Golgi transport by COPI. CREB I is activated by proteolysis upon stress, translocates to the nucleus and activates expression of ARF4 resulting in Golgi stress-induced apoptosis and disruption of the Golgi. The functional compartmentalization of the Golgi determines targeted activity of stress responsive pathways on the *cis* or *trans* face of the Golgi [58,74].

Transport System and Vesicle Pools

Directed and regulated transport between different compartments of the endomembrane system and organelles of a cell, but also endo- and exocytosis are operated through coated vesicles, which underlie a steady circle of budding and fusion reactions at endomembranes of organelles and the cellular plasma membrane. Vesicles are formed by strictly regulated self-assembly of proteins, budding, and finally scission from the membrane in a GTP-dependent manner. Budding is an energy consuming process that occurs on the surface of membranes when cytoplasmic coat protein (COP) complexes assemble on the membrane surface. Within eukaryotes COPI buds vesicles from the Golgi, COPII functions at the ER, and the clathrin/adaptin system is operating at plasma membranes and endosomes. Taken together, at present three main coats are known: (1) COPI, (2) COPII, and (3) clathrin/adaptin.

While all of them exhibit similar function and comparable ancestry, severe differences exist regarding structure and assembly to the target and of the load concentrated within the specialized vesicle [75–77]. COPI vesicles may be understood as "Golgi vesicles." They shuttle within the Golgi and from the Golgi back to the ER. COPII vesicles on the other hand export proteins from the ER to their target. Third, clathrin-coated vesicles (CCVs) provide the late secretory pathway and the endocytic pathway at the plasma membrane of cells.

Generally, vesicle forming is an energy consuming process and participating proteins can be functionally divided into two subgroups:

first, adaptor-proteins and second, cage forming, that is, "coating" proteins. Vesicle formation is tightly regulated by GTP-binding proteins and activated by a GTPase that is stimulated by guanine exchange factors. The protein complex is anchored to the membrane by an amphipathic alpha-helix, recruits coat-proteins to form a complex, which further interacts with recognition and targeting sequences of possible cargo structures. Thus, specific cargo can be concentrated within a vesicle.

Generally, formation of these vesicle-forming protein-complexes differs between the three mentioned coats and can be observed (1) at the endomembrane itself or (2) within the cytosol, which leads to subsequent recruitment to the endomembrane and following vesicle formation.

During COPI-guided vesicle formation adaptor and cage-forming complexes are associated as a single heptameric complex, which is recruited afterwards to the membrane. Briefly, an ARF family G protein and coatomer, a 550 kDa cytoplasmic complex of seven COPs: α-, β-, β'-, γ-, δ-, ϵ-, and ζ-COP is formed and is then recruited to the membrane as a solid complex. By exchanging GDP for GTP on ARF (by guanine nucleotide exchange factor; GEF), budding is initiated. The early vesicle recruits cargo through coatamer-ARF-GTP interaction and forms spherical cages, which are coated by COPI thereafter [78–80].

During vesicle formation by COPII and clathrin, the adaptor proteins are first bound to the membrane. Subsequently, cage complexes are polymerized to form the coated vesicle. The adaptor complexes consist of AP1–5, AP180, and the Golgi-localizing, γ-adaptin ear containing, ARF-binding (GGA) proteins for clathrin, and Sec23–24, for COPII [77]. In contrast, endocytosis does not require GTPases but is initiated by phoshoinisitol-dependent recruitment of AP2 adaptor complexes to the membrane [81].

Lately, vesicular transport originating from the mt has been identified. Small vesicles derived from mt (MDVs) carry outer mitochondrial membrane protein MAPL to peroxisomes and furthermore, subpopulations of MDV targeted to the endosome and Golgi have been identified as well [82–84]. Thus, besides direct contact between mt and other organelles to interact and to transfer ions, metabolites and proteins, vesicular transport facilitates directional transfer [85,86]. MDV are sized between 70–150 nm. The "budding" and scission is independent from mitochondrial fission protein DrpI. MDV directed to lysosomes and Golgi have been shown to be enriched with oxidized proteins, the purpose of vesicles delivering to peroxisomes is not understood so far [83,87]. Based on the stress induced to mitochondria, selective incorporation of cargo into MDV is distinguishable. ROS originating from metabolic stress by xanthine oxidase/xanthine induce genesis of MDVs carrying the outer membrane pore voltage-dependent anion channel (VDAC). Otherwise oxidative stress resulting from dysfunction of the respiratory chain has been shown to lead to MDVs carrying an oxidized Complex III subunit (core2) without VDAC enrichment [88]. However, attribution of cargo, aggregation, or oligomerization in MDVs seems to depend on the structure damaged or oxidized first and may affect each mitochondrial structure. The generation of MDVs destined for lysosomes requires the protein kinase PINK1 and the cytosolic ubiquitin E3 ligase Parkin [89]. More recent data suggest that MDVs may be a first-line defense mechanisms to mt enabling export of damaged proteins to avoid mitochondrial dysfunction or failure without activation of autophagy [84].

CELL DEATH: NECROSIS, APOPTOSIS, NECROPTOSIS

Cell death may be defined as an irreversible loss of plasma membrane integrity [90]. A number of different types of cell death can be distinguished including programmed cell death, which is mediated by a highly regulated intracellular program, or necrosis, which occurs as a result of a cellular injury induced by toxins, mechanical trauma, heat, or infections. Apoptosis and autophagic cell death are both independent forms of programmed cell death. In addition, a novel form of necrosis, called necroptosis, has recently been described as an alternate form of programmed cell death.

Apoptosis

The term "apoptosis" has been introduced by Kerr et al. to describe a mechanism of controlled cell deletion characterized by specific morphologic aspects [91]. These typical features include cellular shrinkage, condensation and margination of nuclear chromatin, DNA fragmentation, and cytoplasmic vacuolization. Apoptosis results in the controlled cellular breakdown into apoptotic bodies, which are recognized and engulfed by surrounding cells and phagocytes. Apoptosis does not trigger an inflammatory response and depends on the sequential activation of a family of cysteine proteases, the caspases. Apoptotic caspases can be further subdivided into initiator (caspase-2, -8, -9, -10) and executioner (caspase-3, -6, -7) caspases. Apoptosis can be triggered through two major different biochemical routes: the "intrinsic" pathway or the "extrinsic" pathway, which result in the activation of specific caspases but differ in the primary contribution of mitochondria. The intrinsic pathway acts through mitochondria, and is controlled by the Bcl-2 family of proteins. The antiapoptotic Bcl-2 family members, which reside in the outer mitochondrial membrane, maintain mitochondrial integrity by preventing the proapoptotic Bcl-2 family members such as Bax and Bak from causing mitochondrial damage. During cytotoxic insults, these antiapoptotic effects are antagonized and Bax and/or Bak are oligomerized, resulting in the formation of a channel through which mitochondrial cytochrome c is released into the cytosol. Upon release from mitochondria, cytochrome c binds with Apaf-1 and ATP to form an activation complex with caspase-9, called the apoptosome. Thus, activated caspase-9 cleaves and activates the executioner caspases-3, -6, and -7, which are crucial for the execution of apoptotic cell death. The extrinsic pathway of apoptosis on the other hand involves stimulation of death receptors belonging to the TNFR family such as TNF receptor-1 (TNFR1), CD95 (also called Fas and APO-1), death receptor 3 (DR3), TNF-related apoptosis-inducing ligand receptor-1 (TRAIL-R1; also called DR4), and TRAIL-R2 (also called DR5) by their respective ligands TNF-alpha, Fas ligand (FasL), or TRAIL. FasL induces the formation of a death-inducing signaling complex (DISC), which contains the Fas-associated death domain (FADD), and the initiator caspases-8 and/or -10. Binding of TNF-alpha to TNFR1 leads to the sequential formation of two complexes. Complex I consists of TNFR1, TNFR-associated death domain (TRADD), TRAF2, RIP1, cIAP1, and cIAP. Endocytosis of TNFR1 is followed by the formation of Complex II, which includes TRADD, FADD, and caspase-8 and/or -10. Activation of caspase-8 and -10 leads to activation of the downstream executioner caspases-3, -6, and -7. Caspase-8 also activates the intrinsic apoptotic pathway through cleavage of BID thus amplifying the death receptor-induced cell death program.

Autophagic Cell Death

Although physiologic levels of autophagy are essential for the maintenance of cellular homeostasis and to promote survival, excessive levels of autophagy are able to induce autophagic cell death. Autophagic cell death is morphologically defined as a type of cell death that occurs in the presence of massive autophagic vacuolization of the cytoplasm but in the absence of chromatin condensation and without association to phagocytes [92,93]. Caspase activation and DNA fragmentation occur very late (if at all) in autophagic

cell death. A prominent morphologic change observed in this type of cell death is the appearance of double-membrane-enclosed vesicles that engulf portions of cytoplasm or organelles such as mitochondria. These vesicles fuse with lysosomes and deliver their content for degradation by lysosomal enzymes of the same cell. The finding that inhibition of autophagic activity by knocking down autophagic proteins such as Atg5, Atg7, and Beclin-1 attenuates cell death and supports the existence of a nonapoptotic, caspase-independent type of programmed cell death. These Atg genes (autophagy-related genes) are involved in vesicle nucleation and expansion, autophagic vesicle fusion to late endosome/lysosome, and degradation of sequestered cytoplasmic material. However, there has also been a controversy on the existence of autophagic cell death as a novel, independent pattern of cell death. Some studies suggested that autophagy was simply present in dying cells while others, including studies utilizing genetic inactivation of autophagic genes, suggested that autophagy had actually caused cell death.

Necrosis

Necrosis was long regarded as an unprogrammed death of cells and living tissues, which does not follow a highly regulated intracellular program such as the apoptotic signal transduction pathway. It has therefore often been defined in a negative manner as death lacking the characteristics of apoptosis or autophagic cell death. Indeed, unlike apoptosis, necrotic cell death is not the result of one or two well-described signaling cascades but is the consequence of an extensive cross talk between various molecular events. Six characteristic morphologic patterns of necrosis are distinguished in pathology: coagulative necrosis, caseous necrosis, liquefactive necrosis, fat necrosis, fibrinoid necrosis, and gangrenous necrosis. Necrosis typically results in a loss of cell membrane integrity and an uncontrolled release of products of cell death into the extracellular space, which initiates an inflammatory response. Typical morphologic features of necrotic cell death, which occur in most but not in all eukaryotic cells, include mitochondrial swelling, lysosome rupture, and plasma membrane rupture. While apoptotic cells are taken up completely by phagocytes, necrotic cells are internalized by a macropinocytotic mechanism, meaning that only parts of the cell are taken up by phagocytes. Necrosis can be induced by injury, infection, heat, cancer, infarction, toxins, and inflammation. Necrosis is accompanied by the release of special enzymes, which are stored by lysosomes and capable of digesting cell components. However, there is also evidence indicating that the postulated features attributed to apoptosis but not to necrosis such as the absence of inflammation, being a genetically controlled, energy-dependent method of cellular deletion, which follows a coordinated, predictable, and predetermined pathway; chromatin condensation and the nucleosomal DNA laddering may not be exclusive characteristics of apoptosis. Apoptosis and necrosis may actually represent the two extremes of a continuum. Many insults induce apoptosis at lower doses and necrosis at higher doses.

Necroptosis

Necroptosis [94], an alternate form of programmed cell death, can serve as a cell-death backup when apoptosis signaling is blocked by endogenous or exogenous factors such as viruses or mutations. Necroptosis shares some molecular players with apoptotic cell death. However, necroptosis does not depend on caspase activity, but on the activity of the receptor interacting protein kinase 1 (RIPK1). RIPK1 functions as the initiator of the pathway, while several downstream kinases (most notable RIPK3) serve as the executioners. Necroptosis can be activated by members of the TNF family (through TNFR1, TNFR2, TRAILR1, and TRAILR2), FasL, toll-like receptors, LPS, genotoxic stress, ionizing

radiation, or calcium overload. TNF alpha, for example, can activate TNFR1, resulting in the recruitment of RIPK1 and other proteins to form Complex I. Subsequently, Complex II b is formed, which includes RIP1, receptor-interacting protein 3 (RIP3) kinase, caspase-8, and FADD and leads to necroptosis. In the absence of active caspase-8, RIPK1, and RIPK3 auto- and transphosphorylate each other, leading to the formation of a microfilament-like complex called the necrosome. The necrosome then activates the pronecroptotic protein MLKL allowing it to permeabilize plasma membranes and organelles. As in all forms of necrotic cell death, cells undergo necroptosis rupture, and leak their contents such as damage-associated molecular patterns (DAMPs) into the intercellular space, resulting in the recruitment of immune cells.

References

[1] Christian BE, Spremulli LL. Mechanism of protein biosynthesis in mammalian mitochondria. Biochim Biophys Acta 2012;1819(9–10):1035–54.
[2] Heinrich PC, Müller M, Graeve L. In: Heinrich PC, Müller M, Graeve L, editors. Löffler/Petrides Biochemie und Pathobiochemie, 9th ed., vol. 9. Berlin: Heidelberg: Springer Lehrbuch; 2014. p. 225–51.
[3] Boengler K, et al. Mitochondrial respiration and membrane potential after low-flow ischemia are not affected by ischemic preconditioning. J Mol Cell Cardiol 2007;43(5):610–5.
[4] Lesnefsky EJ, et al. Myocardial ischemia decreases oxidative phosphorylation through cytochrome oxidase in subsarcolemmal mitochondria. Am J Physiol 1997; 273(3 Pt 2):H1544–54.
[5] Paradies G, et al. Decrease in mitochondrial complex I activity in ischemic/reperfused rat heart: involvement of reactive oxygen species and cardiolipin. Circ Res 2004;94(1):53–9.
[6] Guan KL, Xiong Y. Regulation of intermediary metabolism by protein acetylation. Trends Biochem Sci 2011;36(2):108–16.
[7] Lu Z, et al. The emerging characterization of lysine residue deacetylation on the modulation of mitochondrial function and cardiovascular biology. Circ Res 2009; 105(9):830–41.
[8] Sack MN. Caloric excess or restriction mediated modulation of metabolic enzyme acetylation-proposed effects on cardiac growth and function. Biochim Biophys Acta 2011;1813(7):1279–85.

[9] Kim SC, et al. Substrate and functional diversity of lysine acetylation revealed by a proteomics survey. Mol Cell 2006;23(4):607–18.

[10] Schwer B, et al. Calorie restriction alters mitochondrial protein acetylation. Aging Cell 2009;8(5):604–6.

[11] Wang Q, et al. Acetylation of metabolic enzymes coordinates carbon source utilization and metabolic flux. Science 2010;327(5968):1004–7.

[12] Zhao S, et al. Regulation of cellular metabolism by protein lysine acetylation. Science 2010;327(5968):1000–4.

[13] Ahn BH, et al. A role for the mitochondrial deacetylase Sirt3 in regulating energy homeostasis. Proc Natl Acad Sci USA 2008;105(38):14447–52.

[14] Cimen H, et al. Regulation of succinate dehydrogenase activity by SIRT3 in mammalian mitochondria. Biochemistry 2010;49(2):304–11.

[15] Finley LW, et al. Succinate dehydrogenase is a direct target of sirtuin 3 deacetylase activity. PLoS ONE 2011;6(8):e23295.

[16] Onyango P, et al. SIRT3, a human SIR2 homologue, is an NAD-dependent deacetylase localized to mitochondria. Proc Natl Acad Sci USA 2002;99(21):13653–8.

[17] Michishita E, et al. Evolutionarily conserved and nonconserved cellular localizations and functions of human SIRT proteins. Mol Biol Cell 2005;16(10):4623–35.

[18] Schwer B, et al. Reversible lysine acetylation controls the activity of the mitochondrial enzyme acetyl-CoA synthetase 2. Proc Natl Acad Sci USA 2006;103(27):10224–9.

[19] Haigis MC, Yankner BA. The aging stress response. Mol Cell 2010;40(2):333–44.

[20] Lomb DJ, Laurent G, Haigis MC. Sirtuins regulate key aspects of lipid metabolism. Biochim Biophys Acta 2010;1804(8):1652–7.

[21] Verdin E, et al. Sirtuin regulation of mitochondria: energy production, apoptosis, and signaling. Trends Biochem Sci 2010;35(12):669–75.

[22] Droge W. Free radicals in the physiological control of cell function. Physiol Rev 2002;82(1):47–95.

[23] Balaban RS, Nemoto S, Finkel T. Mitochondria, oxidants, and aging. Cell 2005;120(4):483–95.

[24] Murphy MP. How mitochondria produce reactive oxygen species. Biochem J 2009;417(1):1–13.

[25] Giorgio V, et al. Dimers of mitochondrial ATP synthase form the permeability transition pore. Proc Natl Acad Sci USA 2013;110(15):5887–92.

[26] Griffiths EJ. Mitochondria and heart disease. Adv Exp Med Biol 2012;942:249–67.

[27] Di Lisa F, et al. The mitochondrial permeability transition pore and cyclophilin D in cardioprotection. Biochim Biophys Acta 2011;1813(7):1316–22.

[28] Zou H, et al. An APAF-1.cytochrome c multimeric complex is a functional apoptosome that activates procaspase-9. J Biol Chem 1999;274(17):11549–56.

[29] Youle RJ, van der Bliek AM. Mitochondrial fission, fusion, and stress. Science 2012;337(6098):1062–5.

[30] Westermann B. Mitochondrial fusion and fission in cell life and death. Nat Rev Mol Cell Biol 2010;11(12):872–84.

[31] Otera H, Mihara K. Molecular mechanisms and physiologic functions of mitochondrial dynamics. J Biochem 2011;149(3):241–51.

[32] Campello S, Scorrano L. Mitochondrial shape changes: orchestrating cell pathophysiology. EMBO Rep 2010;11(9):678–84.

[33] Ong SB, Hausenloy DJ. Mitochondrial morphology and cardiovascular disease. Cardiovasc Res 2010;88(1):16–29.

[34] Ong SB, et al. Inhibiting mitochondrial fission protects the heart against ischemia/reperfusion injury. Circulation 2010;121(18):2012–22.

[35] Nakada K, et al. Inter-mitochondrial complementation: mitochondria-specific system preventing mice from expression of disease phenotypes by mutant mtDNA. Nat Med 2001;7(8):934–40.

[36] Chen H, et al. Mitochondrial fusion is required for mtDNA stability in skeletal muscle and tolerance of mtDNA mutations. Cell 2010;141(2):280–9.

[37] Chen Q, et al. Isolating the segment of the mitochondrial electron transport chain responsible for mitochondrial damage during cardiac ischemia. Biochem Biophys Res Commun 2010;397(4):656–60.

[38] Kubli DA, Gustafsson AB. Mitochondria and mitophagy: the yin and yang of cell death control. Circ Res 2012;111(9):1208–21.

[39] Kanki T, Klionsky DJ, Okamoto K. Mitochondria autophagy in yeast. Antioxid Redox Signal 2011;14(10):1989–2001.

[40] Kim I, Rodriguez-Enriquez S, Lemasters JJ. Selective degradation of mitochondria by mitophagy. Arch Biochem Biophys 2007;462(2):245–53.

[41] Lemasters JJ. Selective mitochondrial autophagy, or mitophagy, as a targeted defense against oxidative stress, mitochondrial dysfunction, and aging. Rejuvenation Res 2005;8(1):3–5.

[42] Priault M, et al. Impairing the bioenergetic status and the biogenesis of mitochondria triggers mitophagy in yeast. Cell Death Differ 2005;12(12):1613–21.

[43] Twig G, et al. Fission and selective fusion govern mitochondrial segregation and elimination by autophagy. EMBO J 2008;27(2):433–46.

[44] Nakai A, et al. The role of autophagy in cardiomyocytes in the basal state and in response to hemodynamic stress. Nat Med 2007;13(5):619–24.

[45] Gottlieb RA, Carreira RS. Autophagy in health and disease. 5. Mitophagy as a way of life. Am J Physiol Cell Physiol 2010;299(2):C203–10.

[46] Gottlieb RA, Mentzer RM. Autophagy during cardiac stress: joys and frustrations of autophagy. Annu Rev Physiol 2010;72:45–59.

[47] Sheng R, et al. Autophagy activation is associated with neuroprotection in a rat model of focal cerebral ischemic preconditioning. Autophagy 2010;6(4):482–94.

[48] Lopez-Lluch G, et al. Mitochondrial biogenesis and healthy aging. Exp Gerontol 2008;43(9):813–9.

[49] Goffart S, Wiesner RJ. Regulation and co-ordination of nuclear gene expression during mitochondrial biogenesis. Exp Physiol 2003;88(1):33–40.

[50] Gleyzer N, Vercauteren K, Scarpulla RC. Control of mitochondrial transcription specificity factors (TFB1M and TFB2M) by nuclear respiratory factors (NRF-1 and NRF-2) and PGC-1 family coactivators. Mol Cell Biol 2005;25(4):1354–66.

[51] Schreiber SN, et al. The estrogen-related receptor alpha (ERRalpha) functions in PPARgamma coactivator 1alpha (PGC-1alpha)-induced mitochondrial biogenesis. Proc Natl Acad Sci USA 2004;101(17):6472–7.

[52] Safdar A, et al. Exercise increases mitochondrial PGC-1alpha content and promotes nuclear-mitochondrial cross-talk to coordinate mitochondrial biogenesis. J Biol Chem 2011;286(12):10605–17.

[53] Zhu LL, et al. PGC-1alpha coactivates estrogen-related receptor-alpha to induce the expression of glucokinase. Am J Physiol Endocrinol Metab 2010;298(6):E1210–8.

[54] Arany Z, et al. Transcriptional coactivator PGC-1 alpha controls the energy state and contractile function of cardiac muscle. Cell Metab 2005;1(4):259–71.

[55] Leone TC, et al. PGC-1alpha deficiency causes multisystem energy metabolic derangements: muscle dysfunction, abnormal weight control and hepatic steatosis. PLoS Biol 2005;3(4):e101.

[56] Arany Z, et al. Transverse aortic constriction leads to accelerated heart failure in mice lacking PPAR-gamma coactivator 1alpha. Proc Natl Acad Sci USA 2006; 103(26):10086–91.

[57] Lehman JJ, et al. Peroxisome proliferator-activated receptor gamma coactivator-1 promotes cardiac mitochondrial biogenesis. J Clin Invest 2000;106(7): 847–56.

[58] Sasaki K, Yoshida H. Organelle autoregulation-stress responses in the ER, Golgi, mitochondria and lysosome. J Biochem 2015;157(4):185–95.

[59] Altman BJ, Dang CV. Normal and cancer cell metabolism: lymphocytes and lymphoma. FEBS J 2012; 279(15):2598–609.

[60] Thorens B, Mueckler M. Glucose transporters in the 21st century. Am J Physiol Endocrinol Metab 2010;298(2): E141–5.

[61] Woods A, et al. LKB1 is the upstream kinase in the AMP-activated protein kinase cascade. Curr Biol 2003; 13(22):2004–8.

[62] Woods A, et al. Ca2+/calmodulin-dependent protein kinase kinase-beta acts upstream of AMP-activated protein kinase in mammalian cells. Cell Metab 2005;2(1): 21–33.

[63] Lizcano JM, et al. LKB1 is a master kinase that activates 13 kinases of the AMPK subfamily, including MARK/PAR-1. EMBO J 2004;23(4):833–43.

[64] Hawley SA, et al. Calmodulin-dependent protein kinase kinase-beta is an alternative upstream kinase for AMP-activated protein kinase. Cell Metab 2005;2(1):9–19.

[65] Momcilovic M, Hong SP, Carlson M. Mammalian TAK1 activates Snf1 protein kinase in yeast and phosphorylates AMP-activated protein kinase in vitro. J Biol Chem 2006;281(35):25336–43.

[66] Gwinn DM, et al. AMPK phosphorylation of raptor mediates a metabolic checkpoint. Mol Cell 2008;30(2): 214–26.

[67] Inoki K, et al. Rheb GTPase is a direct target of TSC2 GAP activity and regulates mTOR signaling. Genes Dev 2003;17(15):1829–34.

[68] Inoki K, Zhu T, Guan KL. TSC2 mediates cellular energy response to control cell growth and survival. Cell 2003;115(5):577–90.

[69] Kusnadi EP, et al. Regulation of rDNA transcription in response to growth factors, nutrients and energy. Gene 2015;556(1):27–34.

[70] Koc EC, Koc H. Regulation of mammalian mitochondrial translation by post-translational modifications. Biochim Biophys Acta 2012;1819(9–10):1055–66.

[71] Koc EC, et al. A proteomics approach to the identification of mammalian mitochondrial small subunit ribosomal proteins. J Biol Chem 2000;275(42):32585–91.

[72] Koc EC, et al. Identification of four proteins from the small subunit of the mammalian mitochondrial ribosome using a proteomics approach. Protein Sci 2001; 10(3):471–81.

[73] Miyata S, et al. The endoplasmic reticulum-resident chaperone heat shock protein 47 protects the Golgi apparatus from the effects of O-glycosylation inhibition. PLoS ONE 2013;8(7):e69732.

[74] Reiling JH, et al. A CREB3-ARF4 signalling pathway mediates the response to Golgi stress and susceptibility to pathogens. Nat Cell Biol 2013;15(12):1473–85.

[75] Bonifacino JS, Glick BS. The mechanisms of vesicle budding and fusion. Cell 2004;116(2):153–66.

[76] Lee C, Goldberg J. Structure of coatomer cage proteins and the relationship among COPI, COPII, and clathrin vesicle coats. Cell 2010;142(1):123–32.

[77] Faini M, et al. Vesicle coats: structure, function, and general principles of assembly. Trends Cell Biol 2013;23(6): 279–88.

[78] Waters MG, Serafini T, Rothman JE. 'Coatomer': a cytosolic protein complex containing subunits of non-clathrin-coated Golgi transport vesicles. Nature 1991; 349(6306):248–51.

[79] Serafini T, et al. ADP-ribosylation factor is a subunit of the coat of Golgi-derived COP-coated vesicles: a novel role for a GTP-binding protein. Cell 1991;67(2): 239–53.

[80] Peyroche A, Paris S, Jackson CL. Nucleotide exchange on ARF mediated by yeast Gea1 protein. Nature 1996; 384(6608):479–81.

[81] Gaidarov I, et al. A functional phosphatidylinositol 3,4,5-trisphosphate/phosphoinositide binding domain in the clathrin adaptor AP-2 alpha subunit. Implications for the endocytic pathway. J Biol Chem 1996;271(34): 20922–9.

[82] Sugiura A, et al. A new pathway for mitochondrial quality control: mitochondrial-derived vesicles. EMBO J 2014;33(19):2142–56.

[83] Neuspiel M, et al. Cargo-selected transport from the mitochondria to peroxisomes is mediated by vesicular carriers. Curr Biol 2008;18(2):102–8.

[84] Soubannier V, et al. A vesicular transport pathway shuttles cargo from mitochondria to lysosomes. Curr Biol 2012;22(2):135–41.

[85] Shiao YJ, Lupo G, Vance JE. Evidence that phosphatidylserine is imported into mitochondria via a mitochondria-associated membrane and that the majority of mitochondrial phosphatidylethanolamine is derived from decarboxylation of phosphatidylserine. J Biol Chem 1995;270(19):11190–8.

[86] Rizzuto R, et al. Close contacts with the endoplasmic reticulum as determinants of mitochondrial Ca2+ responses. Science 1998;280(5370):1763–6.

[87] Braschi E, Zunino R, McBride HM. MAPL is a new mitochondrial SUMO E3 ligase that regulates mitochondrial fission. EMBO Rep 2009;10(7):748–54.

[88] Soubannier V, et al. Reconstitution of mitochondria derived vesicle formation demonstrates selective enrichment of oxidized cargo. PLoS ONE 2012;7(12): e52830.

[89] McLelland GL, et al. Parkin and PINK1 function in a vesicular trafficking pathway regulating mitochondrial quality control. EMBO J 2014;33(4):282–95.

[90] Kroemer G, et al. Classification of cell death: recommendations of the Nomenclature Committee on Cell Death. Cell Death Differ 2005;12(Suppl. 2):1463–7.

[91] Kerr JF, Wyllie AH, Currie AR. Apoptosis: a basic biological phenomenon with wide-ranging implications in tissue kinetics. Br J Cancer 1972;26(4):239–57.

[92] Debnath J, Baehrecke EH, Kroemer G. Does autophagy contribute to cell death? Autophagy 2005;1(2):66–74.

[93] Baehrecke EH. Autophagy: dual roles in life and death? Nat Rev Mol Cell Biol 2005;6(6):505–10.

[94] Degterev A, et al. Chemical inhibitor of nonapoptotic cell death with therapeutic potential for ischemic brain injury. Nat Chem Biol 2005;1(2):112–9.

Metabolic Pathways and Cycles

Christina Werner, Torsten Doenst, Michael Schwarzer

Department of Cardiothoracic Surgery, Jena University Hospital,
Friedrich Schiller University of Jena, Jena, Germany

INTRODUCTION

Metabolic pathways and cycles are reaction chains where chemical products become the substrate for the next step. All substrates are chemically transformed in reactions that belong to either pathways (if the reactions are aligned in linear fashion) or metabolic cycles (if the moieties of the reactions are preserved). The term substrate oxidation is used for substrate degradation ultimately leading to CO_2 production. However, such an oxidative pathway may be interrupted for several reasons, such as in the case of glucose oxidation, where limited oxygen results in the production of lactate rather than water and CO_2. Some but not all substrates can also be converted to building blocks for biosynthetic processes. This chapter will describe the main metabolic pathways and cycles of cardiac substrate metabolism. It will also address their interactions and their contributions to meet cardiac energy requirements.

LIPIDS AND FATTY ACIDS

Lipids are the major source of energy (60–90%) for cardiac contraction [1]. Almost the entire nutritional fat consists of one to three fatty acids esterified with glycerol. Lipase dissociates glycerol and prepares free fatty acids for cellular uptake. Cellular uptake of fatty acids is passive or mediated by fatty acid translocase (FAT/CD36) and cardiac-specific fatty acid binding protein (H-FABP) [2]. Once inside the cell, fatty acids are esterified to coenzyme A (CoA) in an ATP-requiring reaction that is catalyzed by fatty acid thiokinase. After this activation step, 70–90% of the fatty acids undergo immediate oxidation. The rest are stored in the intracellular triglyceride pool [3].

In order to be oxidized, fatty acids must be transported into the mitochondria. Short-chain fatty acids and ketone bodies do not require a specific transport system but enter the mitochondria by diffusion. Since the majority of fatty acids are long chained, a specific transport mechanism, the carnitine-palmitoyl transferase system, is required. This system consists of two enzymes, carnitine-palmitoyl transferase I and II (CPT-I/II) and transfers the impermeable acyl-CoA to the mitochondrial matrix by temporarily converting it to acyl-carnitine. CPT-I catalyzes the reaction from acyl-CoA to acyl-carnitine at the outer mitochondrial membrane and CPT-II reverts the reaction at the inner mitochondrial membrane. Acyl-CoA inside the mitochondrial matrix enters fatty acid oxidation immediately. CPT-I activity is therefore seen as a rate-limiting step for fatty acid oxidation [4]. Fatty acid breakdown is driven by β-oxidation. It mainly takes place inside the mitochondria, but to a much smaller extent, may also occur in peroxisomes.

FATTY ACID OXIDATION

The principle of β-oxidation is the stepwise dissociation of two carbon atoms from fatty acids and the coenzyme A supported formation of acetyl-CoA, NADH, and FADH$_2$ [5]. The full spiral of β-oxidation is shown in Fig. 4.1 and its energetic balance is given in Table 4.1.

First, acyl-CoA is oxidized by acyl-CoA dehydrogenase and FAD$^+$ and results in α,β-unsaturated acyl-CoA and FADH$_2$. This is the first of four reactions in the β-oxidation chain. The α,β-unsaturated acyl-CoA is hydrated by enoyltransferase to β-hydroxy-acyl-CoA. The

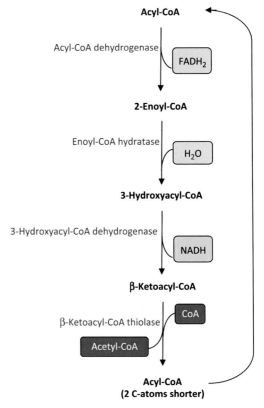

FIGURE 4.1 Reaction steps of fatty acid oxidation (β-oxidation).

NAD$^+$-dependent β-hydroxyacyl-CoA dehydrogenase oxidizes β-hydroxyacyl-CoA to β-ketoacyl-CoA and NADH. This reaction is the second oxidation step and has given β-oxidation pathway its name. The next step can be seen as

TABLE 4.1 Energetic Balance of Complete Fatty Acid Oxidation Using Stearyl-CoA as Example (18 C-Atoms Fatty Acid). First Step: ß-Oxidation and Second Step: Oxidation of Acetyl-CoA in the Citric Acid Cycle

Metabolic step	Reducing equivalents		Resulting ATP after oxidation of		ATP output from citric acid cycle
	NADH	FADH$_2$	NADH	FADH$_2$	
Stearyl-CoA → 9 Acetyl-CoA	8	8	21.6	12.8	–
9 Acetyl-CoA → 18 CO$_2$ + 18 H$_2$O	27	9	72.7	14.4	9
			Sum 130.7		

the last and at the same time the first step of one cycle of the β-oxidation spiral. Another coenzyme A is added to β-ketoacyl-CoA dissociating one molecule of acetyl-CoA. The remaining acyl-CoA has two carbon atoms less than the initial fatty acid. Fatty acids with an even number of carbon atoms (practically all naturally occurring fatty acids) run through several passages of β-oxidation until all pairs of carbon atoms are converted to acetyl-CoA. Acetyl-CoA directly enters the citric acid cycle (CAC).

Fatty acids with odd numbers of carbon atoms only occur in some plants and have rather low importance for cardiac-energy production. However, they can be catabolized to acetyl-CoA as described earlier. The last step leaves the activated three-carbon fatty acid propionyl-CoA. Propionyl-CoA carboxylase catalyzes the ATP-dependent reaction of propionyl-CoA to 2-methylmalonyl-CoA. Activated carboxybiotin serves as the carbon donor for this carboxylation. 2-methylmalonyl-CoA is converted into succinyl-CoA by an isomerase. Finally, succinyl-CoA is able to enter the CAC. This anaplerotic reaction serves as replenishment of the cycle intermediates rather than for energy generation. For more detailed information on anaplerotic reactions the reader is referred to the following text.

β-Oxidation is mainly controlled by the presence of activated fatty acids in the cardiomyocyte. The mechanisms of regulation will be discussed in Chapter 5. In times of fasting with low substrate supply and glucose restriction, the heart is able to use ketone bodies as substrate for energy generation. In the liver, acetyl-CoA is transformed to the ketone bodies acetoacetic acid and β-hydroxybutyric acid. They can enter the cardiomyocyte and mitochondria directly through monocarboxylate transporters or diffusion and are regenerated to acetyl-CoA by β-ketoacyl-CoA transferase. They contribute to substrate competition by the same mechanisms as fatty acids through inhibition of pyruvate dehydrogenase activity by acetyl-CoA [6].

GLUCOSE UTILIZATION

The second important substrate for cardiac energy metabolism is glucose. Glucose metabolism involves the different pathways of glycolysis, glycogen synthesis, and the two alternate pathways pentose phosphate pathway (PPP) and hexosamine biosynthesis pathway (HBP). Figure 4.2 gives an overview of the interactions of the PPP, the HBP, and glycolysis.

Glucose is first transported across the plasma membrane and then phosphorylated by hexokinase. These two steps are defined as glucose uptake and this phosphorylation step is irreversible since there is no glucose-6-phosphatase in the heart (e.g., as it is in the liver). If glucose is ultimately converted to water and CO_2, the term "glucose oxidation" is used. Full oxidation of glucose represents 10–40% of cardiac energy production.

GLUCOSE UPTAKE AND GLYCOGEN

Glucose transport in myocytes is driven by the translocation of monosaccharide transporters (GLUT-4 and GLUT-1) to the sarcolemma. Insulin-mediated GLUT-4 is the major glucose transporter in cardiac and skeletal muscle [7]. In contrast, GLUT-1 shows a much lower glucose uptake rate and rather mediates general glucose uptake in most tissues. Additionally, intracellular glycogen stores are another potential source of glucose. Cardiac glycogen stores are small compared to other tissues such as liver or skeletal muscle. There is a rapid turnover of glucose to glycogen for storage and of glycogen to glucose as substrate in glycolysis (see Chapter 5). Cardiomyocytes therefore present with quite stable glycogen concentration. However, high extracellular glucose concentrations increase the glycogen pool [8]. In contrast, elevated amounts of AMP, inorganic phosphate, and a fall in ATP

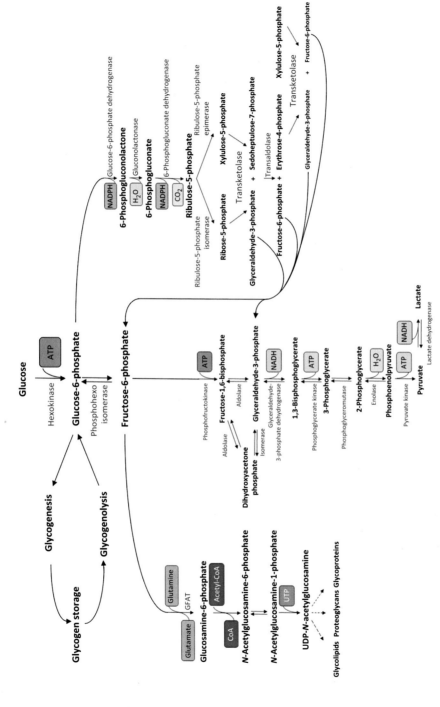

FIGURE 4.2   Interactions of the pentose phosphate pathway, the hexosamine biosynthesis pathway, and glycolysis.

activate glycogenolysis and result in enhanced substrate supply for glycolysis [9].

GLYCOLYSIS

Glycolytic enzymes are located in the sarcoplasm and are associated with the sarcoplasmic reticulum [10,11]. They convert glucose-6-phosphate and nicotinamide adenine dinucleotides (NAD^+) to pyruvate and NADH by producing two molecules of ATP. Table 4.2 shows the key chemical reactions of glycolysis and their energetic efficiency [12].

A scheme of the full glycolysis pathway is shown in Fig. 4.3. First, sarcoplasmic, free glucose is activated to glucose-6-phosphate by hexokinase and transformed to fructose-6-phosphate by phosphohexoisomerase. Hexokinase has a high affinity to glucose but is also able to utilize fructose or galactose as substrates. The first step of glycolysis is catalyzed by phosphofructokinase (PFK) and converts fructose-6-phosphate to fructose-1,6-bisphosphate. The two initial phosphorylations by hexokinase and PFK require ATP and are therefore irreversible. PFK is one of the key regulators in glycolysis. It is activated by fructose-2,6-bisphosphate and AMP and inhibited by increased formation of citrate and ATP during oxidative substrate phosphorylation in the CAC. In the next step, the enzyme aldolase divides the hexose fructose-1,6-bisphosphate in two trioses, dihydroxyacetone phosphate and glyceraldehyde-3-phosphate, which are in a dynamic equilibrium driven by triosephosphate isomerase. Only glyceraldehyde-3-phosphate underlies further reaction and is oxidized and phosphorylated by the NAD^+-specific glyceraldehyde-3-phosphate dehydrogenase (GAPDH) to 1,3-bisphosphoglycerate and NADH. The following reaction of 1,3-bisphosphoglycerat to 3-phosphoglycerate, catalyzed by phosphoglycerate kinase, transfers inorganic phosphate to ADP and results in the first substrate chain phosphorylation generating ATP. The enzyme mutase transforms 3-phosphoglycerate to 2-phosphoglycerate, which is dehydrated by enolase under formation of energy-rich phosphoenolpyruvate and water. Phosphoenolpyruvate is dephosphorylated to pyruvate by pyruvate kinase. Within this reaction step, inorganic phosphate is transferred to ADP for a second time during glycolysis and results in the production of the second molecule ATP. This reaction is again irreversible due to the high energetic gradient between phosphoenolpyruvate and pyruvate.

Glycolytic breakdown of glucose ends with the formation of pyruvate. There are four different reactions pyruvate is involved in afterwards: decarboxylation to acetyl-CoA, reduction to lactate, carboxylation to oxaloacetate or malate or transamination with glutamate to alanine. While acetyl-CoA and lactate are further metabolized within the process of ATP generation, the formation of oxaloacetate, malate, or alanine have

TABLE 4.2 Catalytic Efficiency and ATP Production of Glycolysis

Enzyme	Metabolic step	ATP output
Hexokinase	Glucose + ATP → Glucose-6-phosphate + ADP	−1
PFK	Fructose-6-phosphate + ATP → Fructose-1,6-bisphosphate + ADP	−1
Phosphoglycerate kinase	1,3-Bisphosphoglycerate + ADP → 3-Phosphoglycerate + ATP	2
Pyruvate kinase	Phosphoenolpyruvate + ADP → Pyruvate + ATP	2
		Sum 2

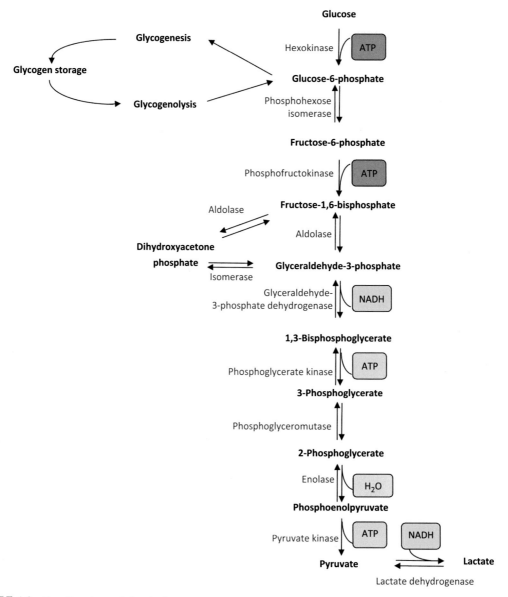

FIGURE 4.3   **Reaction steps of glycolysis.**

anaplerotic function (see later in the chapter) and are favored in case of a lack of CAC intermediates. Under aerobic conditions, pyruvate is transported to the inner mitochondrial membrane by the mitochondrial pyruvate carrier MCP1/2 [13,14]. The following irreversible decarboxyl-

ation and oxidation of pyruvate to acetyl-CoA is supported by the pyruvate dehydrogenase complex (PDH), a highly regulated multienzyme complex with coenzyme A and thiamine pyrophosphate as coenzymes. PDH serves as a key regulator and catalyzes the irreversible step in carbohydrate decarboxylation or glucose oxidation. The regulation of this enzyme is described in detail in Chapter 5. Briefly, PDH activity is limited by its product acetyl-CoA as well as by NADH generated in the CAC. The enzyme activity is also inhibited by phosphorylation through pyruvate dehydrogenase kinase (PDK). Elevated acetyl-CoA levels and increased conversion of ADP to ATP leads to activation of PDK and hence to phosphorylation and inactivation of PDH. In contrast, elevated concentrations of pyruvate, ADP, and pyrophosphate lead to activation of pyruvate dehydrogenase phosphatase, resulting in dephosphorylation of PDH thereby activating it [15].

Under ischemic conditions, pyruvate is not decarboxylated but reduced to lactate by lactate dehydrogenase (LDH) and NADH that was built in the previous GAPDH reaction. Therefore increasing amounts of lactate act as feedback inhibitor of glycolysis at the level of GAPDH. In the absence of oxygen, the glycolytic production of lactate is often referred to as "anaerobic glycolysis" although the reaction steps do not differ from those in the presence of oxygen. The fate of the end product differs – instead of oxidation to acetyl-CoA, pyruvate is reduced to lactate. Importantly, lactate is a normal (and even preferred) substrate for cardiac energy metabolism under aerobic conditions. Especially in times of high skeletal muscle activity (e.g., during exercise), lactate utilization by the heart increases proportionally to its appearance in plasma through increased skeletal muscle contraction [16]. Therefore, LDH oxidizes lactate to pyruvate in a NAD^+-dependent reaction.

Increasing amounts of NADH and/or ATP act as negative allosteric effectors controlling the flux through glycolysis [1,2]. Furthermore, metabolites of the CAC are able to regulate glycolysis. Higher rates of fatty acid oxidation lead to accumulation of citric acid, the first intermediate of CAC. This acts as a negative allosteric inhibitor of the glycolytic enzyme PFK and is hence able to slow glycolytic breakdown [17].

Besides glycolysis and glycogen synthesis, there are two additional pathways using glucose as substrate: the pentose phosphate pathway and the hexosamine biosynthesis pathway. They are not necessary for the generation of ATP but promote formation of metabolically relevant metabolites for DNA synthesis and antioxidative defense as well as mechanisms of cell signaling.

PENTOSE PHOSPHATE PATHWAY

The pentose phosphate pathway (PPP) is an alternative way of glucose use. It consists of an aerobic and an anaerobic part. For that reason, the PPP can act as a pathway or a cycle both at the same time. Its aerobic part leads to ribulose-5-phosphate, carbon dioxide (CO_2), and reduced nicotinamide adenine dinucleotide phosphate (NADPH). One molecule of ribulose-5-phosphate and two molecules of NADPH are produced out of one molecule of glucose. Subsequently, following anaerobic transformation of ribulose-5-phosphate delivers no energy but new glucose-6-phosphate. This can reenter other glycolytic pathways such as glycolysis or the HBP (see later in the chapter). In total, a series of PPP reactions cycle 6 molecules of glucose-6-phosphate to 5 molecules of glucose-6-phosphate, 12 NADPH and 6 CO_2 [18]. Figure 4.4 shows a scheme of all reactions within the PPP. NADPH is mainly used for fatty acid synthesis, pyruvate oxidation to malate, and the reduction of glutathione. Ribulose-5-phosphate, the product of the aerobic part of PPP is easily converted to ribose-5-phosphate, which is used for synthesis of nucleotides and nucleic acids. Hence, the PPP links carbohydrate and fatty acid metabolism, anaplerosis, nucleotide synthesis, and antioxidative defense depending on the individual need of a cell's metabolism.

The PPP as well as glycolysis and the hexosamine biosynthesis pathway use glucose-6-phosphate. This substrate is oxidized twice by

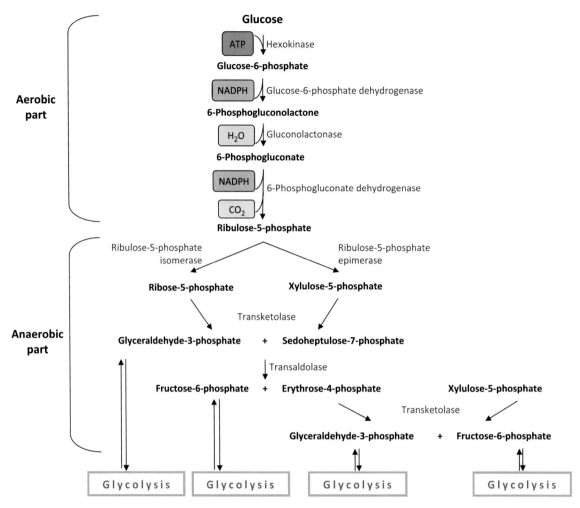

FIGURE 4.4   **Reaction steps of the pentose phosphate pathway.**

the NADP[+]-dependent glucose-6-phosphate dehydrogenase to 6-phospho-glucon-δ-lacton as intermediate and by gluconolactonase to 6-phosphogluconate. In the next step, 6-phosphogluconate is converted to ribulose-5-phosphate by NADP[+]-dependent 6-phosphogluconate dehydrogenase. Hereby, 3-keto 6-phosphogluconate occurs as an unstable intermediate. These last three reactions of the PPP result in two molecules of NADPH. The following anaerobic part of PPP allows

the conversion of ribulose-5-phosphate to intermediates of glycolysis. So the PPP may rather be seen as a cycle instead of a linear pathway. Isomerization and epimerization of ribulose-5-phosphate allow the formation of ribose-5-phosphate and xylolose-5-phosphate. The enzyme transketolase catalyzes their reaction to seduheptulose-7-phosphate and glycerinaldehyde-3-phosphate. The latter two build erythrose-4-phosphate and fructose-6-phosphate, catalyzed by transaldolase. A final reaction of erythrose-4-phosphate and another molecule of xylolose-5-phosphate form fructose-6-phosphate and glycerinaldehyde-3-phosphate, which may directly enter glycolysis.

HEXOSAMINE BIOSYNTHESIS PATHWAY

The hexosamine biosynthesis pathway (HBP) is another branch for glucose utilization. In the normal working healthy heart, there is only a small flux of glucose into PPP and HBP [19]. Metabolite supply of these alternate pathways is enhanced when glycolysis is limited as it occurs with increasing amounts of free fatty acids. The HBP has importance in glucose metabolism and is activated not only by increasing but also rapidly decreasing glucose concentrations or changing intracellular calcium levels [20]. A detailed overview of all reaction steps within the HBP is given in Fig. 4.5.

Initially, glucose-6-phosphate is converted to fructose-6-phosphate, which is not utilized by

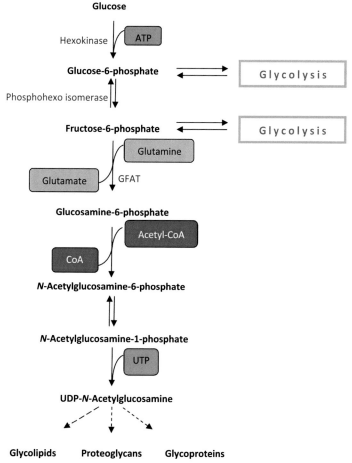

FIGURE 4.5   Reaction steps of the hexosamine biosynthesis pathway.

PFK as in glycolysis but is rather reacting with the amino-donor glutamine to glucosamine-6-phosphate. This reaction is driven by the rate-limiting enzyme glucosamine-6-phosphate synthase/glutamine-fructose-6-phosphate aminotransferase (GFAT). Due to its short half-life of less than an hour GFAT mainly seems to be regulated by its gene expression. It is also controlled by allosteric inhibition and/or covalent modification [21]. In the next step, acetyl-CoA is transferred to glucosamine-6-phosphate resulting in N-acetylglucosamine-6-phosphate. After isomerization N-acetylglucosamine-1-phosphate and UTP react to the final product UDP-N-acetylglucosamine (UDP-GlcNAc). The end product UDP-GlcNAc plays an important role in reversible posttranslational protein modification and may have special impact in signal transduction [22]. The linkage of UDP-GlcNAc to proteins is called glycosylation and takes place in the Golgi apparatus. It is mediated by O-GlcNAc transferase (OGT) that catalyzes the O-linked binding of UDP-GlcNAc to mainly hydroxyl oxygen of serine and threonine residues but also of tyrosine, hydroxylysine, or hydroxyproline side chains of proteins [23]. Glycosylation of proteins leads to formation of various protein alterations such as proteoglycans and glycoproteins. The removal of O-GlcNAc from proteins is driven by O-GlcNAcase (OGA). In contrast to protein modifications, such as phosphorylations, glycosylation is mediated by only two enzymes OGT and OGA. For that reason, it is less specific than phosphorylation, which in turn is mediated by dozens of particular kinases and phosphatases [24]. Cellular O-GlcNAc levels are increased by stressors such as heat, free radicals, and ultraviolet (UV) radiation [25]. Levels are chronically elevated in response to diabetes, impaired Ca^{2+} handling, and in heart failure [26]. A special focus of recent work was on cellular signaling pathways and transcriptional regulation. However, there is also growing evidence that O-GlcNAcylation may play a role in epigenetic regulation [20,24].

AMINO ACIDS

Besides fatty acids and glucose, the heart is also able to use amino acids as substrates for energy production [27], although amino acids are the primary building blocks for protein synthesis. For use as substrate, they are either taken up by the cell or generated by protein degradation. The major site of amino acid degradation in mammals is the liver. However, the branched chain amino acids (BCAA) valine, leucine, and isoleucine are also metabolized in the heart muscle. First, the amino group is removed in a deamination reaction that is driven by transaminases. The carbon skeletons of the 20 amino acids are transformed into only 7 molecules: acetyl-CoA, acetoacetyl-CoA (ketogenic amino acids), and pyruvate, α-ketoglutarate, succinyl-CoA, fumarate, and oxaloacetate (glucogenic amino acids) [28]. These carbon skeletons of the resulting α-ketoacids enter metabolic pathways as precursors to glucose or CAC intermediates (see later in the chapter).

The BCAA valine, leucine, and isoleucine are of major interest with regard to regulation of cardiac metabolism and the development of common diseases such as insulin resistance, diabetes, and cardiovascular diseases [29]. The BCAA leucine, and the metabolites of leucine, valine, and isoleucine have been shown to inhibit protein degradation and accelerated protein synthesis in the heart and may play a role in the regulation of protein turnover [30].

CITRIC ACID CYCLE

The CAC marks the center of interconnected energy providing pathways and cycles. In the mitochondrial matrix, enzymes of the CAC (also known as Krebs cycle or tricarboxylic acid cycle -TCA) produce the reducing equivalents NADH and $FADH_2$. They deliver electrons to complexes of mitochondrial electron-transport chain, which builds up a proton gradient that

TABLE 4.3 Catalytic Efficiency of the Citric Acid Cycle

Metabolic step	Reducing equivalent	ATP output
Isocitrate $\rightarrow$ α-Ketoglutarate	$NAD^+ \rightarrow NADH + H^+$	2.7
α-Ketoglutarate $\rightarrow$ Succinyl-CoA	$NAD^+ \rightarrow NADH + H^+$	2.7
Succinyl-CoA $\rightarrow$ Succinate	Substrate chain phosphorylation	
	$GDP + P \rightarrow GTP$	1.0
Succinate $\rightarrow$ Fumarate	$FAD \rightarrow FADH_2$	1.6
Malate $\rightarrow$ Oxaloacetate	$NAD^+ \rightarrow NADH + H^+$	2.7
		Sum 10.7

drives ATP production. This key process links glycolysis, fatty acid oxidation, and also amino acid oxidation. In contrast to glycolysis and fatty acid oxidation, which can be described as linear pathways, cycles such as the CAC are energetically more efficient. ATP output of a full cycle is shown in Table 4.3 [31,32]. The centrality of the Krebs cycle for metabolism is remarkable. It not only connects all catabolic substrate oxidation pathways to the respiratory chain, but also represents the source of biosynthetic products for many anabolic processes [33]. Thus, although the cycle does not lose intermediates (i.e., moieties) by its own reactions, it still needs to be replenished continuously (anaplerosis, see later in the chapter).

The CAC stepwise catabolizes one molecule of acetyl-CoA, the high-energetic end product of glycolysis and β-oxidation into two molecules of carbon dioxide and the reducing equivalents NADH and FADH$_2$. A schematic overview of the full CAC is shown in Fig. 4.6. Initially, acetyl-CoA reacts with the cycle intermediate oxaloacetate. This reaction is driven by citrate synthase, which is the key enzyme of the cycle and may also be seen as a marker of mitochondrial activity. The product citrate is dehydrated to the unstable intermediate *cis*-aconitate, which is subsequently hydrated to isocitrate. Both reactions are catalyzed by aconitase. In the third step of the CAC, NAD$^+$-dependent isocitrate dehydrogenase transforms isocitrate to oxalosuccinate

with the result of the first reducing equivalent NADH. Decarboxylation of oxalosuccinate leads to formation of α-ketoglutarate and CO$_2$. In the next step, α-ketoglutarate dehydrogenase catalyzes the oxidation and carboxylation of α-ketoglutarate to succinate by formation of succinyl-CoA as an unstable intermediate. This reaction is comparable to the pyruvate dehydrogenase reaction and leads to formation of one molecule of high-energetic GTP. Efficiency of α-ketoglutarate dehydrogenase is crucial and controls the substrate flux through the complete CAC [34]. Further oxidation of succinate by FAD$^+$-dependent succinate dehydrogenase results in the reducing equivalent FADH$_2$ and fumarate, which is then hydrogenated to malate by fumarase. In the "last" step of the CAC, the NAD$^+$-dependent malate dehydrogenase converts malate to oxaloacetate and another NADH. Consequently, oxaloacetate may react with another molecule of acetyl-CoA and start a new round of the cycle.

The CAC as the pivotal element of energy metabolism is strictly regulated. The cycle is triggered by ADP, inorganic phosphate, and calcium while it is inhibited by NADH and ATP. Additionally, most CAC enzymes are regulated by their educts and products in a kinetic manner. For example, succinate dehydrogenase is inhibited by oxaloacetate but activated by succinate. Although the amount of the intermediates of the CAC is strictly regulated, they also

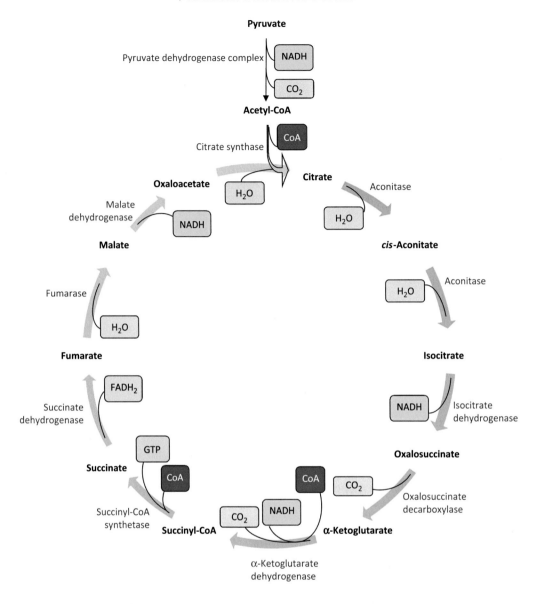

FIGURE 4.6   **Reactions steps of the citric acid cycle.**

represent the basic building blocks for many synthesis processes. The pool of CAC intermediates is therefore continuously "in change" and needs to be replenished if synthesis processes have prevalence – a process that is called anaplerosis.

ANAPLEROSIS

Anaplerosis was first mentioned by Kornberg in 1966 [35,36]. He described the mechanism as pathways to recover a metabolic cycle. In conditions of catabolic oxidation concentrations of CAC intermediates are very low ($10^{-5}-10^{-4}$ mol/L) [1]. No cycle intermediates are either removed or built. However, besides substrate oxidation, intermediates of the CAC act as starting points of anabolic pathways. The intermediates serve for fatty acid biosynthesis (citrate), heme biosynthesis (succinyl-CoA), gluconeogenesis (oxaloacetate), and biosynthesis of nonessential amino acids (α-ketoglutarate and oxaloacetate) [37]. Depletion of CAC intermediates leads to a rapid decline of contractile function of the heart [15]. It has been shown that impairment of anaplerotic pathways rapidly causes contractile dysfunction [19,38]. Changes in anaplerotic pathways have recently been implicated in heart disease. As a mechanism to maintain steady state, anaplerosis becomes necessary in case of loss of carbon intermediates of the CAC [39,40]. In contrast to skeletal muscle, where concentrations of intermediates of the CAC are not totally stable and for example increase immediately during exercise [41–43], strong interactions between metabolites and enzymes maintain a steady state in the heart. Here, metabolites such as pyruvate that are able to undergo various reactions and that act as substrate for different enzymes are essential. They allow the flexibility of moving from ATP generation (conversion of pyruvate to acetyl-CoA) to the replenishment of cycle intermediates. Pyruvate as a link between glycolysis and the CAC plays an important role in anaplerosis. The most typical anaplerotic reaction is the carboxylation of pyruvate to oxaloacetate, which is catalyzed by the ATP-dependent pyruvate carboxylase. This reaction is initiated by elevated amounts of acetyl-CoA, the allosteric activator, as well as by exercise [44,45] and ensures a sufficient supply of oxaloacetate to run the CAC. Another possible carboxylation of pyruvate produces malate by the $NADP^+$-dependent malic enzyme. This reaction might be more important in the opposite direction (decarboxylation of malate to pyruvate) because during anaplerotic reactions it may

come to a lack of NADPH which is necessary for fatty acid synthesis and defense of reactive oxygen species. A second focus of anaplerosis is on transamination reactions driven by transaminases. An example of such transaminations is the generation of α-ketoglutarate from the amino acid glutamate. Also oxaloacetate can be refilled. It is generated by aspartate [39]. Although odd chain fatty acids are of lower importance for the generation of ATP, they may serve for anaplerotic reactions. Here, the CAC intermediate succinyl-CoA can be regenerated from propionyl-CoA, the three-carbon residue of odd chain fatty acid oxidation. Most notably, during muscle contraction another anaplerotic reaction is important: the conversion of aspartate to ammonia and fumarate in a GTP-consuming reaction [46]. Increasing amounts of fumarate are able to replenish CAC intermediates, especially in conditions of high efflux such as exercise and fasting.

Many of the anaplerotic reactions discussed earlier, only lead to small increases in CAC intermediates, especially the net synthesis of malate from pyruvate is kinetically very unfavorable [47]. For that reason, transamination reactions [41,45] in parallel with the generation of fumarate by the purine nucleotide cycle might be the most important anaplerotic reactions with regard to the amount of intermediate that is replenished in the CAC.

THE RESPIRATORY CHAIN

In a final step, all the energy so far converted and not used for anabolic processes is transferred to the respiratory chain. Here, the reducing equivalents (NADH to Complex I and FADH to Complex II) are handing off their electrons and protons to be passed on to oxygen, resulting in water production. The energy released is used to generate a proton gradient across the inner mitochondrial membrane. The release of this gradient through Complex V (the F_0F_1-ATPase)

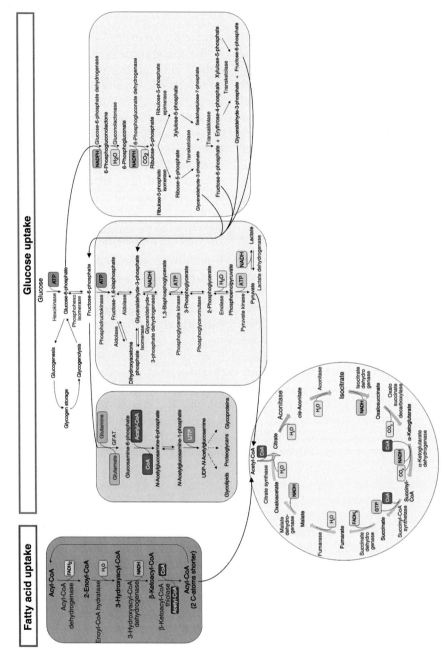

FIGURE 4.7   Complexity of cardiac metabolic pathways and cycles.

is used to generate ATP from ADP and inorganic phosphate. The function of the respiratory chain is described in detail together with its assessment in Chapter 3 of this book.

PUTTING IT ALL TOGETHER

The functional interaction of all metabolic pathways and cycles described in this chapter is summarized in Fig. 4.7. The pathways and cycles function for the permanent supply of energy (i.e., glycogen breakdown, glycolysis, β-oxidation), the generation of metabolites for synthesis processes, and replenishment of the CAC (i.e., PPP and anaplerosis), and metabolic regulation (i.e., HBP). The high-energy demand of cardiac contractile function is primarily covered by oxidative phosphorylation of acetyl-CoA derived from β-oxidation and to a lesser amount from glycolysis. Within the CAC, acetyl-CoA results in the formation NADH and $FADH_2$ delivering their protons to Complexes I and II of mitochondrial electron-transport chain to finally run ATP synthase. The PPP generates ribose to build DNA and RNA elements and NADPH that is an important player in cellular defense of reactive oxygen species. The hexosamine pathway generates O-GlcNAc that leads to glycosylated proteins, which contribute to the regulation of the complex interplay of all metabolic pathways and cycles.

SUMMARY

Cardiac metabolism is a highly concerted plethora of chemical reactions leading to the conversion of substrates for energy production in the form of ATP to sustain cell function and allow contraction and also for the synthesis of building blocks to allow for growth, repair, and regeneration. Knowing the key pathways and cycles of metabolism is a basic requirement for the understanding of cardiac metabolism and the investigation of disease. Here, we have illustrated a summary of individual reactions and some regulatory aspects for the key pathways and cycles focusing on cardiac energy metabolism. We have not described the respiratory chain in this chapter, as it is described in detail in Chapter 3.

References

[1] Stanley WC, Recchia FA, Lopaschuk GD. Myocardial substrate metabolism in the normal and failing heart. Physiol Rev 2005;85(3):1093–129.

[2] Sharma N, Okere IC, Duda MK, Chess DJ, O'Shea KM, Stanley WC. Potential impact of carbohydrate and fat intake on pathological left ventricular hypertrophy. Cardiovasc Res 2007;73(2):257–68.

[3] Lopaschuk GD, Belke DD, Gamble J, Itoi T, Schonekess BO. Regulation of fatty acid oxidation in the mammalian heart in health and disease. Biochim Biophys Acta 1994;1213(3):263–76.

[4] Lewandowski ED, Kudej RK, White LT, O'Donnell JM, Vatner SF. Mitochondrial preference for short chain fatty acid oxidation during coronary artery constriction. Circulation 2002;105(3):367–72.

[5] Berg JM, Tymoczko JL, Stryer L. Biochemistry. 5th ed. New York: W. H. Freeman; 2002. Section 22.2, The Utilization of Fatty Acids as Fuel Requires Three Stages of Processing.

[6] Kodde IF, van der Stok J, Smolenski RT, de Jong JW. Metabolic and genetic regulation of cardiac energy substrate preference. Comp Biochem Physiol A Mol Integr Physiol 2007;146(1):26–39.

[7] Abel ED. Glucose transport in the heart. Front Biosci 2004;9:201–15.

[8] Depre C, Vanoverschelde JL, Taegtmeyer H. Glucose for the heart. Circulation 1999;99(4):578–88.

[9] Depre C, Rider MH, Hue L. Mechanisms of control of heart glycolysis. Eur J Biochem 1998;258(2):277–90.

[10] Pierce GN, Philipson KD. Binding of glycolytic enzymes to cardiac sarcolemmal and sarcoplasmic reticular membranes. J Biol Chem 1985;260(11):6862–70.

[11] Weiss J, Hiltbrand B. Functional compartmentation of glycolytic versus oxidative metabolism in isolated rabbit heart. J Clin Invest 1985;75(2):436–47.

[12] Berg JM, Tymoczko JL, Stryer L. Biochemistry. 5th ed. New York: W. H. Freeman; 2002. Section 16.1 Glycolysis is an Energy-Conversion Pathway in Many Organisms.

[13] Bricker DK, Taylor EB, Schell JC, Orsak T, Boutron

A, Chen Y-C, Cox JE, Cardon CM, Van Vranken JG, Dephoure N, Redin C, Boudina S, Gygi SP, Brivet M, Thummel CS, Rutter J. A mitochondrial pyruvate carrier required for pyruvate uptake in yeast, drosophila, and humans. Science 2012;337(6090):96–100.

[14] Herzig S, Raemy E, Montessuit S, Veuthey J-L, Zamboni N, Westermann B, Kunji ERS, Martinou J-C. Identification and functional expression of the mitochondrial pyruvate carrier. Science 2012;337(6090):93–6.

[15] Hansford RG, Cohen L. Relative importance of pyruvate dehydrogenase interconversion and feed-back inhibition in the effect of fatty acids on pyruvate oxidation by rat heart mitochondria. Arch Biochem Biophys 1978; 191(1):65–81.

[16] Drake AJ, Haines JR, Noble MI. Preferential uptake of lactate by the normal myocardium in dogs. Cardiovasc Res 1980;14(2):65–72.

[17] Garland PB, Randle PJ, Newsholme EA. Citrate as an intermediary in the inhibition of phosphofructokinase in rat heart muscle by fatty acids, ketone bodies, pyruvate, diabetes and starvation. Nature 1963;200(4902):169–70.

[18] Baldwin JE, Krebs H. The evolution of metabolic cycles. Nature 1981;291(5814):381–2.

[19] Doenst T, Nguyen TD, Abel ED. Cardiac metabolism in heart failure: implications beyond ATP production. Circ Res 2013;113(6):709–24.

[20] Chatham JC, Marchase RB. Protein O-GlcNAcylation: a critical regulator of the cellular response to stress. Curr Signal Transd Ther 2010;5(1):49–59.

[21] Karason K, Sjöström L, Wallentin I, Peltonen M. Impact of blood pressure and insulin on the relationship between body fat and left ventricular structure. Eur Heart J 2003;24(16):1500–5.

[22] Kreppel LK, Blomberg MA, Hart GW. Dynamic glycosylation of nuclear and cytosolic proteins. Cloning and characterization of a unique O-GlcNAc transferase with multiple tetratricopeptide repeats. J Biol Chem 1997;272(14):9308–15.

[23] Torres CR, Hart GW. Topography and polypeptide distribution of terminal N-acetylglucosamine residues on the surfaces of intact lymphocytes. Evidence for O-linked GlcNAc. J Biol Chem 1984;259(5):3308–17.

[24] McLarty JL, Marsh SA, Chatham JC. Post-translational protein modification by O-linked N-acetyl-glucosamine: its role in mediating the adverse effects of diabetes on the heart. Life Sci 2012;92:621–7.

[25] Zachara NE, Hart GW. O-GlcNAc a sensor of cellular state: the role of nucleocytoplasmic glycosylation in modulating cellular function in response to nutrition and stress. Biochim Biophys Acta 2004;1673(1–2): 13–28.

[26] Lunde IG, Aronsen JM, Kvaløy H, Qvigstad E, Sjaastad I, Tønnessen T, Christensen G, Grønning-Wang LM, Carlson CR. Cardiac O-GlcNAc signaling is increased in hypertrophy and heart failure. Physiol Genomics 2012; 44:162–72.

[27] Chatham JC. Lactate – the forgotten fuel. J Physiol 2002;542(Pt 2):333.

[28] Berg JM, Tymoczko JL, Stryer L. Biochemistry. 5th ed. New York: W H Freeman; 2002. Section 23.5, Carbon Atoms of Degraded Amino Acids Emerge as Major Metabolic Intermediates.

[29] Newgard CB. Interplay between lipids and branched-chain amino acids in development of insulin resistance. Cell Metab 2012;15(5):606–14.

[30] Chua B, Siehl DL, Morgan HE. Effect of leucine and metabolites of branched chain amino acids on protein turnover in heart. J Biol Chem 1979;254(17):8358–62.

[31] Brandt U, Der Citratzyklus – Abbau von Acetyl-CoA zu CO_2 und H_2O. In: Heinrich PC, Müller M, Graeve L, editors. Löffler/Petrides Biochemie und Pathobiochemie. Berlin, Heidelberg: Springer; 2014. p. 226–34.

[32] Berg JM, Tymoczko JL, Stryer L. Biochemistry. 5th ed. New York: W. H. Freeman; 2002. Section 17.1, The Citric Acid Cycle Oxidizes Two-Carbon Units. The Citric Acid Cycle Oxidizes Two-Carbon Units.

[33] Doenst T, Nguyen TD, Abel ED. Cardiac metabolism in heart failure: implications beyond ATP production. Circ Res 2013;113(6):709–24.

[34] Cooney GJ, Taegtmeyer H, Newsholme EA. Tricarboxylic acid cycle flux and enzyme activities in the isolated working rat heart. Biochem J 1981;200(3):701–3.

[35] Kornberg H. Anaplerotic sequences and their role in metabolism. Essays Biochem 1966;2:1–31.

[36] Owen OE, Kalhan SC, Hanson RW. The key role of anaplerosis and cataplerosis for citric acid cycle function. J Biol Chem 2002;277(34):30409–12.

[37] Gray LR, Tompkins SC, Taylor EB. Regulation of pyruvate metabolism and human disease. Cell Mol Life Sci 2014;71(14):2577–604.

[38] Russell RR, Taegtmeyer H. Changes in citric acid cycle flux and anaplerosis antedate the functional decline in isolated rat hearts utilizing acetoacetate. J Clin Invest 1991;87(2):384–90.

[39] Des Rosiers C, Labarthe F, Lloyd SG, Chatham JC. Cardiac anaplerosis in health and disease: food for thought. Cardiovasc Res 2011;90:210–9.

[40] Wessells RJ, Fitzgerald E, Cypser JR, Tatar M, Bodmer R. Insulin regulation of heart function in aging fruit flies. Nat Genet 2004;36(12):1275–81.

[41] Sahlin K, Katz A, Broberg S. Tricarboxylic acid cycle intermediates in human muscle during prolonged exercise. Am J Physiol 1990;259(5 Pt 1):C834–41.

[42] Russell RR, Taegtmeyer H. Pyruvate carboxylation prevents the decline in contractile function of rat hearts oxidizing acetoacetate. Am J Physiol 1991;261:H1756–62.

[43] Gibala MJ, MacLean DA, Graham TE, Saltin B. Tricarboxylic acid cycle intermediate pool size and estimated

cycle flux in human muscle during exercise. Am J Physiol 1998;275(2):E235–42.

[44] Lancha AH Jr, Recco MB, Curi R. Pyruvate carboxylase activity in the heart and skeletal muscles of the rat. Evidence for a stimulating effect of exercise. Biochem Mol Biol Int 1994;32(3):483–9.

[45] Gibala MJ, MacLean DA, Graham TE, Saltin B. Anaplerotic processes in human skeletal muscle during brief dynamic exercise. J Physiol 1997;502(3):703–13.

[46] Lowenstein JM. The purine nucleotide cycle revised. Int J Sports Med 1990;11(S 2):S37–46.

[47] Lowenstein J, Tornheim K. Ammonia production in muscle: the purine nucleotide cycle. Science 1971;171(3969):397–400.

Principles in the Regulation of Cardiac Metabolism

Louis Hue*, Christophe Beauloye[†,**], Luc Bertrand[†]

*Université catholique de Louvain, de Duve Institute, Protein Phosphorylation Unit, Brussels, Belgium; [†]Université catholique de Louvain, Institut de Recherche Expérimentale et Clinique, Pole of Cardiovascular Research, Brussels, Belgium; **Université catholique de Louvain, Cliniques Universitaires Saint Luc, Division of Cardiology, Cardiovascular Intensive Care, Brussels, Belgium

INTRODUCTION

The heart continuously adapts its activity to ensure appropriate blood supply for the energy needs of the body. It has to cope with changes in workload, substrate availability, hormones, and nutrients. Adaptation is indeed needed to achieve energy homeostasis, which, by definition, implies physiological and biochemical control mechanisms. Through these mechanisms, the cardiac pump produces and consumes energy in proportion to the work to be carried out. From a metabolic point of view, the heart is characterized by its ability to match energy production with demand, and by its metabolic flexibility, which allows for adjustment of its metabolism to changes in substrate quality and quantity. Any imbalance can be deleterious and demonstrates that both energy production and demand have to be under control.

To adapt its metabolism, the heart resorts to several control mechanisms that affect metabolic fluxes by modulating the activity of key enzymes and transporters. This chapter presents a brief description of the biochemical mechanisms responsible for metabolic adaptation. Metabolic and signaling pathways are defined, and the mechanisms of extrinsic control by hormones and nervous stimulation involving signal transduction systems are briefly described. Finally, we review the mechanisms applied to modulate the activity of enzymes and the flux through pathways.

PATHWAYS

A *metabolic pathway* transforms matter and energy. Its direction is driven by reactions that are displaced far from equilibrium, that is, with a negative change in free energy, which, among other things, depends on the actual concentration of reactants and products. A pathway is an entity governed by its own rules (global and local parameters), in such a way that the control is distributed through the pathway. According to *metabolic control analysis*, the control is unevenly shared between the enzymes and transporters of the pathway, and the extent to which a particular enzyme in the pathway controls the overall flux can be evaluated and is expressed by its "*flux control coefficient.*" In fact, a pathway may often contain several so-called "*key*" or "*control*" steps, which exert most of the control. The reader may find an invaluable introduction to metabolic control analysis in the book written by Fell [1].

Ideally, control mechanisms should allow for large changes in metabolic fluxes through the pathway with minimal changes in the concentrations of intermediate metabolites. This *intrinsic metabolic homeostasis*, that is, within a metabolic pathway, is essential to avoid large variations in osmotic pressure and ionic composition that could endanger cell integrity [1,2]. Changes in the overall flux through a pathway are generally imposed from outside by "*extrinsic*" signals, such as hormones or neurotransmitters that modify the activity of key enzymes or transporters. Hormones bind to their specific receptor proteins that are located either on the cell surface (insulin, catecholamines) or inside the cell (thyroid and steroid hormones) and are involved in rapid metabolic adaptations via signal transduction systems or exert long-term effects through changes in gene expression.

Signal Transduction and Signaling Pathway

The binding of a hormone to its membrane receptor triggers a cascade of events that take place within the cell and eventually modulate metabolic fluxes. The link between hormone binding and its intracellular effects is called *signal transduction*. Generally it requires three components: a sensor (the hormone-specific receptor), a transducer (G-proteins or some other receptor-linked enzymes) and an effector that produces a signal molecule also called second messenger, which activates a downstream cascade of events. The G-proteins function as molecular switches. The transition between their inactive GDP-bound to the active GTP-bound state is immediate and irreversible, and is triggered by conformational changes that affect nucleotide binding. The opposite transition is also irreversible and depends on the intrinsic GTPase activity, which acts as an automatic turn-off mechanism. These built-in mechanisms ensure that G-proteins can only be in either one of the two states, "on" or "off" [3].

A *signaling pathway* conveys information and generates specific signal molecules. A signal molecule can be a small molecular-mass ligand (cyclic nucleotides, phosphoinositides, ions, metabolic intermediate, nitric oxide, etc.). In contrast to the metabolic pathway, the control of a signaling pathway aims at obtaining large changes in the concentration of the signal molecule, keeping the flux of matter and energy at a minimal level [4]. This is possible because the concentration of the signal molecule remains low in absolute values (micromolar range or even below for signaling proteins). Interestingly, many signals are produced by so-called signaling cascades, which allow for signal amplification through successive enzymatic reactions because one molecule of the enzyme can transform many molecules of its substrate [5–8]. The amplification resulting from hormonal signaling cascades is remarkable. For example, a small change in insulin concentration (in the nanomolar range) may eventually change the concentration of circulating glucose (in the millimolar range).

AMP is an interesting metabolic intermediate, which may act as an amplifier. AMP is part of a metabolic pathway and at the same time can act as a signal molecule. It is the substrate of a small number of reactions but it binds to several enzymes and modifies their activities, as is the case for AMP-activated protein kinase (AMPK), a protein kinase that affects the activity of several key enzymes controlling metabolism [9–13]. The interconversion of the three adenine nucleotides is catalyzed by the very active adenylate kinase, which is close to equilibrium. Assuming that the equilibrium constant equals 1, then $[AMP] = [ADP]^2/[ATP]$. This means that any small change in the opposite direction of ATP and ADP results in a much larger change in AMP. For example, a decrease in ATP from 5 mM to 4.5 mM, with a concomitant increase in ADP from 0.5 mM to 1 mM, results approximately in a more than fourfold increase in AMP from 0.05 mM to 0.22 mM. Thanks to adenylate kinase, AMP qualifies as a signal molecule that amplifies opposite changes in ATP and ADP and thus signals any change in adenine nucleotide levels, that is, the energy state of the cell, to AMP-sensitive enzymes, such as AMPK [1,2].

Reactive oxygen species (ROS) are chemically reactive molecules that are normally produced by oxygen metabolism in the mitochondria and elsewhere. Their concentration may however drastically increase under certain conditions and become responsible of oxidative stress. It is now accepted that at low concentration, ROS are signal molecules that regulate kinase-driven pathways and are responsible of the redox signaling involved in cardiac hypertrophy. At higher concentrations, they are definitely toxic and destructive [14–17].

Changes in the concentration of the signal are usually transient and depend on the difference between its rate of synthesis and degradation, or for ions for the rate of channels and pumps. For cyclic AMP, it is the difference between the activities of adenylate cyclase and phosphodiesterases; for Ca^{2+}, it depends on the activities of several different channels and pumps. The generation of these signals are in turn under the control of other extrinsic upstream events, such as, hormones that may impose coordinated changes in different organs. The efficiency of a signal relies on the extent of the changes in activity of key enzymes or transporters that may affect the overall flux through the pathway.

CONTROL OF ENZYME ACTIVITY

The rate of an enzyme-catalyzed reaction is directly proportional to the concentration of the enzyme and follows saturation kinetics for its substrates.

Long-Term Control

The long-term adaptation to the environment induces changes in the enzyme content of a tissue. This depends on the turnover of the enzyme, a continuous process, which in turn depends on the rate of its synthesis and degradation. There is a wide range of turnover rates, with enzymes having half-lives ranging from minutes to days or even weeks.

The control of the anabolic part is mainly exerted at the level of gene transcription into mRNAs, and of mRNA translation into proteins.

First, gene expression is controlled by transcription factors, which bind to and control the transcription machinery. There are many ligand-modulated transcription factors, which mediate the effects of steroid and thyroid hormones, lipids, and xenobiotics [18]. In the liver, insulin favors glucose utilization and inhibits fatty acid oxidation through the action of the sterol regulatory element binding protein-1c (SREBP1c) [19]. Interestingly, nutrients are also able to regulate gene expression. The carbohydrate response element binding protein (ChREBP) controls carbohydrate responsive genes in response to glucose in liver and adipose tissue [20,21], whereas the peroxisome proliferator-activated receptors (PPARs) control fatty acid oxidation, with RXRalpha as the predominant heterodimeric partner [22–24]. In addition, PPARs and the estrogen-related receptors (ERR alpha and gamma), together with the peroxisome-activated receptor gamma coactivator (PGC-1alpha), are the main regulators of mitochondrial biogenesis and coordinate cardiac metabolism [25–27]. PPARs are particularly important in the regulation of nutrient utilization in the heart. Finally, oxygen availability also affects the expression of several genes through hypoxia-induced factor, which favors the anaerobic utilization of glucose [28].

Second, the rate of mRNA translation into proteins, that is, the overall rate of protein synthesis, is regulated by factors that control the processes of initiation, elongation, and termination, and which are also under hormonal control. For example, the stimulation of protein synthesis by insulin is mediated by the insulin-signaling pathway that involves the activation of the phosphatidylinositol 3-kinase/protein kinase B (PKB/AKT) axis and leads to an overall stimulation of the translation process by activation of the nutrient-sensitive kinase, the mammalian target of rapamycin (mTOR), and of initiation and elongation factors [29,30]. Amino acids, especially leucine, also control the rate of protein synthesis by the activation of mTOR [31–33].

Interestingly, the new class of nonprotein coding RNAs opens a new era in metabolic control and especially offers a new insight on the control of various steps of protein synthesis [34]. For example, the microRNA cluster miRNA199a-214 actively represses cardiac PPAR expression under hemodynamic stress, and so may facilitate the metabolic shift from fatty acid oxidation toward reliance on glucose metabolism [35].

Mechanisms of protein degradation are lysosomal or nonlysosomal, they may require ATP and be selective or not. The long-lived enzymes, the so-called "house-keeping" enzymes, are in excess and are generally degraded by lysosomal proteases. The short-lived enzymes are less abundant and usually degraded by nonlysosomal processes. The main extralysosomal protein degradation pathway is specific and under the control of ubiquitin, a well-conserved 76 residue protein. The tagging of a lysine residue with several ubiq-uitin molecules in an ATP-dependent three-step pathway marks this protein out for degradation by the proteasome, by the so-called ubiquitin-proteasome system (UPS) [36,37]. In addition, proteome homeostasis, also termed *proteostasis*, results from the subtle balance between intracellular proteolytic processes, autophagy, and hormone action, and is also controlled by AMPK and mTOR [38–41]. Abnormal autophagic fluxes, either excessive or insufficient, may lead to heart disease [42–44].

Short-Term Control of Enzyme Activity: Effect of Substrates and Ligands

The saturation kinetics of an enzyme goes from first-order at low-substrate concentration (directly proportional to substrate concentration) to zero order rate (constant rate) at saturating concentration of substrate, tending toward V_{max}, a limiting value. As explained in the last section of this chapter, these saturating conditions may be required as flux-generating steps to maintain a steady flux through the pathway.

The effect of substrate concentration on the reaction rate in a simple "Michaelian" enzyme is relatively modest [1,2]. To increase its activity, say from 10% to 90% of V_{max}, a more than 80-fold increase in substrate concentration is required. Such a large change in substrate concentration is seldom observed and could perturb the osmotic and ionic state of the cell. However conformational changes by reversible binding of ligands to allosteric enzymes allow for much larger change in activity. The same eight- to ninefold increase in activity could be achieved with a three- or fourfold increase in substrate concentration depending on the extent of cooperativity [1,2]. This substrate response is much more satisfactory and applies to a large number of key enzymes. In addition, metabolic signals may act at different sites in a coordinate manner (see Fig. 5.1 and the last section of this chapter).

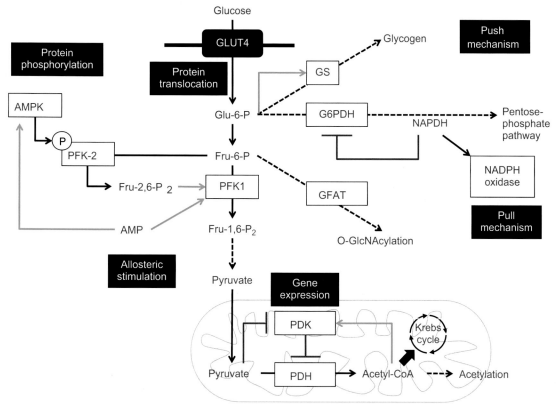

FIGURE 5.1 **Multistep coordinated control of glucose metabolism in the heart.** The fate of glucose is regulated by several mechanisms, including allosteric interaction, gene expression, protein phosphorylation, protein translocation, pull mechanism, and push mechanism as detailed in the text. Glucose metabolism may lead to acetylation and O-GlcNAcylation of proteins. Abbreviations: AMPK, AMP-activated protein kinase; Fru-1,6-P$_2$, fructose-1,6-bisphosphate; Fru-2,6-P$_2$, fructose-2,6-bisphosphate; G6PDH, glucose-6-phosphate dehydrogenase; GFAT, glutamine fructose-6-phosphate amido-transferase; GLUT4, glucose transporter 4; PDH, pyruvate dehydrogenase; PDK, pyruvate dehydrogenase kinase; PFK1, 6-phosphofructo-1-kinase; PFK2, 6-phosphofructo-2-kinase.

Compartmentation

A large number of metabolites are distributed between several subcellular compartments where their concentrations may differ [1]. This is the case for adenine nucleotides and acetyl-CoA whose mitochondrial and cytosolic concentrations differ. In addition, what matters to enzymes, is not the total concentration of substrates and ligands, but their free concentrations. Binding of metabolites to abundant proteins result in chemi-

cal compartmentation that decreases their actual free concentration. For example, the total concentration of AMP (about 0.2 mM in normoxic tissues) greatly exceeds its free concentration. AMP is distributed between the cytosol and mitochondria and binds to several abundant proteins such as glycogen phosphorylase and adenylate kinase and, in perfused hearts, its free concentration calculated from P-NMR spectroscopy is at least 10 times lower than the total concentrations [45].

Short-Term Control of Enzyme Activity: Covalent Modification by Phosphorylation

Phosphorylation, the most prevalent reversible covalent modification, is catalyzed by protein kinases, which transfer the terminal phosphoryl group of ATP to the hydroxyl group of serine or threonine residues (protein Ser/Thr kinase), or of tyrosine (protein Tyr kinase). Dephosphorylation is catalyzed by protein phosphatases, which hydrolyze the phosphate group. The different classes of protein kinases and protein phosphatases are listed in Tables 5.1 and 5.2.

The term interconversion cycle is preferred to reversible phosphorylation because the reaction of phosphoryl transfer is displaced far from equilibrium. In the interconversion cycle, separate converter enzymes act on their respective substrates, thus forming a "cascade." Such a simple "cascade" acts as a metabolic integrator, which integrates changes in the concentrations of ligands that affect the interconvertible enzymes or either or both of the converter enzymes, thus translating the change in effector concentration into a metabolic adjustment in the steady state level of activity of the interconvertible enzyme. The cascade is also characterized by its capacity for signal amplification, which transforms small changes in the concentration of allosteric modifiers into much larger changes in activity of the interconvertible enzyme. This property and the sensitivity to a large number of allosteric modifiers are further enhanced in multicyclic cascades [5–8].

The interconversion of key metabolic enzymes by (de)phosphorylation is a common, rapid, and efficient mechanism for reversibly controlling their activity. The phosphorylation state of interconvertible enzymes, and hence their activation state, depend on the dynamic equilibrium between the intervening protein kinases and protein phosphatases. All intermediary levels of activation states can exist so that the activity and the resulting flux can be very finely tuned.

TABLE 5.1   Some Protein Kinases Involved in Metabolism

Group	Protein kinase	Signal/ligand	Metabolism
AGC	PKA	Cyclic AMP	Glucose, lipid, prot.
	PKB/AKT	Insulin	Glucose, lipid, prot.
	PKC family	DAG, Ca^{2+}	NADPH oxidase
	Ribosomal S6 kinase	Insulin	Protein synthesis
CaMK	Ca^{2+}/calmodulin PK	Ca^{2+} and CM	Multifunctional
	Phosphorylase kinase	Idem	Glycogen
	Myosin light chain kinase	Idem	Muscle contraction
	Elongation factor 2-kinase	Idem	Protein synthesis
	AMPK	AMP	Glucose, lipid, prot.
CMGC	Mitogen-activated PK	Growth factors	Transcription
	Glycogen synthase kinase 3	Constitutively active	Glycogen synthase
PTK	Insulin receptor TK	Insulin	Glucose, lipid, prot.
Others	PDK, mitochondrial	Pyruvate	Pyruvate oxidation

CM, calmodulin; DAG, diacylglycerol; PDK, pyruvate dehydrogenase kinase; PKA, cyclic AMP-dependent protein kinase; PKB, protein kinase B or AKT; PKC, protein kinase C; prot., protein; PTK, protein tyrosine kinase.

TABLE 5.2 Some Protein Ser/Thr Phosphatases Involved in Metabolism

Type	Targets	Specific properties
PP1	Glycogen (PP-1G)	Inhibited by phosphorylated inhibitor1, okadaic acid or microcystin
PP2A	Fatty acid synthesis Glycolysis	Inhibited by okadaic acid or microcystin
PP2B (calcineurin)	NFAT	Ca^{2+}/calmodulin-dependent
PP2C	AMPK	Mg^{2+}-dependent

PP, protein phosphatase; PP-1G, glycogen-targeting subunit of PP1; NFAT, nuclear factor of activated T cells.

As regards, the effects of (de)phosphorylation on enzyme activity, (de)phosphorylation may affect the kinetic properties of the (de)phosphorylated enzymes, with all possible changes (V_{max}, K_m, affinity for allosteric modifiers) resulting in either activation or inactivation. Otherwise, it appears that the kinetic properties of a large number of phosphorylated enzymes are not modified by phosphorylation (silent phosphorylation). The exact role of this silent phosphorylation remains poorly understood, even if changes in cell localization, stability, or possible protein-interaction could be expected.

Protein Kinases

The protein kinases represent a large family of proteins, which have been classified into several groups that share several structural and functional properties (Table 5.1) [46–48]. The 3D structure of the catalytic subunit of all protein kinases is remarkably conserved [48]. For most of these protein kinases, their K_m for ATP is relatively low (<0.2 mM), that is, more than 10 times lower than the concentration of cellular ATP under normoxic conditions [49]. This remarkable property ensures that protein kinases, and especially AMPK, can still work when more

than 90% of ATP is lost, as occurs under ischemic conditions.

The following protein Ser/Thr kinases control key metabolic enzymes.

Cyclic-AMP-dependent protein kinase (PKA) is involved in the control of carbohydrate, lipid, and protein metabolism. Cyclic AMP binds to the two regulatory subunits and so liberates the catalytic subunits, which can then phosphorylate the PKA target [46,47,50,51].

AMPK is involved in energy homeostasis [9–13]. It senses the energy status of the cell and is activated by energy imbalance resulting from decreased ATP production or increased ATP consumption, and has been called the fuel gauge of the cell [52]. It is stimulated by AMP and ADP, and inhibited by ATP. Any increase in the AMP/ATP ratio of the cell promotes AMPK phosphorylation and activation by upstream kinases. Once activated, AMPK aims at restoring the energy charge of the cell by inhibiting the biosynthetic ATP-consuming pathways and by stimulating the catabolic ATP-generating pathways (glucose uptake, glycolysis, fatty acid oxidation). The interest in AMPK also stems from its insulin-independent stimulation of glucose transport in muscle and from the notion that AMPK mediates the antidiabetic effect of biguanides.

Insulin receptor tyrosine kinase and insulin signaling [30,53]. The binding of insulin to its receptor induces the autophosphorylation of three tyrosine residues, which in turn stimulates the tyrosine kinase activity of the receptor toward its two substrates, IRS 1 and 2, which are required for the metabolic response to insulin. Tyrosine phosphorylation of both IRS creates binding sites for SH2 domain proteins, including the regulatory subunits of phosphatidylinositol 3-kinase, which produces PIP3 (phosphatidylinositol-3,4,5-trisphosphate) and eventually recruits the protein serine/

threonine kinase PKB, also called AKT, which is activated by phosphorylation on its T308 by PDK1. Activated PKB/AKT mediates most of the metabolic effects of insulin. Interestingly, phosphorylation of certain Ser/Thr residues of IRS1/2 inhibits the insulin receptor tyrosine kinase activity and could be involved in insulin resistance.

Multiple and hierarchical phosphorylations. Multiple phosphorylations by different protein kinases on the same enzyme are relatively frequent and may lead to interesting interactions, such as sequential and hierarchical phosphorylation. A classical example is glycogen synthase, in which nine different phosphorylation sites are found in three clusters [54]. Phosphorylation of the N-terminal site 2 by several protein kinases, including PKA, phosphorylase kinase, and AMPK, inactivates the enzyme and allows phosphorylation of the adjacent site 2a by the constitutively active casein kinase 1, which further inactivates the enzyme. Phosphorylation of site 5 in the central domain allows sequential phosphorylation of the adjacent sites 3a, 3b, 3c, and 4 by glycogen synthase kinase 3, which also inactivates the enzyme. Thus it appears that sites 2, 2a, 3a, and 3b control the activity, whereas the other sites are involved in the hierarchical control of phosphorylation. Another example is the cross talk between AKT/PKB and AMPK. The effect of insulin to antagonize AMPK by ischemia is mediated by a hierarchical mechanism whereby phosphorylation of an Ser residue of AMPK by AKT/PKB reduces subsequent phosphorylation of the activating threonine by LKB1 and the resulting activation of AMPK [55].

Protein Phosphatases

The family of protein phosphatases involved in the dephosphorylation of serine and threonine have rather broad specificities and have been classified by Cohen and coworkers into four main groups depending on their sensitivity to inhibitors and their requirements for divalent cations (see Table 5.2) [56–58]. Their specificity depends on their association with regulatory subunits that target the phosphatases to cellular structures. For example, PP1 is related to glycogen metabolism through PP-1G, its glycogen targeting subunit. PP1 is inhibited by Inhibitor-1, only when phosphorylated by PKA, and by okadaic acid and microcystin. Calcineurin, a Ca^{2+}-dependent protein phosphatase, dephosphorylates nuclear factor of activated T cells (NFAT), a transcription factor present in the cytosol, which then migrates to the nucleus where it activates the transcription of hypertrophic genes in the heart [59].

Protein tyrosine phosphatases (PTP) contain a conserved catalytic domain that differs from the catalytic domain of the protein Ser/Thr phosphatases [60,61]. They are very active and efficiently antagonize the tyrosine kinases so that the tyrosine phosphorylation state of their substrates is usually rather low, not exceeding 10%. PTP1B, an isoform of protein tyrosine phosphatases, controls insulin and leptin action.

Pyruvate Dehydrogenase: A Case in Point

Pyruvate dehydrogenase (PDH) catalyzes an irreversible and no return metabolic step because its substrate pyruvate is gluconeogenic or anaplerotic, whereas its product acetyl-CoA is not [62–65]. The control of its activity is complex and involves control by its substrates and products, covalent modification by (de)phosphorylation and long-term adaptation. PDH is feedback inhibited by its products and is inactivated by phosphorylation of three serine residues in the alpha-subunit of E1, one of the three components of the PDH complex. Four different PDH kinases (PDK) are known. Site 1 phosphorylation by all PDKs is inactivating, with PDK2 being the most active kinase on this site. Phosphorylation of site 2 (mainly by PDK4) and Site 3 (by PDK1 only) introduces hierarchical control by retarding site 1 dephosphorylation, thus keeping PDH in its inactive state. This is expected to occur in the heart, which expresses PDK1 on top of the other PDKs. In addition,

PDH substrates and products also control PDK activity. Pyruvate inhibits, whereas acetyl-CoA stimulates, PDK. However, the relative insensitivity of PDK4 for pyruvate maintains heart PDH in its inactive phosphorylated state, after prolonged starvation. Moreover, upregulation of PDK in response to high-fat diet, starvation, or insulin deficiency, keeps glucose oxidation at a low level, whereas fatty acid oxidation is increased, thus mimicking "metabolic inflexibility," a characteristic metabolic feature of insulin resistance [64,66–69]. Conversely, blood glucose in starved PDK4-deficient mice is lower than in the controls, probably because the active PDH diverts pyruvate, a gluconeogenic substrate, into acetyl-CoA [70]. These findings underline the pivotal role of PDH in the control of glucose and lipid metabolism.

Short-Term Control of Enzyme Activity: Other Types of Covalent Modifications

There are several types of posttranslational covalent modifications of proteins (PTM) other than phosphorylation. Recently, two types of PTMs, acetylation and O-GlcNAcylation, were particularly studied in the heart. First, acetylation can occur on the N-terminus or on the side chain of lysine residues. The N-terminal acetylation of most, if not all, cytosolic proteins is rather common and does not seem to affect their properties. However, acetylation of proteins on lysine could affect their properties, as is the case for several mitochondrial proteins [71]. In the heart, acetylation levels of lysine are modulated by Sirtuins, which are deacetylases [71–73]. Protein acetylation could also increase as a result of an increased concentration of acetyl-CoA [74]. Such an increase occurs when the oxidation of fatty acids or ketone bodies increase and could participate in the fuel selection in the heart.

Second, intracellular protein O-GlcN-Acylation by addition of *N*-acetyl glucosamine to the hydroxyl group of serine or threonine residues is another interesting case of interconver-

sion cycle [75,76]. It is catalyzed by O-GlcNAc transferase and O-GlcNAcase and seems to exhibit a diurnal variation in mouse hearts [77] and to mediate adverse effects in diabetes [78,79], although it could protect against certain stresses [80]. However, the effect of hyperglycemia to increase protein glycosylation has not been confirmed in isolated adult cardiomyocytes [81]. Increase in O-GlcNAcylation has also been reported in cardiovascular diseases such as heart failure and hypertrophy [75,82]. The exact role of such PTMs in the pathogenesis of these diseases is however not clear and needs further investigation. Interestingly, there is often a site-specific recognition for both phosphorylation and glycosylation, such that the dynamic glycosylation/phosphorylation is mutually exclusive and indirectly controls the activation state of enzymes [83,84].

CONTROL OF THE OVERALL FLUX THROUGH A PATHWAY

To generate large changes in flux, the following mechanisms are used (see Fig. 5.1).

Flux Reversal

The largest changes in flux result from covalent modification inducing opposite changes in the activity of enzymes catalyzing opposite reactions. They can reverse the direction of flux, if the changes in activity are large enough [4]. For example, PKA activates glycogen phosphorylase but inactivates glycogen synthase. The net result is flux reversal. The same applies to the control of lipolysis, fatty acid oxidation, and lipogenesis.

Flux-Generating Steps

To generate a lasting change in the overall flux through a pathway, a signal should activate a *"flux-generating step"* [85]. That is a reaction whose rate remains unaffected when the substrate concentration decreases but remains almost saturated to maintain a steady flux through the pathway. For example, activation of glycogen breakdown, be it hormonal or hypoxia-induced, provides glycolysis with a steady flow of glucosyl units until the glycogen stores are exhausted. Similarly, the stimulation of glucose uptake, by insulin or exercise, which increases the translocation of GLUT4 to the plasma membrane, allows the large blood glucose pool to supply the heart with glucose at a constant rate because the concentration of blood glucose remains almost saturating. Under these conditions, the limiting step is glucose transport and not its phosphorylation because of the high affinity of hexokinase for glucose.

Control by Supply: Push Mechanism

The stimulation of glucose uptake and oxidation, as well as the stimulation of glycogen synthesis, by insulin is a good example of a push mechanism that involves *feed-forward stimulation* and *multisite coordinated control* of key metabolic steps. We consider here three examples. First, for glucose uptake and oxidation, the insulin-induced activation of PKB/AKT leads to an increased recruitment of GLUT4 to the plasma membrane, and to a phosphorylation and activation of 6-phosphofructo-2-kinase, the enzyme responsible for the synthesis of fructose-2,6-bisphosphate, which is a potent allosteric activator of PFK1 (feed-forward stimulation). Insulin also activates PDH [86], a key step in mitochondrial glucose oxidation because of the inhibition of PDK by pyruvate and the activation of PDH phosphatase [51,65,87]. Second, for glycogen synthesis, the push mechanism involves a double activation of glycogen synthase as a result of the increased concentration of glucose-6-phosphate, an allosteric activator of glycogen synthase and inhibitor of glycogen synthase kinase (feed-forward stimulation), following the increased transport and phosphorylation

of glucose, and of the dephosphorylation and activation of glycogen synthase [88]. Finally, for the stimulation of glycolysis by oxygen deprivation, AMP orchestrates the metabolic response by stimulating glycogen phosphorylase, 6-phosphofructo1-kinase (PFK1), and AMPK, which in turn stimulates PFK2 and the recruitment of GLUT4 [89,90]. PFK1 is a multimodulated allosteric enzyme, which is maximally stimulated under hypoxic conditions when the concentration of allosteric activators increase (fructose-6-phosphate, AMP, Fructose-2,6-bisphosphate), whereas those of inhibitors decrease (ATP) [91]. However, the stimulation of PFK1 alone is unable to enhance the glycolytic flux. The overall flux can only increase if the glucose-6-phosphate supply is also increased. Therefore, GLUT4 recruitment and the stimulation of glycogen breakdown are intrinsic parts of the multisite concerted mechanism by which glycolysis is stimulated. In conclusion, a multisite coordination is required to obtain large changes in overall flux and to avoid accumulation of metabolic intermediates.

Control by Demand: Pull Mechanism

Flux control by demand, combined with feedback inhibition, is observed in many biosynthetic pathways, in which the consumption of the final product by another pathway decreases its concentration and so relieves the *feedback inhibition* that it exerts on the first step of the supply pathway. It is a ligand-gated mechanism that controls the access to a pathway. These feedback inhibition loops allow for relatively large changes in flux with minimal changes in the concentrations of metabolic intermediates [1,2,92] and have been described for many biosynthetic pathways, such as the synthesis of certain amino acids, cholesterol, purine nucleotides, etc.

Pull mechanisms are also involved in the control of flux through *metabolic crossroads*. For example, the metabolic fate of glucose-6-phosphate under anabolic conditions is driven by demand via a pull mechanism: glycogen synthesis by activation of glycogen synthase [88], glycolysis by stimulation of PFK1 by fructose-2,6-bisphosphate [87,91], or pentose-phosphate pathway by deinhibition of glucose-6-phosphate dehydrogenase by NADPH-consumption [93]. At the pyruvate crossroads, the metabolic fate of pyruvate mainly depends on the activation state of PDH, which controls its oxidation.

The lack of feedback inhibition may have dramatic consequences, as is the case for the synthesis of ketone bodies. Their blood concentrations result from the hormonal stimulation of lipolysis in adipose tissue, fatty acid oxidation, and ketogenesis in the liver and from their consumption in extrahepatic tissues. In the absence of insulin, the overproduction together with the decreased consumption of ketone bodies causes ketoacidosis, a life-threatening consequence of the imbalance [4].

It is not clear why certain metabolic pathways contain feedback inhibition, whereas others do not. It could be that in certain metabolic pathways, large changes in the concentration of the product are required to supply the consuming pathway, which has a low affinity for this product [1,2,4].

Substrate Interaction and Competition

Feedback inhibition also applies to the well-known glucose-sparing effect of alternative oxidizable fuels discovered by Randle and coworkers in the 1960s, which results from an inhibition of glycolysis [65,94]. Fatty acid, ketone bodies, or even lactate, the product of glycolysis, exert this inhibition, which is unevenly exerted at several levels (multisite coordinated control), glucose transport being less inhibited than PFK1 and less than PDH. These inhibition loops may reroute glucose toward glycogen, and pyruvate toward oxaloacetate, which, through this anaplerotic diversion, may replenish the Krebs cycle intermediates. Shulman and coworkers have questioned

the primordial inhibition of PDH and have put forward an alternative explanation for the skeletal muscle. They have shown that, during fatty acid oxidation, the first steps of glucose metabolism, namely glucose transport by GLUT4, are more inhibited than the mitochondrial oxidation of pyruvate [95]. Remarkably, the common initial trigger, whether reported by Randle, Shulman, or by others, is a mitochondrial event related to fatty acid oxidation that eventually decreases glucose uptake [94–96]. This intricate cross talk gives mitochondria a pivotal role in the control of glucose metabolism.

Interestingly, the opposite, namely the inhibition of fatty acid oxidation, by glucose and insulin also holds true [65]. The mechanism implies an inhibition of fatty acid oxidation by malonyl-CoA. This compound inhibits carnitine palmitoyltransferase (CPT1), which controls the entry and oxidation of fatty acids in the mitochondria [97,98]. The accumulation of malonyl-CoA results from the accumulation of acetyl-CoA in the cytosol and its carboxylation into malonyl-Co by acetyl-CoA carboxylase. Here again, there is a cross talk between insulin and AMPK because the latter phosphorylates and inactivates acetyl-CoA carboxylase. Therefore it appears that the combined action of several key metabolites together with covalent modification is able to change the substrate availability to the heart. This metabolic flexibility is decreased or even lost in insulin resistant hearts.

Protein–Protein Interactions

They include, among others, compartmentation, anchoring, docking, adaptor proteins, interactions with the cytoskeleton, lipid rafts, scaffold formation, etc. The number of metabolic and signaling pathways involving protein interactions as well as the number of conserved structures implied in these interactions is enormous and increasing by the day. If the *qualitative* effect of protein interaction is rather well documented, a *quantitative* analysis of this type of interaction still awaits formal mathematical description and explains why the precise assessment of their importance cannot be taken into account.

CONCLUSIONS

The principles that govern the regulation of cardiac metabolism are general and relatively simple. Metabolic adaptation to changes in energy demand or supply is signaled by hormones or nutrients and mediated by changes in the activity of the key enzymes in the pathway. The changes are immediate or delayed depending on the signals and they have to be coordinated in order to affect the overall flux through the pathway.

References

[1] Fell D. Understanding the control of metabolism. London, UK: Portland Press; 1997.

[2] Cornish-Bowden A. Fundamentals of enzyme kinetics. London, UK: Portland Press; 1995.

[3] Wittinghofer A, Vetter IR. Structure-function relationships of the G domain, a canonical switch motif. Annu Rev Biochem 2011;80:943–71.

[4] Hue L. From control to regulation. In: Cornish-Bowden AJ, Cardenas ML, editors. Technological and medical implications of metabolic control analysis. Dordrecht, The Netherlands: Kluwer Academic Publisher; 2000. p. 329–38.

[5] Stadtman ER, Chock PB. Superiority of interconvertible enzyme cascades in metabolic regulation: analysis of monocyclic systems. Proc Natl Acad Sci USA 1977;74:2761–5.

[6] Chock PB, Stadtman ER. Superiority of interconvertible enzyme cascades in metabolite regulation: analysis of multicyclic systems. Proc Natl Acad Sci USA 1977;74:2766–70.

[7] Goldbeter A, Koshland DE Jr. Ultrasensitivity in biochemical systems controlled by covalent modification. Interplay between zero-order and multistep effects. J Biol Chem 1984;259:14441–7.

[8] Cardenas ML, Cornish-Bowden A. Characteristics necessary for an interconvertible enzyme cascade to generate a highly sensitive response to an effector. Biochem J 1989;257:339–45.

[9] Hue L, Rider MH. The AMP-activated protein kinase: more than an energy sensor. Essays Biochem 2007;43: 121–37.

[10] Beauloye C, Bertrand L, Horman S, Hue L. AMPK activation, a preventive therapeutic target in the transition from cardiac injury to heart failure. Cardiovasc Res 2011;90:224–33.

[11] Zaha VG, Young LH. AMP-activated protein kinase regulation and biological actions in the heart. Circ Res 2012;111:800–14.

[12] Hardie DG. AMPK: positive and negative regulation, and its role in whole-body energy homeostasis. Curr Opin Cell Biol 2014;33C:1–7.

[13] Kim TT, Dyck JR. Is AMPK the savior of the failing heart? Trends Endocrinol Metab 2015;26:40–8.

[14] Gough DR, Cotter TG. Hydrogen peroxide: a Jekyll and Hyde signalling molecule. Cell Death Dis 2011;2:e213.

[15] Nickel A, Kohlhaas M, Maack C. Mitochondrial reactive oxygen species production and elimination. J Mol Cell Cardiol 2014;73:26–33.

[16] Reczek CR, Chandel NS. ROS-dependent signal transduction. Curr Opin Cell Biol 2014;33C:8–13.

[17] Sag CM, Santos CX, Shah AM. Redox regulation of cardiac hypertrophy. J Mol Cell Cardiol 2014;73:103–11.

[18] Fan W, Downes M, Atkins A, Yu R, Evans RM. Nuclear receptors and AMPK: resetting metabolism. Cold Spring Harb Symp Quant Biol 2011;76:17–22.

[19] Foufelle F, Ferre P. New perspectives in the regulation of hepatic glycolytic and lipogenic genes by insulin and glucose: a role for the transcription factor sterol regulatory element binding protein-1c. Biochem J 2002;366:377–91.

[20] Uyeda K, Repa JJ. Carbohydrate response element binding protein, ChREBP, a transcription factor coupling hepatic glucose utilization and lipid synthesis. Cell Metab 2006;4:107–10.

[21] Postic C, Dentin R, Denechaud PD, Girard J. ChREBP, a transcriptional regulator of glucose and lipid metabolism. Annu Rev Nutr 2007;27:179–92.

[22] Desvergne B, Michalik L, Wahli W. Transcriptional regulation of metabolism. Physiol Rev 2006;86:465–514.

[23] Finck BN. The PPAR regulatory system in cardiac physiology and disease. Cardiovasc Res 2007;73:269–77.

[24] Ahmadian M, Suh JM, Hah N, Liddle C, Atkins AR, Downes M, Evans RM. PPARgamma signaling and metabolism: the good, the bad and the future. Nat Med 2013;19:557–66.

[25] Schreiber SN, Emter R, Hock MB, Knutti D, Cardenas J, Podvinec M, Oakeley EJ, Kralli A. The estrogen-related receptor alpha (ERRalpha) functions in PPARgamma coactivator 1alpha (PGC-1alpha)-induced mitochondrial biogenesis. Proc Natl Acad Sci USA 2004;101:6472–7.

[26] Ventura-Clapier R, Garnier A, Veksler V. Transcriptional control of mitochondrial biogenesis: the central role of PGC-1alpha. Cardiovasc Res 2008;79:208–17.

[27] Wang T, McDonald C, Petrenko NB, Leblanc M, Wang T, Giguere V, Evans RM, Patel VV, Pei L. ERRalpha and

ERRgamma are essential coordinators of cardiac metabolism and function. Mol Cell Biol 2015;35:1281–98.

[28] Semenza GL. Hypoxia-inducible factor 1 and cardiovascular disease. Annu Rev Physiol 2014;76:39–56.

[29] Proud CG. Regulation of protein synthesis by insulin. Biochem Soc Trans 2006;34:213–6.

[30] Bertrand L, Horman S, Beauloye C, Vanoverschelde JL. Insulin signalling in the heart. Cardiovasc Res 2008;79:238–48.

[31] Proud CG. Control of the translational machinery by amino acids. Am J Clin Nutr 2014;99:231S–6S.

[32] Averous J, Lambert-Langlais S, Carraro V, Gourbeyre O, Parry L, B'Chir W, Muranishi Y, Jousse C, Bruhat A, Maurin AC, Proud CG, Fafournoux P. Requirement for lysosomal localization of mTOR for its activation differs between leucine and other amino acids. Cell Signal 2014;26:1918–27.

[33] Jewell JL, Kim YC, Russell RC, Yu FX, Park HW, Plouffe SW, Tagliabracci VS, Guan KL. Metabolism. Differential regulation of mTORC1 by leucine and glutamine. Science 2015;347:194–8.

[34] Thum T, Condorelli G. Long noncoding RNAs and MicroRNAs in cardiovascular pathophysiology. Circ Res 2015;116:751–62.

[35] el Azzouzi H, Leptidis S, Dirkx E, Hoeks J, van Bree B, Brand K, McClellan EA, Poels E, Sluimer JC, van den Hoogenhof MM, Armand AS, Yin X, Langley S, Bourajjaj M, Olieslagers S, Krishnan J, Vooijs M, Kurihara H, Stubbs A, Pinto YM, Krek W, Mayr M, da Costa Martins PA, Schrauwen P, De Windt LJ. The hypoxia-inducible microRNA cluster miR-199a approximately 214 targets myocardial PPARdelta and impairs mitochondrial fatty acid oxidation. Cell Metab 2013;18:341–54.

[36] Portbury AL, Ronnebaum SM, Zungu M, Patterson C, Willis MS. Back to your heart: ubiquitin proteasome system-regulated signal transduction. J Mol Cell Cardiol 2012;52:526–37.

[37] Cui Z, Scruggs SB, Gilda JE, Ping P, Gomes AV. Regulation of cardiac proteasomes by ubiquitination, SUMOylation, and beyond. J Mol Cell Cardiol 2014;71:32–42.

[38] Ronnebaum SM, Patterson C, Schisler JC. Minireview: hey U(PS): metabolic and proteolytic homeostasis linked via AMPK and the ubiquitin proteasome system. Mol Endocrinol 2014;28:1602–15.

[39] Russell RC, Yuan HX, Guan KL. Autophagy regulation by nutrient signaling. Cell Res 2014;24:42–57.

[40] Kubli DA, Gustafsson AB. Cardiomyocyte health: adapting to metabolic changes through autophagy. Trends Endocrinol Metab 2014;25:156–64.

[41] Li Y, Chen C, Yao F, Su Q, Liu D, Xue R, Dai G, Fang R, Zeng J, Chen Y, Huang H, Ma Y, Li W, Zhang L, Liu C, Dong Y. AMPK inhibits cardiac hypertrophy by promoting autophagy via mTORC1. Arch Biochem Biophys 2014;558:79–86.

[42] Schiaffino S, Dyar KA, Ciciliot S, Blaauw B, Sandri M. Mechanisms regulating skeletal muscle growth and atrophy. FEBS J 2013;280:4294–314.

[43] Sandri M, Robbins J. Proteotoxicity: an underappreciated pathology in cardiac disease. J Mol Cell Cardiol 2014;71:3–10.

[44] Lavandero S, Chiong M, Rothermel BA, Hill JA. Autophagy in cardiovascular biology. J Clin Invest 2015;125:55–64.

[45] Frederich M, Balschi JA. The relationship between AMP-activated protein kinase activity and AMP concentration in the isolated perfused rat heart. J Biol Chem 2002;277:1928–32.

[46] Hardie G, Hanks S. The protein kinases facts book. San Diego: Academic Press; 1995.

[47] Cohen P. The role of protein phosphorylation in the hormonal control of enzyme activity. Eur J Biochem 1985;151:439–48.

[48] Endicott JA, Noble ME, Johnson LN. The structural basis for control of eukaryotic protein kinases. Annu Rev Biochem 2012;81:587–613.

[49] Knight ZA, Shokat KM. Features of selective kinase inhibitors. Chem Biol 2005;12:621–37.

[50] Depre C, Ponchaut S, Deprez J, Maisin L, Hue L. Cyclic AMP suppresses the inhibition of glycolysis by alternative oxidizable substrates in the heart. J Clin Invest 1998;101:390–7.

[51] Depre C, Rider MH, Hue L. Mechanisms of control of heart glycolysis. Eur J Biochem 1998;258:277–90.

[52] Hardie DG, Carling D. The AMP-activated protein kinase – fuel gauge of the mammalian cell? Eur J Biochem 1997;246:259–73.

[53] Copps KD, White MF. Regulation of insulin sensitivity by serine/threonine phosphorylation of insulin receptor substrate proteins IRS1 and IRS2. Diabetologia 2012;55:2565–82.

[54] Roach PJ. Control of glycogen synthase by hierarchal protein phosphorylation. FASEB J 1990;4:2961–8.

[55] Horman S, Vertommen D, Heath R, Neumann D, Mouton V, Woods A, Schlattner U, Wallimann T, Carling D, Hue L, Rider MH. Insulin antagonizes ischemia-induced Thr172 phosphorylation of AMP-activated protein kinase alpha-subunits in heart via hierarchical phosphorylation of Ser485/491. J Biol Chem 2006;281:5335–40.

[56] Cohen P. The structure and regulation of protein phosphatases. Annu Rev Biochem 1989;58:453–508.

[57] Cohen PT. Novel protein serine/threonine phosphatases: variety is the spice of life. Trends Biochem Sci 1997;22:245–51.

[58] Dawson JF, Holmes CF. Molecular mechanisms underlying inhibition of protein phosphatases by marine toxins. Front Biosci 1999;4:D646–58.

[59] Tham YK, Bernardo BC, Ooi JY, Weeks KL, McMullen JR. Pathophysiology of cardiac hypertrophy and heart

failure: signaling pathways and novel therapeutic targets. Arch Toxicol 2015; 89(9):1401-38.

[60] Tonks NK. Protein tyrosine phosphatases: from genes, to function, to disease. Nat Rev Mol Cell Biol 2006;7:833–46.

[61] Tonks NK. Protein tyrosine phosphatases – from housekeeping enzymes to master regulators of signal transduction. FEBS J 2013;280:346–78.

[62] Holness MJ, Sugden MC. Regulation of pyruvate dehydrogenase complex activity by reversible phosphorylation. Biochem Soc Trans 2003;31:1143–51.

[63] Sugden MC, Holness MJ. Mechanisms underlying regulation of the expression and activities of the mammalian pyruvate dehydrogenase kinases. Arch Physiol Biochem 2006;112:139–49.

[64] Sugden MC, Zariwala MG, Holness MJ. PPARs and the orchestration of metabolic fuel selection. Pharmacol Res 2009;60:141–50.

[65] Hue L, Taegtmeyer H. The Randle cycle revisited: a new head for an old hat. Am J Physiol Endocrinol Metab 2009;297:E578–91.

[66] Zhao G, Jeoung NH, Burgess SC, Rosaaen-Stowe KA, Inagaki T, Latif S, Shelton JM, McAnally J, Bassel-Duby R, Harris RA, Richardson JA, Kliewer SA. Overexpression of pyruvate dehydrogenase kinase 4 in heart perturbs metabolism and exacerbates calcineurin-induced cardiomyopathy. Am J Physiol Heart Circ Physiol 2008;294:H936–43.

[67] Chambers KT, Leone TC, Sambandam N, Kovacs A, Wagg CS, Lopaschuk GD, Finck BN, Kelly DP. Chronic inhibition of pyruvate dehydrogenase in heart triggers an adaptive metabolic response. J Biol Chem 2011;286:11155–62.

[68] Houten SM, Chegary M, Te Brinke H, Wijnen WJ, Glatz JF, Luiken JJ, Wijburg FA, Wanders RJ. Pyruvate dehydrogenase kinase 4 expression is synergistically induced by AMP-activated protein kinase and fatty acids. Cell Mol Life Sci 2009;66:1283–94.

[69] Jeong JY, Jeoung NH, Park KG, Lee IK. Transcriptional regulation of pyruvate dehydrogenase kinase. Diabetes Metab J 2012;36:328–35.

[70] Jeoung NH, Harris RA. Pyruvate dehydrogenase kinase-4 deficiency lowers blood glucose and improves glucose tolerance in diet-induced obese mice. Am J Physiol Endocrinol Metab 2008;295:E46–54.

[71] Papanicolaou KN, O'Rourke B, Foster DB. Metabolism leaves its mark on the powerhouse: recent progress in post-translational modifications of lysine in mitochondria. Front Physiol 2014;5:301.

[72] Sack MN. Emerging characterization of the role of SIRT3-mediated mitochondrial protein deacetylation in the heart. Am J Physiol Heart Circ Physiol 2011;301:H2191–7.

[73] Rardin MJ, He W, Nishida Y, Newman JC, Carrico C, Danielson SR, Guo A, Gut P, Sahu AK, Li B, Uppala R,

Fitch M, Riiff T, Zhu L, Zhou J, Mulhern D, Stevens RD, Ilkayeva OR, Newgard CB, Jacobson MP, Hellerstein M, Goetzman ES, Gibson BW, Verdin E. SIRT5 regulates the mitochondrial lysine succinylome and metabolic networks. Cell Metab 2013;18:920–33.

[74] Abo Alrob O, Lopaschuk GD. Role of CoA and acetyl-CoA in regulating cardiac fatty acid and glucose oxidation. Biochem Soc Trans 2014;42:1043–51.

[75] Dassanayaka S, Jones SP. O-GlcNAc and the cardiovascular system. Pharmacol Ther 2014;142:62–71.

[76] Ma J, Hart GW. O-GlcNAc profiling: from proteins to proteomes. Clin Proteomics 2014;11:8.

[77] Durgan DJ, Pat BM, Laczy B, Bradley JA, Tsai JY, Grenett MH, Ratcliffe WF, Brewer RA, Nagendran J, Villegas-Montoya C, Zou C, Zou L, Johnson RL Jr, Dyck JR, Bray MS, Gamble KL, Chatham JC, Young ME. O-GlcNAcylation, novel post-translational modification linking myocardial metabolism and cardiomyocyte circadian clock. J Biol Chem 2011;286:44606–19.

[78] Erickson JR, Pereira L, Wang L, Han G, Ferguson A, Dao K, Copeland RJ, Despa F, Hart GW, Ripplinger CM, Bers DM. Diabetic hyperglycaemia activates CaMKII and arrhythmias by O-linked glycosylation. Nature 2013;502:372–6.

[79] McLarty JL, Marsh SA, Chatham JC. Post-translational protein modification by O-linked N-acetyl-glucosamine: its role in mediating the adverse effects of diabetes on the heart. Life Sci 2013;92:621–7.

[80] Ngoh GA, Facundo HT, Zafir A, Jones SP. O-GlcNAc signaling in the cardiovascular system. Circ Res 2010;107:171–85.

[81] Balteau M, Tajeddine N, de Meester C, Ginion A, Des Rosiers C, Brady NR, Sommereyns C, Horman S, Vanoverschelde JL, Gailly P, Hue L, Bertrand L, Beauloye C. NADPH oxidase activation by hyperglycaemia in cardiomyocytes is independent of glucose metabolism but requires SGLT1. Cardiovasc Res 2011;92:237–46.

[82] Lunde IG, Aronsen JM, Kvaloy H, Qvigstad E, Sjaastad I, Tonnessen T, Christensen G, Gronning-Wang LM, Carlson CR. Cardiac O-GlcNAc signaling is increased in hypertrophy and heart failure. Physiol Genomics 2012;44:162–72.

[83] Hart GW, Slawson C, Ramirez-Correa G, Lagerlof O. Cross talk between O-GlcNAcylation and phosphorylation: roles in signaling, transcription, and chronic disease. Annu Rev Biochem 2011;80:825–58.

[84] Zachara NE. The roles of O-linked beta-N-acetylglucosamine in cardiovascular physiology and disease. Am J Physiol Heart Circ Physiol 2012;302:H1905–18.

[85] Newsholme EA, Crabtree B. Flux-generating and regulatory steps in metabolic control. Trends Biochem Sci 1981;6:53–6.

[86] McCormack JG, Halestrap AP, Denton RM. Role of calcium ions in regulation of mammalian intramitochondrial metabolism. Physiol Rev 1990;70:391–425.

[87] Depre C, Rider MH, Veitch K, Hue L. Role of fructose 2,6-bisphosphate in the control of heart glycolysis. J Biol Chem 1993;268:13274–9.

[88] Bouskila M, Hunter RW, Ibrahim AF, Delattre L, Peggie M, van Diepen JA, Voshol PJ, Jensen J, Sakamoto K. Allosteric regulation of glycogen synthase controls glycogen synthesis in muscle. Cell Metab 2010;12:456–66.

[89] Marsin AS, Bertrand L, Rider MH, Deprez J, Beauloye C, Vincent MF, Van den Berghe G, Carling D, Hue L. Phosphorylation and activation of heart PFK-2 by AMPK has a role in the stimulation of glycolysis during ischaemia. Curr Biol 2000;10:1247–55.

[90] Russell RR III, Bergeron R, Shulman GI, Young LH. Translocation of myocardial GLUT-4 and increased glucose uptake through activation of AMPK by AICAR. Am J Physiol 1999;277:H643–9.

[91] Van Schaftingen E, Jett MF, Hue L, Hers HG. Control of liver 6-phosphofructokinase by fructose 2,6-bisphosphate and other effectors. Proc Natl Acad Sci USA 1981;78:3483–6.

[92] Hofmeyr JH, Cornish-Bowden A. Quantitative assessment of regulation in metabolic systems. Eur J Biochem 1991;200:223–36.

[93] Eggleston LV, Krebs HA. Regulation of the pentose phosphate cycle. Biochem J 1974;138:425–35.

[94] Randle PJ, Garland PB, Hales CN, Newsholme EA. The glucose fatty-acid cycle. Its role in insulin sensitivity and the metabolic disturbances of diabetes mellitus. Lancet 1963;1:785–9.

[95] Dresner A, Laurent D, Marcucci M, Griffin ME, Dufour S, Cline GW, Slezak LA, Andersen DK, Hundal RS, Rothman DL, Petersen KF, Shulman GI. Effects of free fatty acids on glucose transport and IRS-1-associated phosphatidylinositol 3-kinase activity. J Clin Invest 1999;103:253–9.

[96] Muoio DM, Neufer PD. Lipid-induced mitochondrial stress and insulin action in muscle. Cell Metab 2012;15:595–605.

[97] McGarry JD, Mannaerts GP, Foster DW. A possible role for malonyl-CoA in the regulation of hepatic fatty acid oxidation and ketogenesis. J Clin Invest 1977;60:265–70.

[98] Zammit VA. Carnitine palmitoyltransferase 1: central to cell function. IUBMB Life 2008;60:347–54.

Cardiac Metabolism During Development and Aging

Andrea Schrepper

Department of Cardiothoracic Surgery, Jena University Hospital,
Friedrich Schiller University of Jena, Jena, Germany

Normal cardiac metabolism in healthy adult humans is well described in Chapter 4. From this chapter we know that the major source of ATP in the adult heart is the mitochondrial oxidation of carbohydrates and lipids. The important carbohydrates used are glucose and lactate, whereas free fatty acids (FFAs) (predominantly oleic acid and palmitic acid) constitute the major lipids oxidized. Amino acids are needed to a much lesser extent under normal conditions. The utilization of glucose, lactate, and fatty acid depends on the concentrations of these substrates in the blood, the oxygen availability as well as the demand of energy.

During development, substrate and oxygen availability as well as ATP demand are markedly different compared to the adult heart. Therefore, the relations of the metabolic ATP-producing pathways are different compared to the adult state (Fig. 6.1). They are adapted to the prevailing conditions. The mechanisms involved in these differences are discussed in this chapter.

THE FETAL METABOLISM IS DIFFERENT FROM MATERNAL METABOLISM

The fetus develops in a hypoxic environment and is dependent on maternal metabolism through nutrient supply via umbilical cord. Fetal growth is a complex process that involves the interaction of the mother, the fetus, and the interconnecting placenta. In mammals, the major determinant of intrauterine growth is the placental supply of nutrients to the fetus. At childbirth in humans, the placenta has an 11 m^2 exchange surface. The mother delivers oxygen and nutrients, while the fetus produces fetal waste (urea, bilirubin) and CO_2. Nutrient transport occurs through several processes. Gases, electrolytes, and water, for example, pass by means of diffusion. The placenta is selectively permeable to glucose, but not fructose. Amino acids, vitamins, and iron are transported by specific transporters, but proteins are transported only

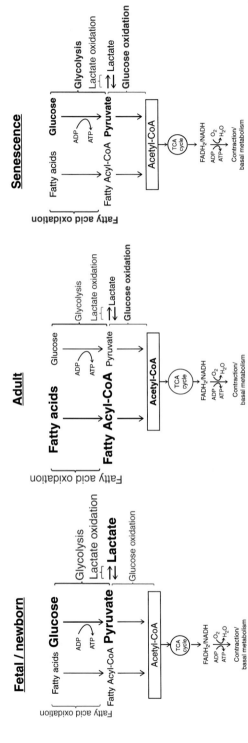

FIGURE 6.1 Overview of the relative priorities of various pathways of energy substrate metabolism that contribute to the production of ATP in fetal/newborn, adult, and old mammals.

slowly by pinocytosis. Before the placenta has sufficiently developed for hemotrophic nutrition of the embryo through umbilical cord, the developing embryo has to be supplied by histiotrophic nutrition. The human trophoblast is highly phagocytic, and has been shown to endocytose maternal erythrocytes and proteins. These glandular secretions contain glycogen, glycoproteins, and lipids [1]. Once placental circulation is established, the fetus receives nutrients through that organ.

Fetal Energy Substrates

In the fetus, glucose and lactate are dominant substrates for growth and development. The fetal heart receives glucose from the maternal circulation and, to a lesser extent, from the fetal liver via glycogenolysis. Endogenous glycogen is an important and readily available source of glucose in both fetal and neonatal hearts. In early gestation, myocardial glycogen contents are high, and start to decline by midgestation [2,3]. Hepatic glycogen synthesis and storage increase during late gestation in most mammals, including man and rodents. The accumulation of glycogen stores, parallels the increase in the activity of the rate-limiting enzyme for glycogen synthesis: glycogen synthase [4]. The fetus prepares for transition mainly in the third trimester by storing glycogen, producing catecholamines, and depositing brown fat. The placenta is a rich source of lactate, resulting in raised concentrations of lactate in the fetal circulation [5]. In general, the lactate concentrations in the umbilical vein are higher than those in the umbilical artery, demonstrating that the fetus as a whole is a net consumer rather than a producer of lactate [6,7]. Although lactate levels are not dramatically raised in the fetal lamb, lactate uptake by the fetal heart is significantly higher than in either newborn or adult hearts [8]. Within hours of birth, blood lactate levels quickly fall to levels seen in the adult [9]. In contrast to lactate, concentrations of plasma-FFAs

are very low in the fetal circulation [9,10]. In humans, a transplacental FFA gradient exists. The conclusion of most investigators is that FFAs are not a major energy source for the fetal heart under normal conditions. However, shortly after birth, blood lactate levels decrease and FFA concentrations increase to levels normally observed in adults [9,11].

The list of amino acids that the fetus must receive from the mother includes all the essential amino acids plus those amino acids that the fetus is not capable of synthesizing in adequate amounts. The concentration of most free amino acids, including some of the essentials, is higher in fetal than in maternal plasma. The higher concentration of essential amino acids in the recipient circulation suggests a process of active transport. The amino acids that the placenta delivers to the umbilical circulation are used by the fetus in part to build new tissues and in part as fuels for catabolic processes [12].

During fetal development, ketone bodies play a subordinated role in the heart. During the last trimester of gestation, maternal lipid metabolism switches to a catabolic condition, as shown by an accelerated breakdown of fat depots. An enhanced adipose tissue lipolytic activity is responsible for the increase in plasma FFA levels seen during the last weeks of gestation in fasted pregnant than in virgin rats (Fig. 6.2) [13]. Besides, and probably due to the limited capability of the placenta for FFA transfer, the level of FFA found in fetal plasma is low. Plasma FFAs are therefore mainly directed to the mother's liver and used for either esterification or oxidation of glycerides and ketone body synthesis. Plasma ketone body levels of pregnant rats increase to values that are much higher than in virgin rats. Despite the fact that the fetus is not able to synthesize ketone bodies [14], the amount of ketone bodies in fetal plasma reach the same level as in the mother since they easily cross the placenta (Fig. 6.2). The fetus therefore benefits from this product of maternal fatty acid metabolism, since ketone bodies may be used not only as

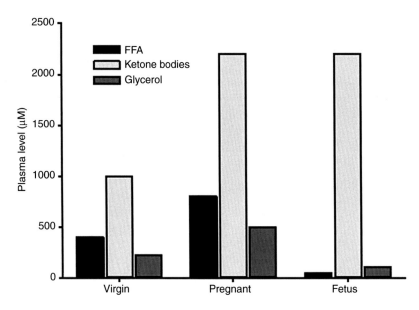

FIGURE 6.2   **Plasma level of FFA, ketone bodies, and glycerol in 24 h fasted virgin and pregnant (20 days) rats and their fetuses.** *Figure adapted from Ref. [13].*

fuels [14] but also as lipogenic substrates [15]. However, since fatty acid metabolism, TCA, and therefore oxidative phosphorylation are not the main energy source in the fetal heart, ketone bodies do not play an essential role in cardiac fetal metabolism.

Fetal Glycolysis

In the absence of adequate mitochondrial oxidative capacity, the primary source of ATP production shifts toward glycolysis. Therefore, the embryonic heart depends almost exclusively on glycolysis for energy production. Early in development, isolated rat and mouse hearts beat even in the absence of oxygen, this activity being sustained by glycolysis. In isolated fetal guinea pig hearts glycolysis can provide 22% of the myocardial ATP production [16]. In addition, a large proportion of glucose passing through glycolysis is converted to lactate, as opposed to being oxidized. This suggests that the ability of mitochondria to oxidize pyruvate is still limited in fetal hearts. This is supported by studies in

rat hearts, in which glucose uptake and lactate production are much higher in fetal hearts compared to newborn hearts [17]. Furthermore, in isolated perfused rabbit hearts obtained immediately following birth, glycolysis provides over 50% of myocardial ATP production, with glucose oxidation providing only 5% of ATP production (Fig. 6.3) [19].

This high glycolytic capacity is probably due to higher enzyme activity in the glycolytic pathway [20]. Furthermore, regulation of the glycolytic pathway in fetal hearts also appears to differ. Phosphofructokinase from fetal rat hearts is less sensitive to inhibition by increasing ATP concentrations, more sensitive to stimulation by increases in fructose 2,6-bisphosphate concentration, and less sensitive to inhibition by citrate [21,22]. As a result, this key enzyme in the glycolytic pathway seems more active in the fetal heart and less affected by inhibition.

Endogenous glycogen is also a potentially important source of glucose for glycolysis in the heart. This endogenous carbon pool can contribute significantly to overall myocardial ATP

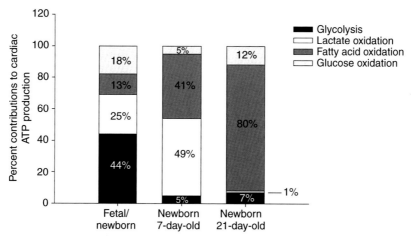

FIGURE 6.3 **Relative contributions of glycolysis, glucose oxidation, lactate oxidation, and fatty acid oxidation to overall cardiac ATP production during the immediate newborn/fetal period and the neonatal period (from 7 days to 21 days) in the isolated working rabbit heart preparation.** *Figure adapted from Ref. [18].*

production. Under conditions of normal oxygen availability, glucose derived from glycogen is not a major source of myocardial ATP. However, during conditions of insufficient oxygen supply, for example, during ischemia or hypoxia, a rapid increase in glycogenolysis occurs. The phosphorolytic cleavage of glycogen is energetically advantageous to the heart during low oxygen availability, because the released sugar is already phosphorylated (G-l-P), which saves an ATP on entering the glycolytic pathway.

Carbohydrate Oxidation in the Fetal Heart

Glucose and lactate are the primary carbohydrates undergoing oxidation in the developing heart. Although glucose oxidation rates are low in the immature heart, lactate oxidation rates can be substantial. In fetal lamb hearts, high lactate extraction has been shown, the circulating levels of which can be substantially higher than seen in adult hearts [9]. This can account for the majority of myocardial oxygen consumption and energy demand [8]. In fetal pig hearts addition of lactate to the perfusate not only accounts for more than 80% of myocardial oxygen consumption, it also

markedly inhibits glucose uptake by the heart [23]. These findings underscore the importance of lactate as a metabolic substrate in the fetal heart. The reason for high rates of lactate oxidation is not well understood. One possibility is that by late gestation, the fetal myocardium develops substantial oxidative capacity. However, at which time point oxidative metabolism becomes important has not been well characterized and seems to be species-dependent. Cox and Gunberg showed, that malonate and 2,4-dinitrophenol addition to fetal rat hearts depress contraction only in the second half of pregnancy (~12 days), indicating a greater reliance on oxidative metabolism beyond this point in development [24]. It should be noted that at this time, the allantoic placenta begins to replace the yolk sac, in the rat, as the site of maternal–fetal gas exchange, resulting in a better-oxygenated environment.

Fatty Acid Oxidation in the Fetal Heart

Late in gestation, the fetal heart has the capacity to utilize fatty acid. In isolated working fetal pig hearts (90% gestation) and neonatal hearts (2 days) substantial palmitate oxidation was observed at both ages but was greater in neonatal hearts. Nevertheless, palmitate inhibited lactate utilization and glucose uptake similarly in fetal and neonatal hearts. Lactate also reduced palmitate uptake and oxidation by 40–60% in both fetal and neonatal hearts. Thus both palmitate and lactate can act as major energy substrates for the immature heart. Both substrates significantly suppress glucose utilization, and each has suppressive effects on the other's metabolism [25]. This means that physiologic concentrations of lactate readily inhibit fatty acid oxidation. When fetal pig hearts were perfused under optimal conditions for fatty acid oxidation (high palmitate, no lactate), it was shown that fatty acid oxidation can account for 30% of oxygen consumption [21]. However, under normal substrate conditions, the actual contribution of fatty acids to myocardial oxygen consumption is probably much lower.

NEONATAL ENERGY METABOLISM

At birth, dramatic changes occur to the newborn, requiring several adjustments to adapt to extrauterine life, such as maintaining normoglycemia. Before birth, fetal glucose levels are maintained by the transplacental transfer of glucose from the mother, reflecting the mother's blood-glucose concentration. After birth, during the first hours of life, blood glucose concentrations drop. This is a critical period between birth and the establishment of suckling when the newborn depends on its own hepatic glycogen stores to maintain blood glucose. Thus, the presence of appropriate hepatic glycogen stores at birth is necessary to survive this crucial transitional period.

Neonatal Glycolysis and Carbohydrate Oxidation

Lactate oxidation and glycolysis are the preferred sources for ATP production in the fetal heart, but during the neonatal period, the transition to aerobic metabolism is completed. Blood lactate levels decrease dramatically after removal of the placenta, and fatty acid concentrations increase to near adult values, reflecting the lipid content of milk. Still, for a time after birth, differences exist in the relative amounts of carbon substrates that can be oxidized, compared with the fully matured heart. The greater glycolytic capacity seen in fetal hearts is still evident in the immediate newborn period. Glucose uptake by 1-day-old isolated rat hearts is greater than that observed in 7-day-old hearts [17]. Direct measurement of glycolysis in isolated rabbit hearts also shows that there is an increased glycolytic capacity in the immediate newborn period. In 1-day-old rabbits, 44% of ATP production from exogenous substrates can be obtained from glycolysis, whereas by 7 days glycolysis provides only 7% of ATP production [19]. However, the rabbit is relatively immature at birth. In contrast, in the perfused piglet heart, glucose oxidation rates can range from 10% to 60% of glycolytic rates [26]. In mature hearts, glucose oxidation rates typically are 20–60% of glycolytic rates [27]. In any given situation, the actual rates of glucose oxidation depend on the availability of competing substrates. FFAs exert their inhibitory effect on glucose oxidation on three major sites: (1) glucose uptake, (2) hexokinase-phosphofructokinase, and (3) pyruvate dehydrogenase [28]. Palmitate can also suppress lactate oxidation (inhibition of PDH) and, conversely, lactate can depress palmitate oxidation (inhibition of acyl-CoA synthetase) [23], depending on the concentrations of each substrate. Furthermore, under certain experimental conditions glucose, lactate, pyruvate, and ketone bodies have all been shown to decrease beta-oxidation [29]. It would appear that glycolysis remains high in the immature heart until fatty acid oxidation takes over as a major source of ATP production.

Neonatal Energy Metabolism – Fatty Acid Oxidation

The transition from carbohydrate to predominately fatty acid metabolism takes place over the first 2 weeks of life. Experiments with isolated working heart perfusions showed that neither piglets [30] nor rabbits [31] immediately after birth can sustain cardiac contractility using fatty acids as the sole exogenous substrate. The presence of glucose and insulin is an absolute must in these hearts. The existing reliance on glycolysis for ATP production directly after birth is a possible reason for the dependence on insulin. After birth the requirement for insulin is rapidly lost, and this is accompanied by a decrease in glycolysis and an increase in fatty acid oxidation [10]. In 1-day-old rabbits, glycolysis from extracellular glucose in isolated working hearts contributes up to 50% of total myocardial ATP production and fatty acid oxidation is low [19]. In 7-day-old hearts glycolytic rates decrease sharply (–60%), with a concomitant increase in fatty acid oxidation. Steady-state palmitate oxidation rates of 7-day-old rabbits were >10 times greater than in 1-day-old rabbit hearts. This suggests that a rapid maturation of the enzymatic pathways involving fatty acid oxidation occurs shortly after birth in the rabbit [19].

It is not well known why fatty acid oxidation rates are very low in the immediate newborn period. A possible reason is inhibition of CPT I by malonyl-CoA [32,33]. Furthermore, several hormones are suspected of being involved in the maturation of fatty acid metabolism. Studies of maternal thyroxine levels on fetal myocardial fatty acid oxidation showed that the offsprings of hypothyroid mothers, revealed no increase in palmitate oxidation. Lopaschuk et al. [34] have explored the effects of insulin on fatty acid oxidation in the developing rabbit heart. Insulin is known to stimulate, and glucagon to inhibit, acetyl-CoA carboxylase (ACC) activity; the enzyme responsible for catalyzing the formation of malonyl-CoA from acetyl-CoA. In the immediate newborn period impaired fatty acid oxidation is referred to insulin's effect on ACC. In the maturing rabbit heart, circulating levels of insulin diminish (and glucagon increase) and malonyl-CoA levels decrease. As a result, inhibition of CPT I by malonyl-CoA is relieved and fatty acid oxidation rates increase [34]. Therefore, it seems that hormonal changes may play an important role in developing cardiac lipid metabolism.

MITOCHONDRIA DURING DEVELOPMENT

Scanning electron microscopic studies have revealed that marked structural changes occur in mitochondria during fetal development [35–37]. Fetal rabbit hearts have smaller mitochondria, with fewer cristae and less dense matrices, and more variable inner membrane configurations compared to cardiac mitochondria from adult animals [36]. Knaapen et al. [35] investigated cardiac segments and intersegmental junctions during rat embryonic development with respect to glycogen stores, mitochondrial volume, and christae density. Glycogen stores increased from day 13 to 17 in all investigated segments. Furthermore, mitochondrial volume and christae density increased continuously from day 11 to 17 (period of investigation). Therefore by late gestation, the number of cristae increase, matrices become denser, and a more consistent pattern of membrane configurations develops [38]. A study by Neary et al. [37] raised suspicion of species-dependent differences. They showed cardiac mitochondria from mice and estimated area occupied by mitochondria in the cell. Mitochondrial volume was relatively constant at fetal stages and increased directly after birth. Furthermore, mitochondria change their shape from small and round before birth to larger and rectangular following birth [37].

This structural maturation is indicative of changes in enzyme activities associated with oxidative metabolism. Enzymes of the electron transport chain and the TCA cycle increase their activity in parallel to these structural changes. Cardiac mitochondria from rats show an increase in complex I and II [39] and also in complex III and IV [40] activity in late fetal development. Goldenthal and coworkers showed in bovine cardiac mitochondria an increase in NADH-dehydrogenase [41], in mitochondria from rats an increase in complexes I, III, IV, and ATPase [42], and in human cardiac mitochondria an increase in citrate synthase activity from birth to young adults [43].

With changes in mitochondrial structure and mitochondrial enzyme activities, changes in mitochondrial respiration also occur. In rabbits, oxygen consumption rates were found to be low in mid-gestation and increase toward birth [44]. The highest rates of oxygen consumption were measured in mitochondria isolated from hearts in the immediate newborn period. In rats, state 3 respiration increased from late fetal stage to the fifth postnatal day [45]. Enhanced oxygen consumption (state 3 respiration), relative to mitochondrial number, has also been documented in fetal and newborn lamb hearts [46]. Presumably, these enhanced rates of oxygen consumption reflect the increased demand on the heart associated with the transition to extrauterine life. These results suggest that oxidative metabolism is functional, and that flux through the electron transport chain may even be enhanced in late gestation.

It has been suggested that the lower mitochondrial content [47,48] and the low activity of mitochondrial enzymes [44,49,50] are a result of the hypoxic uterine environment [48]. Arterial oxygen content in the fetus is lower than in the adult, which may contribute to a low oxidative capacity in the fetal heart [8]. However, whether low oxygen content in the fetal circulation actually results in the fetal heart being in a "relative" hypoxic environment has not been completely resolved. Although arterial oxygen content and myocardial arteriovenous oxygen difference are lower in fetal lambs compared to adult lambs, myocardial oxygen consumption is not different [8,51].

SUBSTRATE METABOLISM IN THE ELDERLY

Normally, the heart displays a high degree of metabolic flexibility and can utilize multiple substrates to generate ATP including fatty acids, glucose, lactate, and ketone bodies. Under normal physiologic conditions, >95% of total ATP is derived from mitochondrial oxidative phosphorylation. In the healthy adult heart, mitochondrial beta-oxidation of fatty acids is the primary source of energy production in the fasted state. Myocardial metabolism is impaired in several cardiac pathologies, such as myocardial ischemia, ventricular hypertrophy, and heart failure. However, little is known about metabolic changes in the elderly, which may contribute to the mentioned pathologies.

In both mouse and rat experimental models of aging, the contribution of fatty acid oxidation to overall myocardial substrate metabolism declines with rising age [52,53]. Using positron emission tomography, Kates et al. could show a decrease in fatty acid oxidation in older humans [54]. However, detailed examination of these studies reveals that overall oxidative metabolism may be depressed as opposed to just fatty acid oxidation. In a model of isolated mouse heart perfusion, Koonen et al. could show that both fatty acid and glucose oxidation are reduced in aged mice [55]. In addition to reduced oxidative metabolism observed in the aged heart, accumulation of lipids within cardiac myocytes has also been observed [55,56]. Lipid accumulation in the heart has been strongly implicated in lipotoxicity and cardiac dysfunction in the setting of obesity and diabetes [57]. Such lipid accumulations result not only from decreased fatty acid oxidation but also from excessive fatty acid uptake into the heart [55]. Reasons for the decline in fatty acid oxidation are not fully understood, reductions in CPT I and carnitine-acylcarnitine translocase activity [53,58,59], reduction of peroxisome proliferator-activated receptor (PPAR)-α [60,61] or/and mitochondrial dysfunction [62–64] may be involved. Similar to fatty acid oxidation, the effect of age on myocardial glucose utilization is also poorly understood. Absolute glucose oxidation rates seem to be markedly diminished [55], while myocardial glucose uptake and glycolysis are increased in the aged heart [52,54]. The heart reverts to a more fetal metabolic phenotype which is commonly observed in the hypertrophied heart [65].

Mitochondria become larger and less numerous with age, accumulating vacuoles, cristae abnormalities, and intramitochondrial paracrystalline inclusions [67]. Mitochondrial respiratory chain enzyme activities decrease [42,67], as well as mitochondrial membrane potential [68,69], on which the production of ATP is dependent, while the amount of oxidative damage to proteins and mtDNA increases, with an associated accumulation in the quantity of mtDNA mutations [68,70]. While the exact causes of mitochondrial dysfunction are not clear, a popular theory proposes that enhanced mitochondrial reactive oxygen species production leads to mitochondrial DNA damage, lipid peroxidation, and mitochondrial dysfunction, creating a vicious cycle of oxidative damage and reduced mitochondrial function that may occur during aging [71,72].

Many studies demonstrated alterations in myocardial substrate metabolism in the elderly. A decline in overall mitochondrial substrate metabolism with reductions in both fatty acid and glucose oxidation in the aged heart has been associated with impaired cardiac performance [54,73]. However, the mechanisms by which these pathophysiologic changes occur have not been completely described. It is also not known if changes in cardiac substrate metabolism are sufficient to impair cardiac performance in the aged heart. Therefore, a better understanding of the metabolic changes that occur in the heart during the normal process of aging could shed light on the pathogenesis of age-related cardiomyopathy and may ultimately lead to improved therapeutic strategies for the treatment of contractile dysfunction in the elderly [73].

References

[1] Mota-Rojas D, Orozco-Gregorio H, Villanueva-Garcia D, Bonilla-Jaime H, Suarez-Bonilla X, Hernandez-Gonzalez R, Roldan-Santiago P, Trujillo-Ortega ME. Foetal and neonatal energy metabolism in pigs and humans: a review. Vet Med 2011;56(5):215–25.

[2] Ascuitto RJ, Ross-Ascuitto NT. Substrate metabolism in the developing heart. Semin Perinatol 1996;20(6):542–63.

[3] Clark CM Jr. Characterization of glucose metabolism in the isolated rat heart during fetal and early neonatal development. Diabetes 1973;22(1):41–9.

[4] Gruppuso PA, Brautigan DL. Induction of hepatic glycogenesis in the fetal rat. Am J Physiol 1989;256(1 Pt 1):E49–54.

[5] Comline RS, Silver M. Some aspects of foetal and uteroplacental metabolism in cows with indwelling umbilical and uterine vascular catheters. J Physiol 1976;260(3):571–86.

[6] Bell AW, Kennaugh JM, Battaglia FC, Makowski EL, Meschia G. Metabolic and circulatory studies of fetal lamb at midgestation. Am J Physiol 1986;250(5 Pt 1):E538–44.

[7] Burd LI, Jones MD Jr, Simmons MA, Makowski EL, Meschia G, Battaglia FC. Placental production and foetal utilisation of lactate and pyruvate. Nature 1975;254(5502):710–1.

[8] Fisher DJ, Heymann MA, Rudolph AM. Myocardial consumption of oxygen and carbohydrates in newborn sheep. Pediatr Res 1981;15(5):843–6.

[9] Medina JM. The role of lactate as an energy substrate for the brain during the early neonatal period. Biol Neonate 1985;48(4):237–44.

[10] Lopaschuk GD, Collins-Nakai RL, Itoi T. Developmental changes in energy substrate use by the heart. Cardiovasc Res 1992;26(12):1172–80.

[11] Phelps RL, Metzger BE, Freinkel N. Carbohydrate metabolism in pregnancy. XVII. Diurnal profiles of plasma glucose, insulin, free fatty acids, triglycerides, cholesterol, and individual amino acids in late normal pregnancy. Am J Obstet Gynecol 1981;140(7):730–6.

[12] Battaglia FC, Meschia G. Principal substrates of fetal metabolism. Physiol Rev 1978;58(2):499–527.

[13] Herrera E. Implications of dietary fatty acids during pregnancy on placental, fetal and postnatal development--a review. Placenta 2002;23(Suppl A):S9–S19.

[14] Shambaugh GE III. Ketone body metabolism in the mother and fetus. Fed Proc 1985;44(7):2347–51.

[15] Edmond J. Ketone bodies as precursors of sterols and fatty acids in the developing rat. J Biol Chem 1974;249(1):72–80.

[16] Rolph TP, Jones CT. Regulation of glycolytic flux in the heart of the fetal guinea pig. J Dev Physiol 1983;5(1):31–49.

[17] Hoerter JA, Opie LH. Perinatal changes in glycolytic function in response to hypoxia in the incubated or perfused rat heart. Biol Neonate 1978;33(3–4):144–61.

[18] Lopaschuk GD, Jaswal JS. Energy metabolic phenotype of the cardiomyocyte during development,

differentiation, and postnatal maturation. J Cardiovasc Pharmacol 2010;56(2):130–40.

[19] Lopaschuk GD, Spafford MA, Marsh DR. Glycolysis is predominant source of myocardial ATP production immediately after birth. Am J Physiol 1991;261(6 Pt 2): H1698–705.

[20] Jones CT, Rolph TP. Metabolism during fetal life: a functional assessment of metabolic development. Physiol Rev 1985;65(2):357–430.

[21] Ascuitto RJ, Ross-Ascuitto NT, Chen V, Downing SE. Ventricular function and fatty acid metabolism in neonatal piglet heart. Am J Physiol 1989;256(1 Pt 2):H9–H15.

[22] Bristow J, Bier DM, Lange LG. Regulation of adult and fetal myocardial phosphofructokinase. Relief of cooperativity and competition between fructose 2,6-bisphosphate, ATP, and citrate. J Biol Chem 1987;262(5):2171–5.

[23] Werner JC, Sicard RE. Lactate metabolism of isolated, perfused fetal, and newborn pig hearts. Pediatr Res 1987;22(5):552–6.

[24] Cox SJ, Gunberg DL. Energy metabolism in isolated rat embryo hearts: effect of metabolic inhibitors. J Embryol Exp Morphol 1972;28(3):591–9.

[25] Werner JC, Sicard RE, Schuler HG. Palmitate oxidation by isolated working fetal and newborn pig hearts. Am J Physiol 1989;256(2 Pt 1):E315–21.

[26] Ascuitto RJ, Joyce JJ, Ross-Ascuitto NT. Mechanical function and substrate oxidation in the neonatal pig heart subjected to pacing-induced tachycardia. Mol Genet Metab 1999;66(3):212–23.

[27] Kobayashi K, Neely JR. Control of maximum rates of glycolysis in rat cardiac muscle. Circ Res 1979;44(2):166–75.

[28] Randle PJ. Regulatory interactions between lipids and carbohydrates: the glucose fatty acid cycle after 35 years. Diabetes Metab Rev 1998;14(4):263–83.

[29] Tripp ME. Developmental cardiac metabolism in health and disease. Pediatr Cardiol 1989;10(3):150–8.

[30] Werner JC, Whitman V, Fripp RR, Schuler HG, Morgan HE. Carbohydrate metabolism in isolated, working newborn pig heart. Am J Physiol 1981;241(5):E364–71.

[31] Lopaschuk GD, Spafford MA. Energy substrate utilization by isolated working hearts from newborn rabbits. Am J Physiol 1990;258(5 Pt 2):H1274–80.

[32] Tomec RJ, Hoppel CL. Carnitine palmitoyltransferase in bovine fetal heart mitochondria. Arch Biochem Biophys 1975;170(2):716–23.

[33] Bartelds B, Takens J, Smid GB, Zammit VA, Prip-Buus C, Kuipers JR, van der Leij FR. Myocardial carnitine palmitoyltransferase I expression and long-chain fatty acid oxidation in fetal and newborn lambs. Am J Physiol Heart Circ Physiol 2004;286(6):H2243–8.

[34] Lopaschuk GD, Witters LA, Itoi T, Barr R, Barr A. Acetyl-CoA carboxylase involvement in the rapid maturation of fatty acid oxidation in the newborn rabbit heart. J Biol Chem 1994;269(41):25871–8.

[35] Knaapen MW, Vrolijk BC, Wenink AC. Ultrastructural changes of the myocardium in the embryonic rat heart. Anat Rec 1997;248(2):233–41.

[36] Sordahl LA, Crow CA, Kraft GH, Schwartz A. Some ultrastructural and biochemical aspects of heart mitochondria associated with development: fetal and cardiomyopathic tissue. J Mol Cell Cardiol 1972;4(1):1–10.

[37] Neary MT, Ng KE, Ludtmann MH, Hall AR, Piotrowska I, Ong SB, Hausenloy DJ, Mohun TJ, Abramov AY, Breckenridge RA. Hypoxia signaling controls postnatal changes in cardiac mitochondrial morphology and function. J Mol Cell Cardiol 2014;74:340–52.

[38] Smith HE, Page E. Ultrastructural changes in rabbit heart mitochondria during the perinatal period. Neonatal transition to aerobic metabolism. Dev Biol 1977;57(1):109–17.

[39] Lang CA. Respiratory enzymes in the heart and liver of the prenatal and postnatal rat. Biochem J 1965;95:365–71.

[40] Schagger H, Noack H, Halangk W, Brandt U, von Jagow G. Cytochrome-c oxidase in developing rat heart. Enzymic properties and amino-terminal sequences suggest identity of the fetal heart and the adult liver isoform. Eur J Biochem 1995;230(1):235–41.

[41] Marin-Garcia J, Ananthakrishnan R, Agrawal N, Goldenthal MJ. Mitochondrial gene expression during bovine cardiac growth and development. J Mol Cell Cardiol 1994;26(8):1029–36.

[42] Marin-Garcia J, Ananthakrishnan R, Goldenthal MJ. Mitochondrial gene expression in rat heart and liver during growth and development. Biochem Cell Biol 1997;75(2):137–42.

[43] Marin-Garcia J, Ananthakrishnan R, Goldenthal MJ. Human mitochondrial function during cardiac growth and development. Mol Cell Biochem 1998;179(1–2):21–6.

[44] Werner JC, Whitman V, Musselman J, Schuler HG. Perinatal changes in mitochondrial respiration of the rabbit heart. Biol Neonate 1982;42(5–6):208–16.

[45] Schonfeld P, Schild L, Bohnensack R. Expression of the ADP/ATP carrier and expansion of the mitochondrial (ATP + ADP) pool contribute to postnatal maturation of the rat heart. Eur J Biochem 1996;241(3):895–900.

[46] Wells RJ, Friedman WF, Sobel BE. Increased oxidative metabolism in the fetal and newborn lamb heart. Am J Physiol 1972;222(6):1488–93.

[47] Goodwin CW, Mela L, Deutsch C, Forster RE, Miller LD, Kelivoria-Papadopoulos M. Development and adaptation of heart mitochondrial respiratory chain function in fetus and in newborn. Adv Exp Med Biol 1976;75:713–9.

[48] Hallman M. Changes in mitochondrial respiratory chain proteins during perinatal development. Evidence of the importance of environmental oxygen tension. Biochim Biophys Acta 1971;253(2):360–72.

[49] Dallman PR, Schwartz HC. Ctyochrome C concentrations during rat and guinea pig development. Pediatrics 1964;33:106–10.

[50] Glatz JF, Veerkamp JH. Postnatal development of palmitate oxidation and mitochondrial enzyme activities in rat cardiac and skeletal muscle. Biochim Biophys Acta 1982;711(2):327–35.

[51] Fisher DJ, Heymann MA, Rudolph AM. Regional myocardial blood flow and oxygen delivery in fetal, newborn, and adult sheep. Am J Physiol 1982;243(5):H729–31.

[52] Abu-Erreish GM, Neely JR, Whitmer JT, Whitman V, Sanadi DR. Fatty acid oxidation by isolated perfused working hearts of aged rats. Am J Physiol 1977;232(3):E258–62.

[53] McMillin JB, Taffet GE, Taegtmeyer H, Hudson EK, Tate CA. Mitochondrial metabolism and substrate competition in the aging Fischer rat heart. Cardiovasc Res 1993;27(12):2222–8.

[54] Kates AM, Herrero P, Dence C, Soto P, Srinivasan M, Delano DG, Ehsani A, Gropler RJ. Impact of aging on substrate metabolism by the human heart. J Am Coll Cardiol 2003;41(2):293–9.

[55] Koonen DP, Febbraio M, Bonnet S, Nagendran J, Young ME, Michelakis ED, Dyck JR. CD36 expression contributes to age-induced cardiomyopathy in mice. Circulation 2007;116(19):2139–47.

[56] van der Meer RW, Rijzewijk LJ, Diamant M, Hammer S, Schar M, Bax JJ, Smit JW, Romijn JA, de Roos A, Lamb HJ. The ageing male heart: myocardial triglyceride content as independent predictor of diastolic function. Eur Heart J 2008;29(12):1516–22.

[57] Wende AR, Abel ED. Lipotoxicity in the heart. Biochim Biophys Acta 2010;1801(3):311–9.

[58] Odiet JA, Boerrigter ME, Wei JY. Carnitine palmitoyl transferase-I activity in the aging mouse heart. Mech Ageing Dev 1995;79(2–3):127–36.

[59] Paradies G, Ruggiero FM, Petrosillo G, Gadaleta MN, Quagliariello E. Carnitine-acylcarnitine translocase activity in cardiac mitochondria from aged rats: the effect of acetyl-L-carnitine. Mech Ageing Dev 1995;84(2):103–12.

[60] Hyyti OM, Ledee D, Ning XH, Ge M, Portman MA. Aging impairs myocardial fatty acid and ketone oxidation and modifies cardiac functional and metabolic responses to insulin in mice. Am J Physiol Heart Circ Physiol 2010;299(3):H868–75.

[61] Rodriguez-Calvo R, Serrano L, Barroso E, Coll T, Palomer X, Camins A, Sanchez RM, Alegret M, Merlos M, Pallas M, Laguna JC, Vazquez-Carrera M. Peroxisome proliferator-activated receptor alpha down-regulation is associated with enhanced ceramide levels in age-associated cardiac hypertrophy. J Gerontol A Biol Sci Med Sci 2007;62(12):1326–36.

[62] Fannin SW, Lesnefsky EJ, Slabe TJ, Hassan MO, Hoppel CL. Aging selectively decreases oxidative capacity in rat heart interfibrillar mitochondria. Arch Biochem Biophys 1999;372(2):399–407.

[63] Navarro A, Boveris A. The mitochondrial energy transduction system and the aging process. Am J Physiol Cell Physiol 2007;292(2):C670–86.

[64] Petersen KF, Befroy D, Dufour S, Dziura J, Ariyan C, Rothman DL, DiPietro L, Cline GW, Shulman GI. Mitochondrial dysfunction in the elderly: possible role in insulin resistance. Science 2003;300(5622):1140–2.

[65] Sambandam N, Lopaschuk GD, Brownsey RW, Allard MF. Energy metabolism in the hypertrophied heart. Heart Fail Rev 2002;7(2):161–73.

[66] Frenzel H, Feimann J. Age-dependent structural changes in the myocardium of rats. A quantitative light- and electron-microscopic study on the right and left chamber wall. Mech Ageing Dev 1984;27(1):29–41.

[67] Kumaran S, Subathra M, Balu M, Panneerselvam C. Age-associated decreased activities of mitochondrial electron transport chain complexes in heart and skeletal muscle: role of L-carnitine. Chem Biol Interact 2004;148(1–2):11–8.

[68] Petrosillo G, Matera M, Moro N, Ruggiero FM, Paradies G. Mitochondrial complex I dysfunction in rat heart with aging: critical role of reactive oxygen species and cardiolipin. Free Radic Biol Med 2009;46(1):88–94.

[69] Cottrell DA, Turnbull DM. Mitochondria and ageing. Curr Opin Clin Nutr Metab Care 2000;3(6):473–8.

[70] Mohamed SA, Hanke T, Erasmi AW, Bechtel MJ, Scharfschwerdt M, Meissner C, Sievers HH, Gosslau A. Mitochondrial DNA deletions and the aging heart. Exp Gerontol 2006;41(5):508–17.

[71] Dai DF, Rabinovitch PS. Cardiac aging in mice and humans: the role of mitochondrial oxidative stress. Trends Cardiovasc Med 2009;19(7):213–20.

[72] Bratic I, Trifunovic A. Mitochondrial energy metabolism and ageing. Biochim Biophys Acta 2010;1797(6–7):961–7.

[73] Dyck JRB, Sung MMY. Energy metabolism in the aging heart. Heart Metab 2011;52:3–7.

Methods to Investigate Cardiac Metabolism

Moritz Osterholt*, Michael Schwarzer†, Torsten Doenst†

*Department of Internal Medicine, Helios Spital Überlingen, Überlingen, Germany;
†Department of Cardiothoracic Surgery, Jena University Hospital, Friedrich Schiller
University of Jena, Jena, Germany

INTRODUCTION

Cardiac metabolism encompasses all biochemical processes that result in the conversion of substrates or intermediates of metabolic pathways and cycles for the purpose of cell function, growth, and contraction. Methods investigating metabolism must therefore address the moieties of metabolic pathways and cycles, flux through them and the activity and regulation of their enzymes. Such investigations can be performed *in vivo*, *in vitro*, or *ex vivo*; in humans, in animals, or in cell culture. In this chapter, we will describe the principles of the main methods used to examine cardiac metabolism in health and disease. As mitochondria have moved into the focus of attention in recent years, we will put a focus on the assessment of mitochondrial function. We will start with the methods of assessing moieties of pathways and cycles, that is, the amounts of RNA, proteins, and metabolites, then continue with the description of methods to assess enzyme activities *in vitro*, and move on to methods

of addressing fluxes and imaging of metabolic activity in intact organs or *in vivo*. This chapter is not meant to describe individual techniques in detail and is not meant to be complete but to provide an overview and to illustrate principles of the currently available methodology.

THE SAMPLE

For the investigation of cellular and specifically metabolic processes, the use (and therefore, the generation) of the right sample is a key feature. Investigations on blood may require separation of the cellular components from plasma or serum. Investigations on organ tissue may require homogenization (blending) of a tissue sample with or without further isolation steps (for instance, differential centrifugation for the isolation of mitochondria). This is often the first critical step for the generation of proper results and the detrimental effects of ignoring this aspect are not compensated by using sophisticated methods. While the various possibilities and techniques to generate a sample are beyond the scope of this article, it is important to be aware that sample preparation may often influence the nature of the obtained results, either directly or indirectly. For instance, the isolation process of differential centrifugation may mechanically affect mitochondria in their function [1]. Thus, the obtained results may not directly reflect function *in vivo*, which should be considered when interpreting the results. Another important aspect is storage of a sample: while some methods require fresh material (e.g., freshly isolated mitochondria that can be kept on ice only for a few hours), other experiments will yield stable results using samples that have been stored at $-80°C$ or in liquid nitrogen for years. It is therefore important to know, which method is suitable for a given setting and which potential alternatives are available. In the case of mitochondrial investigations, alternative methods to assess mitochondrial function have been developed, for instance, with the "saponin-skinned fibers technique" [2,3]. Here, the homogenization step is omitted and the tissue is treated with chemicals that create pores in the cell membrane, such that substrates for an enzymatic assay obtain access to the mitochondria of interest. If the optimal sample for investigation has been determined, its properties can usually be measured reliably.

QUANTIFICATION OF MOIETIES AND INTERMEDIATES

The beginning of metabolism research was characterized biochemically by determining the concentrations of individual components of pathways and cycles. Spectrophotometric assays represented the backbone of metabolic research for many years and are still used today for specific purposes. In general, a specifically designed reaction is initiated inside a cuvette placed in a spectrophotometer. This reaction links the desired moiety to the appearance or disappearance of marker molecules (often nicotinamide adenine dinucleotide) that absorb light at a certain wavelength. Figure 7.1A illustrates the principle of such an assay for the example of glucose-6-phosphate quantification. This principle can also be used to assess enzyme activities, only that the conditions inside the reaction cuvette need to be changed (with substrate provided and measuring absorbance over time; Fig. 7.1B shows the reactions used for assessment of hexokinase activity). Using mainly these methodologies, our metabolic forefathers were able to make important metabolic discoveries, such as the function of the respiratory chain, the citric acid cycle, or the glycolytic pathway [4–8].

Another moiety that can be quantified is RNA, which has been transcribed from the cell's DNA. The initial tools to quantify DNA and RNA were based on electrophoretic separation and hybridization with labeled homologous nucleic acids. They were termed Southern blotting (for DNA) and Northern blotting (for RNA). Later the polymerase chain reaction (PCR) was developed, which allows the rapid quantification of nucleotide chains, that is, the assessment of genes and their transcription. The polymerase chain reaction has also been a prerequisite for the newer array technologies, which use amplification as well as homology hybridization (transcriptomics, etc. described later).

While the metabolic intermediates are organic molecules (such as glucose, pyruvate, or lactate) and RNA is assembled of nucleotides, the enzymes catalyzing the interconversions of these molecules are generally made of protein, that is, an assembly of amino acids. The previously described spectrophotometric assay reveals enzyme activity. In addition, enzyme (protein) abundance may need to be quantified. Assessment of abundance (the quantity) of proteins is mainly performed by electrophoretic separation of a specific sample from enzyme-containing tissue, followed by reaction with an appropriate specific antibody. These antibodies possess properties, which can be detected by fluorescence or

(A)

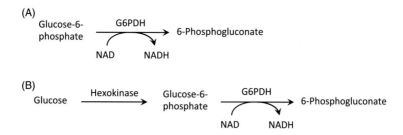

(B)

(C)

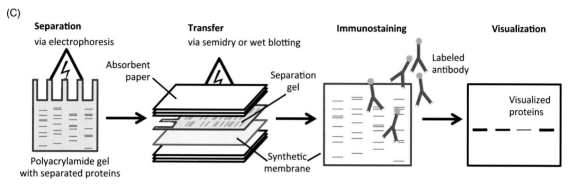

FIGURE 7.1 **Examples for spectrophotometric assays to determine metabolite concentration or enzyme activities as well as western blotting.** (A) Assay to determine the amount of glucose-6-phosphate. Glucose-6-phosphate dehydrogenase (G6PDH) quantitatively transforms glucose-6-phosphate to 6-phosphogluconate and NADH/H. Concentration of the latter can be measured using the increase in absorbance at a wavelength of 340 nm. (B) Indirect assay to determine hexokinase activity. Hexokinase produces glucose-6-phosphate and ADP from glucose and ATP. Here, neither educts nor products can be properly measured directly using spectrophotometry. However, the combination with an excess of G6PDH and NAD quantitatively transfers glucose-6-phosphate to 6-phosphogluconate and NADH/H. The rate of appearance of the latter can easily be monitored at 340 nm to calculate reaction rates. (C) Western blot analysis is used to detect a specific protein in a biological specimen. The technique can be separated into: separation, transfer, immunostaining, and visualization. *Separation*: The sample is solubilized with a detergent and the proteins are then separated by electrophoresis in a polyacrylamide gel. *Transfer*: After electrophoresis, the gel is placed next to a thin, synthetic membrane that has a strong affinity for proteins. The gel and membrane are placed between sheets of absorbent paper in a blotting tank. This arrangement allows buffer to flow across the gel and through the thin membrane. As a result, the proteins in the gel are transferred to the membrane by capillary action. Transfer of the proteins to the membrane may also be accomplished by an electrical current. *Immunostaining*: After the transfer step, the membrane is incubated with an antibody to a specific protein. The antibody may be coupled to an enzyme, which can then be used to detect and quantify the antibody on the membrane. In the example shown, the antibody is coupled to horseradish peroxidase. The membrane is incubated with a substrate that is converted to a luminescent compound after reaction with this enzyme. *Visualization*: A sheet of X-ray film is then placed next to the membrane, which allows visualization of individual proteins or a picture is taken with a luminescent analyzer.

a detectable chemical reaction (luminescence) enabling quantification. This process has been named Western blotting. Since proteins are often regulated by chemical modifications (e.g., phosphorylation), which can also be assessed by Western blotting, this method provides an important tool for the investigation of various signaling cascades within the cell (Fig. 7.1C).

Another sophisticated method to determine proteins is the so-called blue native polyacrylamide gel electrophoresis (BN-PAGE). This technique separates protein complexes [9]. A major advantage of this method is that proteins remain in their native state. Their activity can therefore be assessed *"in-gel"* [10]. In mitochondrial research, the supermolecular assembly of individual respiratory complexes into so-called supercomplexes has gained considerable attention. Instead of a random distribution in the inner mitochondrial membrane (IMM), individual complexes are suggested to be linked to each other [11]. Reduced abundance of such a supercomplex consisting of complexes I, III, and IV was associated with a defect in oxidative phosphorylation in a model of coronary microembolization-induced heart failure in dogs, even in the absence of defects of the individual complexes [12].

More sophisticated techniques for the broad-spectrum quantification of proteins (proteomics), RNA (transcriptomics) and metabolites (metabolomics) have been developed and are described in more detail later. However, as alluded to earlier, the function of an enzyme is not only characterized by its abundance. Thus, elaborate methods to assess enzyme activity and its regulation have been developed. The spectrophotometric principle of enzyme activity determination has been described earlier. The following section addresses the generation and interpretation of assay results with respect to quantification of enzyme activity and their regulation.

MEASURING ENZYME ACTIVITY IN VITRO

Figure 7.2A shows the principle of an enzyme reaction. An enzyme and a substrate form a substrate–enzyme complex, the reaction is catalyzed, and the product separates from the enzyme. The activity of an enzyme is influenced by several factors. First, the amount of substrate; second, the amount of enzyme, as well as by other factors such as the amount of product (e.g., product inhibition), temperature, or the presence of activators or inhibitors. The classic spectrophotometric assay assesses enzyme activity with ample amount of substrate, that is, at substrate saturation. The measured

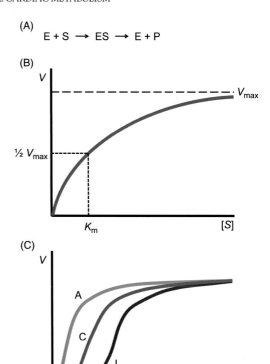

FIGURE 7.2 (A) The basic mechanism of an enzymatic reaction involves two subsequent steps: the enzyme E and the substrate(s) S form a substrate enzyme complex ES. After catalyzation, the product P then separates from the enzyme. It has to be noted that the enzyme is not altered in the course of the reaction. (B) Michaelis–Menten saturation curve showing the relationship between substrate concentration and enzyme activity of an enzyme according to the Michaelis–Menten kinetics. [S], substrate concentration; V, reaction rate; V_{max}, maximal reaction rate; K_m, Michaelis constant. (C) Allosteric regulation of enzymes alters the apparent K_m of the enzyme: with an allosteric inhibitor, the sigmoidal curve is shifted to the right (I), while an allosteric activator shifts the curve to the left (A). The blue curve (C) depicts kinetic without allosteric regulators. Note that V_{max} is not altered due to allosteric regulation.

activity is referred to as the maximal enzyme activity, V_{max}. The classic relationship between enzyme activity and substrate concentration below saturation levels is known as Michaelis–Menten kinetics, where the Michaelis constant

K_m reflects the concentration of substrate at which the enzyme is at half maximal activity (Fig. 7.2B). Thus, varying concentrations of substrate *in vivo* lead to the adaptation of enzyme activity. However, substrate is not the only regulator of enzyme activity and another important regulatory mechanism is that of allosteric regulation. Here, a regulatory site on the enzyme, which differs in location from the catalytic site can exert inhibitory or stimulatory function if occupied by a ligand or cofactor. Allosterically regulated enzymes differ in their catalytic activity/substrate relationship from classic Michaelis–Menten type enzymes by their sigmoidal shape (Fig. 7.2C). Thus, allosteric regulation increases "substrate sensitivity" and allows the cell to respond quickly to alterations in substrate concentrations.

It is important to note that activity measured *in vitro* does not necessarily reflect the actual activity states of the enzymes *in vivo*. The *in vivo* activity of enzymes is difficult to assess because of the complexity of the different regulatory influences and the often-quick changes of substrates and regulators inside the cell. Thus, reports on enzyme activities *in vivo* are often extrapolated from measurements *in vitro*. From a practical standpoint, these limitations are not perceived as a major problem. Knowing the type of regulation of an enzyme from *in vitro* studies together with the knowledge of the amount of enzyme present at a given point of time, usually gives the investigators ample information on the activity status and the potential activity under certain conditions (e.g., if a similarly regulated key enzyme of the respiratory chain is only half abundant in heart failure, an involvement in a pathomechanism is possible). Given the fact that regulation of many enzymes today is known, it is not surprising that novel, broad-spectrum analyses have been developed for the quantification of protein as well as RNA and metabolite concentrations. These technologies are generally referred to as "omics" and are described in principle later.

TRANSCRIPTOMICS AND PROTEOMICS

From gene to phenotype, several distinct steps of molecular control are known and have led to the evolution of distinct scientific fields based on OMICS technology: beginning with the level of genes and their modifications (genomics and epigenomics), followed by gene- and protein-translation and expression (transcriptomics, translatomics, and proteomics) to the assessment of metabolites (metabolomics) [13–16].

The transcriptome is the entirety of RNA in a given cell, with messenger-RNA (mRNA) being the most abundant type. As every RNA molecule is the product of transcription of a part of the DNA, the transcriptome can give insights into the activity of certain genes. When the transcriptome is assessed at large, links between individual genes and their concerted activation/deactivation can be detected. A well-studied example in cardiomyocytes is the return to the fetal gene program, which is found in a variety of pathophysiologic conditions as hypoxia, hypertrophy, and heart failure [17]. For the assessment of the transcriptome, high-throughput methods as microarrays and RNA-sequencing are available [16,18]. The principle of the microarray technology consists of nucleotide sequences attached to a matrix, which are hybridized with a carefully prepared sample (frequently cDNA) to generate a signal. The signal intensity correlates to the expression level of the mRNA in question. The microarray technology allows the assessment of thousands of mRNAs at the same time. Analysis of such assay results require sophisticated bioinformatic knowledge and statistically sound assessment.

It should be noted that the abundance of RNA; however, does not necessarily correlate with the abundance of protein (see also Chapter 9 on diurnal variations), let alone the function of this protein [19]. Therefore, analysis of the transcriptome should be followed by the assessment of the proteome, that is, the entirety of proteins in a biological sample. Using two-dimensional (2D) electrophoresis, proteins can be separated according to their electrical charge and their relative molecular mass. This technique allows greater separation of distinct proteins than conventional one-dimensional electrophoresis. The dynamic range of 2D-electrophoresis can be further increased by using subcellular fractions [19]. Further methods for separation and detection of proteins as chromatography and mass spectrometry (MS) are available to increase the accuracy of proteomic research [15].

Nevertheless, even the concentration of a protein in a system may not provide valid details about its activity/function. To evaluate the activity of metabolic pathways, metabolite profiling is increasingly used.

METABOLOMICS

In 1971, Linus Pauling suggested that measuring the concentration of chemical compounds in biological fluids will prove to be a valid way of diagnosing different diseases, basically by reflecting the functional status of biochemical pathways, marking the birth of metabolomics [20].

The aim of metabolomics is to assess small molecules in biological fluids, tissues, or other specimens. Due to technological advances since Pauling's work, it is now possible to quantify a great variety of substances using high-throughput technology (such as chromatography, nuclear magnetic resonance (NMR) spectroscopy, and MS). As Pauling suggested, one of the advantages of metabolomic research is its proximity to the phenotype of disease, compared to RNA or protein abundance. Changes in the levels of metabolites are more directly and more dynamically related to the development and onset of disease [13].

The number of known metabolites in human specimen spans a large, much more heterogeneous group of compounds than gene transcripts or proteins: metabolites can be anionic or cationic, polar or nonpolar, and their concentration in samples may vary from picomolar to millimolar [21,22]. It is this heterogeneity that requires specific technologies to efficiently identify and reliably quantify all metabolites: typically, a combination of liquid chromatography (LC) or gas chromatography (GC), MS, and NMR spectroscopy is used [23].

Detection and identification by NMR spectroscopy is based on the magnetic properties of hydrogen, carbon, or phosphorus nuclei [24]. In a magnetic field, organic substances containing any of these nuclei develop a specific NMR signal that also depends on neighboring atoms. Thus, metabolites can be distinguished by different NMR spectra and can be further quantified. Compared to chromatography/MS, this technique needs only little preparation of the sample. A disadvantage of NMR spectroscopy is the limited sensitivity for substances of low concentration: small differences in the abundance of already low-concentrated substances between two samples may be overlooked in the presence of NMR signals of higher concentrated metabolites in the specimen [24].

MS is often preceded by chromatographic separation. With chromatography, metabolites are separated by their different chemical properties, facilitating the following step of MS. GC is used for volatile and nonpolar metabolites. LC is applied for nonvolatile metabolites in a liquid solution. During the following step of MS, metabolites can then be differentiated according to their mass-to-charge ratio. Several MS techniques exist, which are reviewed elsewhere in detail [25,26]. GC-MS/LC-MS techniques allow the identification and quantification of metabolites of very low concentrations, making it more sensitive compared to NMR spectroscopy [24].

In general, two distinct approaches are described in metabolomic research [14,21]. With the targeted approach, the focus is on quantification of a small subset of chemically related metabolites (e.g., metabolites of a single metabolic pathway). This is achieved by adding internal standards to the sample before further processing [14,27] and using specific and sensitive detection methods. The second approach is nontargeted, that is, the aim is to detect as many metabolites as possible, which are chemically heterogeneous. As a result, the untargeted approach can deliver spectra with thousands of peaks [23]. These spectra may be used for metabolite profiling, offering the opportunity to find new biomarkers by detecting changes in the metabolome (sometimes referred to as a "disease pattern") [28].

Despite the recent progress in this field, several difficulties are still prevalent: reproducibility between different machines and even different operators is unsatisfactory. Due to a low peak-to-noise ratio especially for low-concentrated metabolites, certain signals can be missed easily [21,28]. It has to be noted that the nontargeted approach provides little insight into pathophysiologic mechanisms. However, for the identification of potential new biomarkers, it has become the method of choice.

ASSESSING METABOLIC FLUX

We had addressed earlier that it is difficult to determine the exact activity status of an individual enzyme *in vivo*. However, several tools exist to determine the activity of entire sequences of enzyme reactions that are aligned in metabolic pathways or cycles. Together with the information from individual enzyme assays, the rate-limiting enzymes within a pathway or cycle can then be determined.

By measuring the concentration of substrates and metabolites in arterial and venous samples, it is possible to calculate rates of substrate metabolism. The underlying principle was first proposed by Adolf Fick in 1870 to measure cardiac output [29], but it can also be applied to evaluate substrate metabolism. Since this technique is relatively crude and is often limited through the inability to quantify coronary flow (especially

in vivo), tracer techniques have been developed. The principle technology behind assessing metabolic fluxes with a tracer is the tracing of a desired molecule through its metabolic fate inside the cell (e.g., radioactively labeled glucose is oxidized and the amount of glucose conversion is measured by quantifying the tracer atom, such as ^{3}H in water or ^{14}C in CO_2). Such tracers can be radioactive (^{3}H, ^{18}F, ^{11}C, or ^{14}C) or visible by NMR (^{1}H or ^{13}C). Depending on the investigational setup and the question desired to investigate, tracers can be used in cell cultures, in isolated organ preparations or even *in vivo*. They can be used for the true or semiquantitative assessment of substrate utilization rates (e.g., glucose uptake or oxidation) or for imaging purposes (e.g., ^{18}F-2-deoxyglucose–positron emission tomography (FDG–PET)). In some cases, not the tracee is labeled but a chemical analog (e.g., 2-deoxyglucose (2-DG) for glucose). Such tracer analogs have advantages and disadvantages.

A prominent example is the use of 2-DG to trace glucose uptake. Figure 7.3 shows the principle function of tracer analogs. In case of glucose uptake, 2-DG is transported and phosphorylated similar to glucose but is not further metabolized. The accumulation of radioactively labeled 2-DG6P can be traced. If 2-DG is labeled with the positron-emitting [^{18}F], the location of the radioactive decay can be determined using PET. FDG–PET has had a tremendous impact on cardiovascular and oncological medicine, because it enabled imaging of glucose metabolism *in vivo*. For a detailed description of radioactive and magnetic resonance imaging *in vivo* and specifically in humans, please see Chapter 14.

In humans, tracer analogs are the most widely used metabolic tracers. However, tracer analogs have weaknesses since they are limited in their ability to truly quantify metabolic processes. In order to relate 2-DG accumulation to glucose uptake, one has to take the kinetic properties of the glucose transporters and of hexokinase into account. When FDG–PET was developed, these differences were accounted for by using

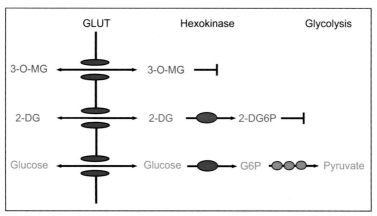

FIGURE 7.3 **Fate of the two glucose tracers 3-O-methylglucose (3-O-MG) and 2-DG in comparison to regular glucose.**
3-O-MG is not phosphorylated by hexokinase, facilitated transport across the cell membrane is possible in both directions
(influx and efflux). 2-DG is phosphorylated by hexokinase after its transport into the cell. The resulting 2-DG-6-phosphate
(2-DG6P); however, is trapped inside the cell. It cannot be transported out of the cell, nor can it be metabolized any further
at significant fractions [30].

a correction factor, named the lumped constant. It received this name because the assumption was made that the evident differences (the glucose transporters prefer the transport of 2-DG and hexokinase prefers the phosphorylation of glucose) were constant and could therefore be "lumped" together. However, we demonstrated that 2-DG cannot be used for the quantification of glucose uptake under nonsteady state conditions because the lumped constant is bound to change with changing metabolic environment [31–33]. As a consequence, investigations *in vivo* using FDG or other 2-DG tracers are mainly semiquantitative by relating the findings to a reference region. Only if true glucose or other metabolite tracers are used, these techniques can be truly quantitative. Although this is possible, due to the fast metabolization of product (e.g., glucose-6-phosphate quickly enters glycolysis or is converted to glycogen), highly sophisticated "pulse-chase" experiments have to be designed.

This difficulty can be overcome by *ex vivo* perfusion of organs. Isolated heart perfusion systems have been established as a powerful tool for metabolic analyses. Several substrate tracers and tracer analogs are available to assess

certain aspects of cardiac metabolism during *ex vivo* perfusion [34]. Table 7.1 shows a list of commonly used tracers and tracer analogs used to assess cardiac substrate metabolism.

Glucose transport may be assessed by measuring the efflux of 3-O-methylglucose, which is imported into the cardiomyocyte, but not further phosphorylated by hexokinase (see also Fig. 7.3). Glucose oxidation rates can be calculated by measuring $^{14}CO_2$ release due to the metabolism of labeled ^{14}C-glucose [34]. Tritium-labeled glucose analogs are established markers of glucose fate. Positioning tritium at different carbon atoms inside the glucose molecule allows the distinction of flux along different pathways: 2-^{3}H-glucose (i.e., glucose labeled with tritium in the C2-position) is used to trace glucose transport and phosphorylation by measuring production of tritium-labeled water (^{3}HOH) [38,39]. ^{3}HOH production during perfusion with 5-^{3}H-glucose has been suggested to measure glycolytic flux [46], although a considerable proportion of ^{3}HOH-production may be due to flux through the pentose phosphate pathway [47].

The rate of fatty acid oxidation is commonly addressed by using radiolabeled fatty acids,

TABLE 7.1 Overview of Commonly Applied Radioactive Glucose and Fatty Acid Tracers and Tracer Analogs in Metabolic Research

Tracer	Metabolic pathway	Measured in	Product measured	References
3-O-[Methyl-^{14}C]-methyl-D-glucose	Glucose transport	Perfusate	^{14}C-3-O-methylglucose efflux	[35,36]
D-[[14]C(U)]-glucose	Glucose oxidation	Perfusate	^{14}C-CO_2	[34]
2[1-^{14}C]-deoxy-D-glucose	Glucose uptake (transport and phosphorylation)	Tissue homogenate	^{14}C-2-deoxyglucose-6-phosphate	[37]
[[14]F]-2-fluoro-2-deoxy-D-glucose	Glucose uptake (transport and phosphorylation)	Tissue	PET signal	[42]
D-[2-^{3}H]-glucose	Glucose uptake (transport and phosphorylation)	Blood/perfusate	^{3}H-H_2O	[38]
D-[3-^{3}H]-glucose	Glucose uptake or glycolysis	Blood/perfusate	^{3}H-H_2O	[39]
D-[5-^{3}H]-glucose	Glycolysis	Blood/perfusate	^{3}H-H_2O	[34,40]
D-[6-^{3}H]-glucose	Glucose oxidation	Blood/perfusate	^{3}H-H_2O	[41]
[9,10-[3]H(N)]-oleic acid	Fatty acid oxidation	Blood/perfusate	^{3}H-H_2O	[43]
[1-^{14}C]-palmitic acid	(Endogenous*) fatty acid oxidation	Perfusate	^{14}C-CO_2	[44]
[9,10-[3]C]-palmitic acid	(Exogenous*) fatty acid oxidation	Perfusate	^{3}H-H_2O	[44]
[1-^{14}C]-octanoate	Fatty acid oxidation (irrespective of carnitine-acylcarnitine-translocase system activity)	Perfusate	^{14}C-CO_2	[45]

* Differentiation between endogenous and exogenous fatty acid oxidation can be achieved by using both tracers in a pulse-chase-experiment. See Ref. [44] for further details.

for example, 9,10-^{3}H-palmitate; 9,10-^{3}H-oleate; or ^{14}C-palmitate. When using tritium-labeled fatty acids, fatty acid oxidation can be determined as the release of [3]HOH, while liberation of [14]CO_2 is measured in experiments with ^{14}C-labeled fatty acids [44]. Oxidation of ^{14}C-octanoate (a medium-chain fatty acid, which is transported into mitochondria independent of carnitine) can also be traced via [14]CO_2 release [45].

A nonradioactive way of obtaining similar information is the use of NMR spectroscopy, which allows the evaluation of intracellular metabolism in vivo and in vitro by tracking NMR signals of nuclei such as [31]phosphorus, [13]carbon, or [1]hydrogen [48]. The visibility of various nuclei allows the detection of several metabolites [49]:

with ^{1}H-NMR spectroscopy, metabolites such as lactate or fatty acids can be detected. ^{31}P-NMR spectroscopy is traditionally used to trace ATP and phosphocreatine (PCr) dynamics and to determine the PCr/ATP ratio, an indicator for the energetic state of the heart. ^{13}C-NMR spectroscopy can be used to evaluate various metabolic pathways as glycolysis, the Krebs cycle, or fatty acid oxidation. Application of this technique to assess metabolism in vivo has proven to be challenging due to its intrinsically low sensitivity: the low abundance of natural [13]C (only 1.1% of total carbon isotopes) results in a low-intensity signal. Therefore, hyperpolarization techniques have been developed to increase the signal from any given ^{13}C-labeled substrate [50].

This method has been used to assess important aspects of cardiac metabolism (e.g., pyruvate dehydrogenase flux and Krebs cycle flux) in the development of heart disease [51–55].

A regular ^{31}P spectrum typically shows six peaks, representing the three ^{31}P-atoms in ATP (α, β, and γ); PCr; 2,3-diphosphoglycerate; and phosphodiesters [49]. Knowing the amount of ATP and PCr, the PCr/ATP ratio can be determined. Whenever ATP synthesis is lower than ATP demand, PCr levels decline, leading to a reduced PCr/ATP ratio. Thus, a reduction of the PCr/ATP ratio is found under circumstances of impaired energy metabolism, for example, heart failure, where it was shown to correlate with the degree of failure and predicted mortality in patients [56,57]. Furthermore, it is also possible to determine actual ATP production rates using ^{31}P saturation transfer NMR spectroscopy [58–60]. Basically, calculation of ATP production rates is based on changes of signal intensities of the γATP- and P_i-peak in the ^{31}P spectrum, representing the incorporation of P_i into ADP to form ATP (for further details, see Refs [60,61]).

The previously described principles of tracer and imaging techniques quickly become highly sophisticated. Therefore, it is not surprising that investigators specializing in one or the other technique quickly search cooperation with other groups to increase the methodological ability to investigate a specific question.

INTEGRATION OF METHODS – MITOCHONDRIAL RESEARCH AS AN EXAMPLE

When addressing a specific question in metabolic research, often a combination of technologies available is necessary. Here we illustrate, how in the context of mitochondrial research such a combination may look like.

Mitochondrial respiration is commonly assessed in isolated mitochondria. In heart muscle, two distinct mitochondrial subpopulations have been described with subsarcolemmal mitochondria (SSM) located underneath the outer cell membrane and interfibrillar mitochondria (IFM) being embedded between the myofibrils. Experiments on isolated mitochondria allow the assessment of differential changes in the activity of IFM and SSM. Besides the eponymous difference in intracellular localization, they exhibit distinct biochemical properties in healthy hearts and react differently during the onset of heart failure [62,63]. Separation of these two fractions can be achieved by differential centrifugation [62]. However, the isolation process itself may result in changes of mitochondrial function and structure [64]. A more elaborated approach is the use of permeabilized muscle fibers. Here, fibers are dissected from whole muscle and treated chemically to permeabilize the cell membrane [65]. In contrast to isolated mitochondria, this approach allows for the assessment of mitochondria while possibly important interactions with other cellular constituents (e.g., the cytoskeleton) are preserved [66]. However, with this method IFM and SSM cannot be assessed separately. Furthermore, as the permeabilization of the cell membrane leads to the equilibration of the cytosol and the incubation solution, relevant intracellular metabolites that regulate mitochondrial function may be washed out. Additionally, under certain circumstances, diffusion may become the limiting step and it has been suggested that SSM are the preferentially assessed mitochondria [67]. Therefore, even with this method, mitochondria cannot be assessed under exact *in vivo* conditions [2].

A key feature of mitochondria is the practically exclusive use of oxygen inside the cell. Thus, the respiratory chain is the hallmark of mitochondria and has been the target of many investigations. Mitochondrial proteomics and metabolomics are helpful for the assessment of the "general equipment" of the mitochondria. More specific technology is used for the assessment of respiratory capacity.

SPECTROPHOTOMETRIC ASSAYS OF MITOCHONDRIAL COMPLEX ACTIVITIES

Enzyme activity of mitochondrial complexes can be measured spectrophotometrically. Results are usually expressed in relation to protein concentration or citrate synthase activity (a marker for mitochondrial abundance under steady state conditions). Mitochondrial complex activities can be assessed in tissue homogenate or in isolated mitochondria. In a given mitochondrial sample, oxygen consumption, respiratory capacity, ATP production, production of reactive oxygen species (ROS), calcium uptake as well as individual complex activities may be measured and is described in detail elsewhere [68–71]. Here, we illustrate the assessment of the respiratory chain beginning with the individual complex activities.

Complex I

Oxidation of NADH/H by complex I leads to a reduction of absorbance at 340 nm. (Please note that in this chapter, NAD is used as a general abbreviation for *nicotinamide adenine dinucleotide*, irrespective of its oxidative state. NAD^+ explicitly denotes the oxidized form and NADH/H the reduced form.) The rate of decline is used to calculate complex I activity. In a direct assay, the oxidation of NADH/H by other (including nonmitochondrial) enzymes leads to considerable background activity. Complex I activity is therefore determined as the rotenone-sensitive change in absorbance [68]. Rotenone is a specific inhibitor of complex I. To further reduce the nonspecific change in absorbance, a coupled assay has been developed, incorporating a second, nonenzymatic step: by adding the dye dichloroindophenol (DCIP), electrons that are transferred from NADH/H to the coenzyme Q_{10} analog decylubiquinone are subsequently delivered to DCIP [72]. As DCIP accepts electrons almost exclusively from decylubiquinol (the reduced form of decylubiquinone), and electrons from other NAD-dependent dehydrogenases are not accepted by decylubiquinone [73], the resulting reduction of DCIP is almost entirely due to complex I activity. A further advantage of this approach is the continuous reoxidation of decylubiquinol to decylubiquinone, as large amounts of decylubiquinol may inhibit complex I activity [74].

Complex II

Similar to complex I, complex II activity can be measured using DCIP as the final electron acceptor [68,75]. To detect noncomplex-II-specific changes in absorbance, complex II inhibitors as thenoyltrifluoroacetone (TTFA) can be used [75,76].

Complex III

Activity of complex III is measured as antimycin-A (a specific complex III inhibitor) sensitive reduction of cytochrome c [75,77]. The reducing agent is decylubiquinol (which has to be prepared from decylubiquinone [68]) and photometrical readings are performed at 550 nm.

Complex IV

Complex IV catalyzes the oxidation of cytochrome c and transfers electrons to molecular oxygen. In analogy to the complex III assay, changes in absorbance are traced at 550 nm. As its product rapidly inhibits cytochrome c oxidase, only little amounts of mitochondrial protein should be used for this assay. By doing so, a pseudolinear rate of change in absorbance can be obtained for several minutes [78,79]. Alternatively, the first-order rate constant can be calculated [80]. To measure the change in absorbance that is not dependent on cytochrome c oxidase, the complex IV-inhibitor cyanide can be used [68].

Complex V

Complex V (ATP synthase) activity can be measured in reverse direction as an oligomycin-sensitive ATPase [68,81]. To do so, lactate dehydrogenase and pyruvate kinase are used as linking enzymes to set up a coupled assay: ADP that is formed during the hydrolysis of ATP is used for the pyruvate kinase reaction, in which phosphoenolpyruvate is converted into pyruvate. Then, pyruvate is reduced to lactate, by lactate dehydrogenase. This reaction requires NADH/H as a cofactor, which is oxidized to NAD^+ in the course of the reaction, a process that can be traced by measuring absorbance at 340 nm.

Combined Complex Assays

In vivo, mitochondrial respiration is a concerted action between the different parts of the electron transport chain. Besides the previously described complexes, the electron carriers coenzyme Q_{10} and cytochrome c are required to transport electrons from one complex to another. Coenzyme Q_{10} transfers electrons from complexes I and II to complex III. From there, they are further transported to complex IV by cytochrome c. For the assessment of individual complex activities (as described earlier), these electron carriers are included in the assays as electron acceptors or donors. However, defects in mitochondrial respirations that are due to changes in the interaction of individual complexes or reduced amounts of endogenous electron carriers may be missed under these conditions. Coenzyme Q_{10} deficiency, for instances, is known to limit mitochondrial function, even in the absence of an isolated defect of the mitochondrial complexes [82] and its deficiency has even been the subject of heart failure therapy in humans [83]. Thus, specific experiments are available to assess the combined action of two complexes, that is, complex I + III activity or complex II + III activity [68,84].

However, when interpreting results of these assays, it is important to consider the relevance of each individual enzyme. For example, flux through complex II and complex III together is more dependent on the slower (and therefore, rate-limiting) activity of complex II. Thus, the assessment of combined complex activities alone does not necessarily allow ruling out a defect of an individual mitochondrial complex [68]. In other words, although a specific disease state may be associated with the impairment of a specific enzyme complex, this reduction in enzyme function may not be functionally relevant if another enzyme is rate limiting. Thus, for the complete assessment of respiratory capacity, mitochondrial respiration needs to be measured as a whole. The standard way of doing this is by polarography.

ASSESSING MITOCHONDRIAL RESPIRATION BY POLAROGRAPHY

Measuring mitochondrial respiration and determining ADP/O ratios has been described by spectrophotometry [85] as well as by polarography [86], with the latter being most commonly used today. During polarography, oxygen concentration in a solution is constantly measured using a Clark-type electrode [87]. The electrode is placed inside a reaction chamber and registers a change in electric current based on changes in oxygen concentration inside the reaction chamber. Owing to its hermetic setup, these changes are due to any reaction consuming (or releasing) oxygen inside the chamber. The rate of mitochondrial respiration, during which molecular oxygen is reduced to water (thus, lowering oxygen concentration), can therefore be measured using cultured cells, tissue homogenate, or isolated mitochondria [88].

Chance and Williams introduced the polarographic approach to measure mitochondrial respiratory rates. They described five steady states of mitochondrial respiration that occur during the experiment [89].

State 1 Respiration

To start state 1 respiration, a fresh preparation of mitochondria is added to the reaction chamber that is filled with an isotonic buffer (containing sufficient amounts of phosphate), which is preincubated at a set temperature and continuously mixed by a small magnetic stirrer. Mitochondrial respiration starts at a slow rate, fueled only by endogenous substrates. While the amount of endogenous substrate declines, reduction of NAD increases. State 1 respiration is limited mostly by the lack of a phosphate acceptor (i.e., ADP).

State 2 Respiration

State 2 respiration is initiated by the addition of ADP. Oxidation of NAD prevails and respiration is ultimately limited by substrate availability. According to Chance and Williams, state 2 respiration is close to zero: with no substrates to fuel the respiratory chain, no electrons are available to be transferred onto molecular oxygen at complex IV. Nevertheless, a small rate of residual oxygen consumption may be found, which is, however, not due to respiration at the site of the electron transport chain. During a presteady-state phase, all endogenous substrate is metabolized in the presence of ADP, leading to a short peak in oxygen consumption, which is not to be confused with the steady state 2 respiration.

State 3 Respiration

In the original publication by Chance and Williams [89], state 3 respiration was initiated by the addition of substrate. Due to excess amounts of substrate and ADP available, mitochondrial respiration is limited mostly by the intrinsic activity of the respiratory chain itself (or the slowest complex of the chain). As ATP generation now exceeds ATP consumption, ADP levels decline again, eventually leading to state 4 respiration. Note that in many studies, state 3 respiration is also termed ADP-stimulated respiration. This at first seems to be confusing regarding the original setup by Chance and Williams, where state 2 respiration is started by the addition of ADP. However, even Chance and Williams suggested examining the effects of ADP on respiration by reversing the transition from state 3 to state 4 by adding ADP a second time. Thus, state 3 can as well be initiated by the addition of ADP to mitochondria already respiring in state 4 (i.e., in the presence of excess substrate).

State 4 Respiration

While substrate is still abundant, lack of a phosphate acceptor again limits mitochondrial respiration during state 4 respiration. Remaining ATPase activity in the preparation leads to slow but continuous degradation of ATP. The resulting small amounts of ADP are then phosphorylated again by ATP synthase. Furthermore, movement of H^+ (uncoupling, described later) through the IMM into the matrix may lead to additional oxygen consumption. Due to the presence of substrate, NAD is found predominantly in the reduced form (i.e., NADH/H).

By adding an exogenous uncoupler (e.g., 2,4-dinitrophenol) during state 4, respiration rates may rise again. This demonstrates the fundamental control of mitochondrial respiration by the electrochemical gradient: when ATP production at complex V is limited during state 4 respiration due to ADP deficiency, the electrochemical potential across the IMM rises to a maximum, at which further transport of protons against this potential would require more energy than can provided by reactions at complexes I, III, and IV. At this point, activity of the electron transport chain is limited. Addition of the uncoupler, however, obviates the necessity of ATP synthesis to reduce the electrochemical gradient (literally uncoupling mitochondrial respiration from ATP synthesis at complex V).

The understanding of respiratory control is important to correctly interpret the results of experiments on mitochondrial respiration: when state 3 respiration is limited by the activity of the ATP phosphorylation apparatus (e.g., a defect of complex V) rather than by the electron transport chain itself, "uncoupled state 4 respiration" may even exceed the level of state 3 respiration.

State 5 Respiration

After a while, the constant mitochondrial respiration leads to an absolute lack of oxygen inside the reaction chamber, naturally ending all respiratory activity.

A marker for the efficiency of mitochondrial ATP generation is the ADP/O ratio, which indicates the number of oxygen atoms needed for mitochondrial respiration to generate one molecule ATP at complex V. The coupling of mitochondrial respiration to ATP production has found a prominent explanation by Peter Mitchell's Nobel prize-winning chemiosmotic hypothesis [90]. Energy from NADH/H and $FADH_2$ is used during mitochondrial respiration to pump protons across the IMM. As the IMM is almost impermeable to protons, a chemical and electrical gradient across the IMM is generated, conserving energy in the so-called proton-motive force. While protons move back into the mitochondrial matrix through complex V, this force is used to generate ATP out of ADP and inorganic phosphate.

According to Mitchell's hypothesis, proton-pumping complexes (I, III, and IV) each pump just enough protons across the membrane for the generation of one ATP. Thus, under ideal conditions, three ATPs are generated per oxygen atom utilized. Indeed, the majority of investigations assessing ADP/O ratios at that time generated values around 3. Later, it was recognized that many other mechanisms may exhaust and influence the proton gradient without resulting in ATP production. Therefore, these mechanisms have been defined as uncoupling mechanisms and they may consist of uncoupling proteins; a constant but small leakage of protons back into the matrix; calcium exchange or under certain conditions, the opening of the mitochondrial permeability transition pore (mPTP) [91].

METHODS TO ASSESS ADDITIONAL ASPECTS OF MITOCHONDRIAL STRUCTURE AND FUNCTION

Various microscopy techniques can be employed to elucidate more details of mitochondrial function and structure. Imaging with high-resolution scanning electron microscopy was used to show structural differences between IFM and SSM in Sprague Dawley rats [62]. Mitochondrial size and density can be measured as well but have been limited by the two-dimensional nature of the method. The three-dimensional reconstruction allows for more precise analysis of mitochondrial morphology [92]. Using confocal microscopy and fluorescent dyes, mitochondrial abundance, morphology, or mitochondrial membrane potential can be detected [71,93]. Such dyes are also frequently used to assess intracellular calcium handling, although many other methods are available [70]. To evaluate intramitochondrial calcium, these dyes can be transformed into cationic molecules by chemical modification, which then preferentially accumulate in the mitochondrial matrix [94].

Opening of the mPTP can be monitored using spectrophotometry as a decrease in light absorbance at 540 nm [95]. In isolated cardiomyocytes, mPTP opening can be measured using fluorescent probes as tetramethylrhodamine-methyl-ester (TMRM) or calcein [96–98].

Another key feature of mitochondrial function is the generation of ROS. In the heart, a significant fraction of ROS are generated at the respiratory chain (see Chapter 3). Direct measurements of different ROS are possible using a variety of techniques: chemiluminescent probes release photons after contact with ROS, which can be detected using scintillation counters. Fluorescent probes as dichlorofluorescin or Amplex Red exhibit their fluorescent properties only after oxidation by ROS. Further, free radicals can be directly assessed by electron spin resonance spectrometry, with the help of so-called spin traps that react with the otherwise short-lived ROS and develop specific spectra. These methods are reviewed in more detail in Ref. [69].

ROS have many effects in the cell, one of them being the oxidative modification (i.e., damage) of cellular structures. The detection of oxidatively modified molecules is a common approach to monitor both production of ROS as well as the amount of damage caused. Malondialdehyde, for example, is a result of lipid peroxidation *in vivo* and can be quantified using ELISA [99]. Alternatively, protein carbonylation (another marker of oxidative stress) can be measured via spectrophotometry or Western blotting [100,101].

SUMMARY

There is a plethora of methodologies available today to investigate cardiac metabolism. The technical abilities by far exceed the information that can be covered by one book, let alone by one book chapter. However, with this chapter, it should have become clear that from individual metabolite or enzyme tests to high-throughput screening technologies, all these techniques have their inherent strengths and limitations. To reliably test a hypothesis, it is therefore important to adapt the technologies to the questions asked and not vice versa. Another important recognition should be that the most sophisticated method will not yield satisfactory results if the sample used for investigation is false or flawed. Consideration of the principles in this chapter should help planning one's own experiments in cardiac metabolism.

References

[1] Chemnitius JM, et al. Rapid preparation of subsarcolemmal and interfibrillar mitochondrial subpopulations from cardiac muscle. Int J Biochem 1993;25(4):589–96.

[2] Hughey CC, et al. Respirometric oxidative phosphorylation assessment in saponin-permeabilized cardiac fibers. J Vis Exp 2011;48:2431.

[3] Schuh RA, et al. Measuring mitochondrial respiration in intact single muscle fibers. Am J Physiol Regul Integr Comp Physiol 2012;302(6):R712–9.

[4] Krebs HA, Johnson WA. The role of citric acid in intermediate metabolism in animal tissues. Enzymologia 1937;4:148–56.

[5] Mitchell P. Keilin's respiratory chain concept and its chemiosmotic consequences. Science 1979;206(4423): 1148–59.

[6] Mitchell P, Moyle J. Stoichiometry of proton translocation through the respiratory chain and adenosine triphosphatase systems of rat liver mitochondria. Nature 1965;208(5006):147–51.

[7] Meyerhof O. The origin of the reaction of harden and young in cell-free alcoholic fermentation. J Biol Chem 1945;157(1):105–20.

[8] Meyerhof O, Junowicz-Kocholaty R. The equilibria of isomerase and aldolase, and the problem of the phosphorylation of glyceraldehyde phosphate. J Biol Chem 1943;149(1):71–92.

[9] Schagger H, von Jagow G. Blue native electrophoresis for isolation of membrane protein complexes in enzymatically active form. Anal Biochem 1991;199(2):223–31.

[10] Nijtmans LGJ, Henderson NS, Holt IJ. Blue Native electrophoresis to study mitochondrial and other protein complexes. Methods 2002;26:327–34.

[11] Schägger H, Pfeiffer K. Supercomplexes in the respiratory chains of yeast and mammalian mitochondria. EMBO J 2000;19:1777–83.

[12] Rosca MG, et al. Cardiac mitochondria in heart failure: decrease in respirasomes and oxidative phosphorylation. Cardiovasc Res 2008;80:30–9.

[13] Shah SH, Kraus WE, Newgard CB. Metabolomic profiling for the identification of novel biomarkers and mechanisms related to common cardiovascular diseases: form and function. Circulation 2012;126(9): 1110–20.

[14] Patti GJ, Yanes O, Siuzdak G. Innovation: metabolomics: the apogee of the omics trilogy. Nat Rev Mol Cell Biol 2012;13:263–9.

[15] Langley SR, et al. Proteomics: from single molecules to biological pathways. Cardiovasc Res 2013;97: 612–22.

[16] Samuel JL, et al. Genomics in cardiac metabolism. Cardiovasc Res 2008;79:218–27.

[17] Taegtmeyer H, Sen S, Vela D. Return to the fetal gene program: a suggested metabolic link to gene expression in the heart. Ann NY Acad Sci 2010;1188:191–8.

[18] Churko JM, et al. Overview of high throughput sequencing technologies to elucidate molecular pathways in cardiovascular diseases. Circ Res 2013;112:1613–23.

[19] Sharma P, Cosme J, Gramolini AO. Recent advances in cardiovascular proteomics. J Proteomics 2013;81:3–14.

[20] Pauling L, et al. Quantitative analysis of urine vapor and breath by gas–liquid partition chromatography. Proc Natl Acad Sci USA 1971;68:2374–6.

[21] Lewis GD, Asnani A, Gerszten RE. Application of metabolomics to cardiovascular biomarker and pathway discovery. J Am Coll Cardiol 2008;52:117–23.

[22] Wishart DS, et al. HMDB 3.0 – The Human Metabolome Database in 2013. Nucleic Acids Res 2013;41(Database): D801–7.

[23] Milne SB, et al. Sum of the parts: mass spectrometry-based metabolomics. Biochemistry 2013;52:3829–40.

[24] Senn T, Hazen SL, Tang WHW. Translating metabolomics to cardiovascular biomarkers. Prog Cardiovasc Dis 2012;55:70–6.

[25] Zhou B, et al. LC-MS-based metabolomics. Mol Biosyst 2012;8:470–81.

[26] Lei Z, Huhman DV, Sumner LW. Mass spectrometry strategies in metabolomics. J Biol Chem 2011;286: 25435–42.

[27] Bain JR, et al. Metabolomics applied to diabetes research: moving from information to knowledge. Diabetes 2009;58:2429–43.

[28] Lee DR, Robert EG. Toward new biomarkers of cardiometabolic diseases. Cell Metab 2013;18(1):43–50.

[29] Fick A. Ueber die Messung des Blutquantums in den Herzventrikeln. Sitzungsberichte der Physikalisch-Medizinischen Gesellschaft zu Wuerzburg; 1870. p. XVI.

[30] Doenst T, Taegtmeyer H. Kinetic differences and similarities among 3 tracers of myocardial glucose uptake. J Nucl Med 2000;41:488–92.

[31] Doenst T, Taegtmeyer H. Profound underestimation of glucose uptake by [18F]2-deoxy-2-fluoroglucose in reperfused rat heart muscle. Circulation 1998;97:2454–62.

[32] Rhodes CG, et al. Variability of the lumped constant for [18F]2-deoxy-2-fluoroglucose and the experimental isolated rat heart model: clinical perspectives for the measurement of myocardial tissue viability in humans. Circulation 1999;99(9):1275–6.

[33] Doenst T, Taegtmeyer H. Complexities underlying the quantitative determination of myocardial glucose uptake with 2-deoxyglucose. J Mol Cell Cardiol 1998;30(8): 1595–604.

[34] Belke DD, et al. Glucose and fatty acid metabolism in the isolated working mouse heart. Am J Physiol 1999;277:R1210–7.

[35] Rovetto MJ, Whitmer JT, Neely JR. Comparison of the effects of anoxia and whole heart ischemia on carbohydrate utilization in isolated working rat hearts. Circ Res 1973;32:699–711.

[36] Zaninetti D, Greco-Perotto R, Jeanrenaud B. Heart glucose transport and transporters in rat heart: regulation by insulin, workload and glucose. Diabetologia 1988;31:108–13.

[37] Russell RW, Young JW. A review of metabolism of labeled glucoses for use in measuring glucose recycling. J Dairy Sci 1990;73:1005–16.

[38] Katz J, Dunn A. Glucose-2-t as a tracer for glucose metabolism. Biochemistry 1967;6(1):1–5.

[39] Katz H, et al. Use of [3-3H]glucose and [6-14C]glucose to measure glucose turnover and glucose metabolism in humans. Am J Physiol 1992;263(1 Pt. 1):E17–22.

[40] Neely JR, et al. The effects of increased heart work on the tricarboxylate cycle and its interactions with glycolysis in the perfused rat heart. Biochem J 1972;128:147–59.

[41] Clark MG, et al. Estimation of the fructose 1,6-diphosphatase-phosphofructokinase substrate cycle and its relationship to gluconeogenesis in rat liver in vivo. J Biol Chem 1974;249:279–90.

[42] Osterholt M, et al. Targeted metabolic imaging to improve the management of heart disease. JACC Cardiovasc Imaging 2012;5:214–26.

[43] Goodwin GW, et al. Energy provision from glycogen, glucose, and fatty acids on adrenergic stimulation of isolated working rat hearts. Am J Physiol 1998;274:H1239–47.

[44] Saddik M, Lopaschuk GD. Myocardial triglyceride turnover and contribution to energy substrate utilization in isolated working rat hearts. J Biol Chem 1991;266:8162–70.

[45] el Alaoui-Talibi Z, et al. Fatty acid oxidation and mechanical performance of volume-overloaded rat hearts. Am J Physiol 1992;262:H1068–74.

[46] Hue L, Hers HG. On the use of (3H, 14C)labelled glucose in the study of the so-called "futile cycles" in liver and muscle. Biochem Biophys Res Commun 1974;58(3):532–9.

[47] Goodwin GW, Cohen DM, Taegtmeyer H. [5-3H] glucose overestimates glycolytic flux in isolated working rat heart: role of the pentose phosphate pathway. Am J Physiol Endocrinol Metab 2001;280(3):E502–8.

[48] Hwang J-H, Choi CS. Use of in vivo magnetic resonance spectroscopy for studying metabolic diseases. Nature 2015;47:e139–8.

[49] Hudsmith LE, Neubauer S. Magnetic resonance spectroscopy in myocardial disease. JACC Cardiovasc Imaging 2009;2:87–96.

[50] Rider OJ, Tyler DJ. Clinical implications of cardiac hyperpolarized magnetic resonance imaging. J Cardiovasc Magn Reson 2013;15:93.

[51] Seymour AM, et al. *In vivo* assessment of cardiac metabolism and function in the abdominal aortic banding model of compensated cardiac hypertrophy. Cardiovasc Res 2015;106(2):249–60.

[52] Dodd MS, et al. Impaired *in vivo* mitochondrial Krebs cycle activity after myocardial infarction assessed using hyperpolarized magnetic resonance spectroscopy. Circ Cardiovasc Imaging 2014;7(6):895–904.

[53] Josan S, et al. *In vivo* investigation of cardiac metabolism in the rat using MRS of hyperpolarized [1-13C] and [2-13C]pyruvate. NMR Biomed 2013;26(12):1680–7.

[54] Ball DR, et al. Metabolic imaging of acute and chronic infarction in the perfused rat heart using hyperpolarised [1-13C]pyruvate. NMR Biomed 2013;26(11):1441–50.

[55] Chen AP, et al. Simultaneous investigation of cardiac pyruvate dehydrogenase flux, Krebs cycle metabolism and pH, using hyperpolarized [1,2-(13)C2]pyruvate *in vivo*. NMR Biomed 2012;25(2):305–11.

[56] Neubauer S, et al. 31P magnetic resonance spectroscopy in dilated cardiomyopathy and coronary artery disease. Altered cardiac high-energy phosphate metabolism in heart failure. Circulation 1992;86:1810–8.

[57] Neubauer S, et al. Myocardial phosphocreatine-to-ATP ratio is a predictor of mortality in patients with dilated cardiomyopathy. Circulation 1997;96:2190–6.

[58] Befroy DE, et al. ^{31}P-magnetization transfer magnetic resonance spectroscopy measurements of *in vivo* metabolism. Diabetes 2012;61:2669–78.

[59] Lebon V, et al. Effect of triiodothyronine on mitochondrial energy coupling in human skeletal muscle. J Clin Invest 2001;108:733–7.

[60] Befroy DE, et al. Assessment of *in vivo* mitochondrial metabolism by magnetic resonance spectroscopy. Methods in enzymology. Elsevier; 2009. p. 373–93. [chapter 21].

[61] Forsén S, Hoffman RA. Study of moderately rapid chemical exchange reactions by means of nuclear magnetic double resonance. J Chem Phys 1963;39:2892.

[62] Palmer JW, Tandler B, Hoppel CL. Biochemical properties of subsarcolemmal and interfibrillar mitochondria isolated from rat cardiac muscle. J Biol Chem 1977;252:8731–9.

[63] Schwarzer M, et al. Pressure overload differentially affects respiratory capacity in interfibrillar and subsarcolemmal mitochondria. Am J Physiol Heart Circ Physiol 2013;304:H529–37.

[64] Picard M, et al. Mitochondrial structure and function are disrupted by standard isolation methods. PLoS ONE 2011;6:e18317.

[65] Kuznetsov AV, et al. Analysis of mitochondrial function *in situ* in permeabilized muscle fibers, tissues and cells. Nat Protoc 2008;3(6):965–76.

[66] Kuznetsov AV, et al. Cytoskeleton and regulation of mitochondrial function: the role of beta-tubulin II. Front Physiol 2013;4:82.

[67] Jepihhina N, et al. Permeabilized rat cardiomyocyte response demonstrates intracellular origin of diffusion obstacles. Biophys J 2011;101(9):2112–21.

[68] Kirby DM, et al. Biochemical assays of respiratory chain complex activity. Methods Cell Biol 2007;80:93–119.

[69] Dikalov SI, Harrison DG. Methods for detection of mitochondrial and cellular reactive oxygen species. Antioxid Redox Signal 2014;20:372–82.

[70] Takahashi A, et al. Measurement of intracellular calcium. Physiol Rev 1999;79:1089–125.

[71] Marin-Garcia J. Methods to study mitochondrial structure and function. US: Springer; 2012. p. 13–27.

[72] Janssen AJM, et al. Spectrophotometric assay for complex I of the respiratory chain in tissue samples and cultured fibroblasts. Clin Chem 2007;53:729–34.

[73] Fischer JC, et al. Estimation of NADH oxidation in human skeletal muscle mitochondria. Clin Chim Acta 1986;155:263–73.

[74] Bénit P, Slama A, Rustin P. Decylubiquinol impedes mitochondrial respiratory chain complex I activity. Mol Cell Biochem 2008;314:45–50.

[75] Krahenbuhl S, et al. Decreased activities of ubiquinol:ferricytochrome c oxidoreductase (complex III) and ferrocytochrome c:oxygen oxidoreductase (complex IV) in liver mitochondria from rats with hydroxycobalamin[c-lactam]-induced methylmalonic aciduria. J Biol Chem 1991;266:20998–1003.

[76] King TE. Preparations of succinate – cytochrome c reductase and the cytochrome b-c1 particle, and reconstitution of succinate-cytochrome c reductase. Methods in Enzymology. New York, London: Academic Press, Elsevier; 1967. p. 216–25.

[77] Heather LC, et al. Critical role of complex III in the early metabolic changes following myocardial infarction. Cardiovasc Res 2010;85:127–36.

[78] Miró O, et al. Cytochrome c oxidase assay in minute amounts of human skeletal muscle using single wavelength spectrophotometers. J Neurosci Methods 1998;80:107–11.

[79] Rustin P, et al. Biochemical and molecular investigations in respiratory chain deficiencies. Clin Chim Acta 1994;228:35–51.

[80] Wharton D, Tzagaloff A. Cytochrome c oxidase from beef heart mitochondria. In: Estabrook R, Pullmann M, editors. Methods in enzymology, vol. X: Oxidation and phosphorylation. New York: Academic Press; 1967. p. 245–250.

[81] Barrientos A. *In vivo* and in organello assessment of OXPHOS activities. Methods 2002;26:307–16.

[82] Quinzii CM, DiMauro S, Hirano M. Human coenzyme Q10 deficiency. Neurochem Res 2007;32:723–7.

[83] Mortensen SA, et al. The effect of coenzyme Q10 on morbidity and mortality in chronic heart failure: results from Q-SYMBIO: a randomized double-blind trial. JACC Heart Fail 2014;2(6):641–9.

[84] Barrientos A, Fontanesi F, Díaz F. Evaluation of the mitochondrial respiratory chain and oxidative phosphorylation system using polarography and spectrophotometric enzyme assays. Curr Protoc Hum Genet [editorial board, Jonathan L. Haines et al., 2009. chapter 19, p. Unit19.3].

[85] Chance B. Spectrophotometry of intracellular respiratory pigments. Science 1954;120(3124):767–75.

[86] Chance B, Williams GR. A simple and rapid assay of oxidative phosphorylation. Nature 1955;175:1120–1.

[87] Clark LC Jr, et al. Continuous recording of blood oxygen tensions by polarography. J Appl Physiol 1953;6(3): 189–93.

[88] Villani G, Attardi G. Polarographic assays of respiratory chain complex activity. Methods Cell Biol 2007;80: 121–33.

[89] Chance B, Williams GR. Respiratory enzymes in oxidative phosphorylation. III. The steady state. J Biol Chem 1955;217:409–27.

[90] Mitchell P. Chemiosmotic coupling of oxidative and photosynthetic phosphorylation. Biol Rev Comb Phil Soc 1966;41:445–502.

[91] Busiello RA, Savarese S, Lombardi A. Mitochondrial uncoupling proteins and energy metabolism. Front Physiol 2015;6:36.

[92] Kalkhoran S, et al. 3D electron microscopy tomography to assess mitochondrial morphology in the adult heart. Heart 2014;100(Suppl. 4):A10.

[93] Cottet-Rousselle C, et al. Cytometric assessment of mitochondria using fluorescent probes. Cytometry 2011;79:405–25.

[94] Dedkova EN, Blatter LA. Measuring mitochondrial function in intact cardiac myocytes. J Mol Cell Cardiol 2012;52:48–61.

[95] Ruiz-Meana M, et al. Mitochondrial Ca^{2+} uptake during simulated ischemia does not affect permeability transition pore opening upon simulated reperfusion. Cardiovasc Res 2006;71:715–24.

[96] Halestrap AP, Pasdois P. The role of the mitochondrial permeability transition pore in heart disease. Biochim Biophys Acta 2009;1787:1402–15.

[97] Hausenloy D. Inhibiting mitochondrial permeability transition pore opening at reperfusion protects against ischaemia–reperfusion injury. Cardiovasc Res 2003;60:617–25.

[98] Bernardi P, et al. Mitochondria and cell death. Mechanistic aspects and methodological issues. Eur J Biochem 1999;264:687–701.

[99] Ho E, et al. Biological markers of oxidative stress: applications to cardiovascular research and practice. Redox Biol 2013;1:483–91.

[100] Levine RL, et al. Determination of carbonyl content in oxidatively modified proteins. Methods Enzymol 1990;186:464–78.

[101] Levine RL, et al. Determination of carbonyl groups in oxidized proteins. Methods Mol Biol 2000;99:15–24.

Models to Investigate Cardiac Metabolism

Michael Schwarzer

Department of Cardiothoracic Surgery, Jena University Hospital,
Friedrich Schiller University of Jena, Jena, Germany

INTRODUCTION

The investigation of cardiac metabolism in humans is limited for ethical reasons. Thus, most fundamental findings have been obtained from animal or cell culture models and have been verified in humans. Furthermore, animal and cell culture models allow for investigation of mechanisms of disease and for genetic manipulation to assess effects of particular genes and proteins. Animal models allow the investigator to assess the influence of a single intervention. The investigator may control in detail for the experimental condition, which is in contrast to the human situation where a plethora of environmental and genetic influences may complicate the situation (e.g., nutritional habits, differences in activity, comorbidities, medication).

To date, the majority of animal experiments to assess cardiac metabolism are carried out in mice or rats. Experiments in these small rodents come at a relatively low cost and with a short generation time. In rats and mice, phenotypes develop rapidly and strains with defined backgrounds are

available. Furthermore rodents are less critical than large animal models regarding regulation of animal experiments. Accordingly, larger number of animals are usually used. In small animals the small size of the heart limits the amount of tissue (especially in the mouse). There are significant physiologic differences coming with the higher heart rate in mice and less in rats compared to humans [1]. Further advantages of murine models are that the murine genome has been sequenced and that genetically modified animals are available allowing for a wide range of cardiomyopathies of different etiology. Investigators, however, should be aware that significant differences in cardiac morphology and physiology have been described between different mouse strains [2]. They should further carefully select transgenic or knockout models as compensation for loss of function may affect the observed phenotypes in such animals [3].

Larger animal models are more similar to the human heart with respect to anatomy and physiology. Furthermore, larger animals are easier to instrument and physiologic analysis is easier [4].

The Scientist's Guide to Cardiac Metabolism
http://dx.doi.org/10.1016/B978-0-12-802394-5.00008-X

However, large animal models are much more expensive than mice or rats and come with reduced molecular biology resources.

We will present animal models for myocardial infarction, hypertension, diabetes and obesity, pressure overload, volume overload, and also for exercise-induced cardiac hypertrophy. However, animal models do not completely mimic the human situation. Thus, while there are a large number of different animal models for different diseases, each of the models has its advantages and limitations. We will thus discuss the strengths and limits of animal models presented in this chapter. Furthermore, for mechanistic analysis, animal models may be insufficient and cell culture systems may be required. The most common models in cell culture are discussed at the end of the chapter.

MODELS OF CARDIAC HYPERTROPHY

Cardiac growth by hyperplasia (increase in the number of cells) seems to be limited to the embryonic, the fetal, and the neonatal phase. Thereafter, the heart compensates for an increase in workload with hypertrophy (increase in cell size). With continued presence, some stimuli for cardiac hypertrophy may eventually lead to heart failure (HF), while others such as exercise do not lead to HF. Sometimes, cardiac hypertrophy induced by the latter (especially exercise) is stated as "physiologic" hypertrophy, while all conditions leading to HF are named "pathologic" hypertrophy. In this chapter, we do not use these terms as cardiac hypertrophy independent of stimulus may be seen as a physiologic response. Furthermore, even exercise-induced hypertrophy may present with pathologic findings [5].

Exercise-Induced Cardiac Hypertrophy

Intense exercise may induce the so-called athlete's heart, a type of cardiac hypertrophy, which does not generally lead to cardiac failure. Cardiac hypertrophy in exercise training depends on intensity of exercise. Depending on the type of exercise, different skeletal muscle groups are involved and different effects on cardiac metabolism may be exerted. In general, it can be distinguished between voluntary exercise (where animals may freely determine their exercise quantity and frequency) and forced exercise (where the experimental protocol determines the type, amount, and time of exercise training). Each mode of training has its advantages and disadvantages. Only treadmill running (a type of forced exercise) allows for uniform and well-controlled workloads. Treadmill running induces functional and structural beneficial adaptive responses in multiple organs [6]. However, treadmill running is labor-intensive and requires special equipment. Furthermore, motivating animals for prolonged exercise may be difficult and animals may experience physical and psychological stress.

Conducting swim training, a relatively large number of animals can be exercised simultaneously and without depending on animal motivation. However, the aqueous environment may add additional stress, as fear of drowning and low water temperatures could lead to hemodynamic and blood pressure changes. Swim training cannot be graded.

Voluntary exercise programs require much less supervision. With voluntary exercise, animals show similarly activated stress response. Exercise is performed in a nonstressful environment. Since the running distances and pattern may vary between animals, results tend to be more heterogeneous and less-detailed data of activity and exercise performance can be obtained.

Controlling, for exercise effects is accomplished by similar animals to remain sedentary. However, it is important to note that this is different from the human situation, where all individuals show different levels of basal activity. Investigating exercise in animal models may be regarded as analysis of two extremes while human behavior will be somewhere in between. Furthermore, animals in their natural environment also carry out physical activity and are not limited to a restricted space. Thus, sedentary animals might be regarded as detrained.

Voluntary Exercise

Voluntary exercise has been described to induce cardiac hypertrophy in mice [7] and rats [8]. Wheel running is a complex behavior and its extent depends on food availability, motivation and reward systems, as well as daily activity patterns [9]. Animals in the wild, display similar wheel-running habits in nature as under laboratory conditions [10]. Wheel running differs in female and male animals [11]. Different diets influence wheel-running activity differently in both sexes [12]. Levels of voluntary daily running peak after 2–4 weeks and declines thereafter. Characterization of the response to cage wheel exercise needs to consider the underlying factors driving cage wheel activity, both in a diseased or normal heart [12].

Treadmill Training

Hypertrophy models using treadmill training have the advantage that exercise intensity and volume may be precisely controlled allowing for uniform and well-controlled workload (duration, inclination, speed, distance) [13]. Only treadmills allow to measure exercise and cardiovascular parameters under controlled conditions [14]. The precise, uniform, and well-controlled exercise leads to low variation in experimental groups [14]. Many different treadmill-running protocols have been developed lasting from weeks to months with running session duration up to hours. Running speed varies between 10 m/min and 97 m/min and the treadmill may be leveled or raised to a maximum of 25° [14]. Treadmill running may be performed as continuous exercise or as interval training with high-intensity running bouts and low-intensity bouts changing.

CONTINUOUS EXERCISE

Most studies in mice and rats have applied continuous treadmill training characterized by fixed or increasing speed, inclination, and duration of the session. In mice such protocols have induced only limited degree of hypertrophy [15] or no hypertrophy [16]. In rats continuous protocols have led to an increase in heart-to-body weight up to 30% [17], but some investigators have found no induction of hypertrophy even after long exercise periods [18].

The reasons for such varying results are unknown. With such protocols, exercise load decreases if the absolute load remains constant as exercise capacity increases. This effect may conceal the effect of exercise [14].

AEROBIC INTERVAL TRAINING

Interval training protocols are increasingly used for the assessment of cardiac hypertrophy. These protocols allow for high-intensity running bouts, accumulating high-intensity exercise time over time. High-intensity exercise (8 min at 90%VO_2 max interspersed with 2 min at 50% VO_2 max) may result in 25–35% increase in left ventricular (LV) weight within 4 weeks of exercise [19]. High-intensity aerobic interval training appears more effective than continuous [20] or moderate-intensity treadmill programs [6,19].

ANAEROBIC EXERCISE TRAINING

Another approach has been described reducing the duration and increasing the speed of running bouts into "anaerobic" intensities (suggesting an increase in VO_2 max and stroke volume) [21,22]. This approach showed only a modest degree of hypertrophy and anaerobic exercise may be causing counterproductive response [14].

Exercise training has rarely been performed in larger animals. Treadmill running of swine with both continuous and interval protocols leads to increased heart-to-body weight [23,24]. A constant load protocol of dog sled pulling increases LV mass index as well as diameter [25] while other protocols report unchanged LV dimensions. In dogs, many exercise protocols are from uncontrolled studies, which may explain the lack of consistency [14]. In rabbits, cardiac hypertrophy has not been firmly established, however treadmill exercise for 12 weeks reportedly leads to signs of cardiac hypertrophy [26].

Swimming Exercise

Swim training is initiated by placing animals in water tanks for a defined period of time. Swim training leads to about 15% cardiac hypertrophy, which can be observed already after 1 week of 1 h per day [27]. Duration of individual training sessions (60 min vs. 90 min), total program length (4 weeks vs. 6 weeks), or frequency of training does not seem to affect the response. Instead, application of external loads during swim exercise (2 or 4% of body weight around the tail) increased hypertrophy [28]. Water temperature influences body temperature and also the hypertrophic response in a nonlinear fashion. In young rats, hypertrophic response was higher at 25°C than at 35°C, while in aged rats the opposite was true. Temperatures between 30°C and 36°C seem not to differ in the induction of hypertrophy but at 38°C no hypertrophy may be found [29]. Furthermore, water movement and depth of the water tank may influence hypertrophic response as well as the number of animals trained in a single tank [30,31].

Strength Training

In animals, strength exercise (also known as resistance training) using climbing or special apparatus is used to observe effects on specific muscles but cardiac effects have been rarely investigated [32,33]. A recent report indicates effects on blood pressure in resistance-trained obese animals compared to untrained obese. An increase in heart-to-body weight has also been observed, which may be more due to a reduction in body weight than to an increase in heart weight [34].

Hypertrophy Induced by Ischemia, Pressure Overload, or Hypertension

Ischemia in myocardial infarction and hypoxia, pressure overload, volume overload, and hypertension may lead to cardiac hypertrophy. However, whether the initiated types of hypertrophy show preserved or deteriorated cardiac functions depends on the strength of the stimulus. Therefore, cardiac hypertrophy in these conditions may eventually lead to HF and the models are discussed in the subsequent sections.

MODELS OF HEART FAILURE

Heart failure (HF) is a clinical syndrome in which myocardial pump function is inadequate for maintaining and supporting an individual's physiologic requirements. The clinical presentation is characterized by pulmonary congestion, dyspnea, and fatigue. Historically, HF was suspected when systolic function was impaired and cardiac hypertrophy with preserved systolic function has been called compensated hypertrophy. Recently, it has become clear that HF symptoms may also be found with isolated diastolic dysfunction, while systolic function remains preserved [35]. This condition is often referred to as diastolic HF [35].

HF may be caused by ischemia and hypoxia (e.g., due to myocardial infarction or low coronary flow), pressure overload (e.g., due to hypertension or aortic stenosis), volume overload (e.g., due to mitral regurgitation or aortocaval fistula), or other causes such as infections, diabetes, and genetic mutations.

Ischemia-Induced Heart Failure

Myocardial Infarction

In patients, myocardial infarction is one of the most common causes for HF. Occlusion of a coronary vessel by an atherosclerotic plaque leads to ischemia downstream of the occlusion.

After a short period of time, afflicted cardiomyocytes die, resulting in irreversible damage to the heart. Size of the infarction and survival differ, depending on time until revascularization [36]. Patient survival has improved with current therapies, however, in most cases patients inevitably develop HF.

In animal models, myocardial infarction may be induced by different approaches. It has to be considered, that different species show great differences in collateral vessels and flow [37]. This is of great importance as these vessels strongly influence the size of the infarct, thus affecting the extent of myocardial damage and alterations in myocardial metabolism. In animal models transmural infarction is frequently accompanied by ventricular fibrillations [38]. There are several methods to induce myocardial infarction: coronary artery ligation (CAL), hydraulic occlusion of vessels, and vessel microembolization.

CORONARY ARTERY LIGATION

CAL was first applied in dogs for the investigation of HF [38]. CAL may be associated with high mortality, which appears to be due to ventricular fibrillation [39]. Myocardial infarction caused by this technique comes with high variability depending on the animal model. Furthermore, not only species but even the strain [40] or the substrain used strongly influence results [41]. Infarct size has been described to be affected further by gender, age, body temperature, and pH of the blood [40]. Thus, these factors need to be carefully controlled. To a certain extent, the size of the infarct may be controlled by the site of coronary ligation [42]. Most coronary ligation models (especially in small animals) are performed as chronic occlusion, which differs from human disease in the sudden onset and lack of reopening of the occluded vessels.

CORONARY ARTERY LIGATION IN SMALL ANIMALS In rats, CAL has been established and infarctions greater than about 45% lead to development of congestive HF after 21 days [43].

Impairment of LV function is directly related to the extent of myocardial loss [43]. Mortality in the rat seems to be strain dependent [44]. Clinical characteristics have been described similar to human CHF [45].

Induction of myocardial infarction in mice is surgically more challenging due to the smaller size of the animal. However, in mice genetically modified models (transgenic and knockout strains) are largely available allowing to assess the influence of genetic factors on HF. After ligation of the left coronary artery, mortality in mice reaches up to 40–50% [46]. To our knowledge, CAL is not used in rabbits, guinea pigs, or syrian hamsters.

CORONARY ARTERY LIGATION IN LARGE ANIMAL MODELS CAL has been developed in dogs [38] and the time course to irreversible myocardial injury has been assessed in dogs with increasing periods of ischemia [47,48]. However, in this model mortality is very high and infarcted areas remain small. The presence of a significant collateral circulation makes it difficult to achieve consistent results and may alter post-MI course [47]. The confounding factors in dogs resulted in increased use of alternative models. Consistent coronary artery anatomy and lack of collateral vessels allow for the creation of infarctions with predictable size and location in pigs and sheep [47,49,50]. Porcine models have been used to study infarct expansion, LV remodeling, and delivery of angiogenic and arteriogenic factors. Stem-cell transplantation affected contractile and bioenergetic function [51].

Ovine models of MI have been used to assess outcomes depending on MI location [52]. In the sheep, progressive LV remodeling and aneurysm formation seem similar to patients with HF. Working with sheep models is limited by zoonotic diseases and the gastrointestinal anatomy, which makes transthoracic imaging difficult, leading to necessity of invasive approaches [47,49,52]. Similar to the sheep model, a model of CAL has been described in the goat [53].

HYDRAULIC OCCLUSION

Hydraulic occlusion and ameroid constriction are methods especially used in large animals [54,55]. Both may be used to induce myocardial infarction as well as coronary stenosis to induce hibernating myocardium. Both methods increase occlusion over time, mimicking coronary artery occlusion found in humans more closely [56].

CORONARY ARTERY MICROEMBOLIZATION

Coronary artery embolization is a frequently used model to induce ischemia in dogs. A single or several subsequent microsphere injections over a 10-week period lead to restriction in coronary flow or to occlusion of small vessels in the downstream area of injection. The model recapitulates ischemic cardiomyopathy through multiple sites of microinfarction and remodeling in contrast to a single focal MI. Size of microspheres, injection site, and amount used determine size and location of ischemic area and grade of flow reduction [57]. Microembolization is further used in swine, cattle, and sheep. Depending on the grade of ischemia it serves not only as a model for myocardial infarction, but may also be regarded as a model for reduced flow as seen in diabetic cardiomyopathy [49,58].

Myocardial Infarction Followed by Reperfusion

In humans, treatment of myocardial infarction by thrombolysis and percutaneous coronary angiography (PTCA) reestablishes coronary flow after occlusion. These treatments have increased survival but at the same time resulted in cardiac dysfunction through ischemia reperfusion (I/R) injury. Similarly, in cardiac surgery ischemia and reperfusion are unavoidable with extracorporeal circulation. Thus, models for the study of I/R injury are important. Two systems for these studies are *ex vivo* isolated perfused hearts and *in vivo* animal models.

Regional I/R in the anaesthetized animal is the standard model and the coronary artery is

ligated with a small tube for safe removal of the ligation. Reperfusion after MI has been described to lead to reduced fibrosis, higher inflammation, and neovascularization in the infarcted area [59]. To reduce the effects of surgical trauma in large animals, occlusion of a coronary vessel with the use of catheter-guided balloon, which is inflated at the desired place for a limited time has been developed [60]. Another method based on the use of a hydraulic occluder with ultrasonic control has been used for I/R in large animals such as pigs [61].

In mice, a "closed chest" model has been developed, by passing the suture around the LAD to the neck without occluding during the surgery. Here I/R may be transiently induced a few days after initial surgery when confounding factors due to surgery have vanished. Myocardial infarction can then be initiated by pulling the suture. Afterward it is released to allow reperfusion [62].

Isolated heart perfusion is a well-suited model to assess the consequences of ischemia and reperfusion. The approach to this model can be found in detail later in this chapter in the section discussing isolated heart perfusion.

Chronic Cardiac Ischemia (Without Infarction)

Chronic myocardial ischemia models have been developed in the pig and the rat. In the pig the same hydraulic occluder as used for progressive occlusion has been used to cause partial coronary artery stenosis under ultrasonic control, providing a model of chronic myocardial hibernation [63]. In this model, changes in glucose handling have been described [64]. In the pig the ameroid constrictor has been used as well. It results in a slowly and gradually increasing stenosis, however, the degree and progress of stenosis cannot be adjusted and eventually results in complete occlusion [65,66].

In the rat model of ischemia a copper wire is ligated with the vessel and removed directly after ligation, resulting in a reduction of the luminal diameter [67]. This model leads to chamber dilation after 45 min and remodeling is found after 5–7 days [67].

Models of Left Ventricular Pressure Overload or Hypertensive Heart Failure

In humans, increased afterload is a frequent cause for the development of HF. Aortic stenosis and hypertension are the most frequent causes for increased LV afterload while pulmonary artery constriction leads to increased right ventricular (RV) afterload. Aortic stenosis is well reflected in animal models of ascending or transverse aortic constriction. Hypertension instead, may be either induced surgically by abdominal aortic constriction or renal wrapping or may be caused by genetic predisposition or induced by infusion of angiotensin II. However, progression and frequency of HF development may vary and depends on banding severity, location, rodent strain, and time course [68–72].

Surgical Models of Left Ventricular Pressure Overload

Pressure overload may be achieved by constricting the aorta. To successfully accomplish this procedure, a certain extent of surgical skill is required, which also influences survival of the animals. Aortic constriction can be performed on the ascending, transverse, or descending aorta. This leads to increased workload, as pressure has to be higher to achieve the same blood flow.

ASCENDING AORTIC CONSTRICTION

Ascending aortic constriction is a frequently used model to induce pressure overload. In this model, a stricture is placed around the ascending aorta, proximal to the brachiocephalic trunk.

In mice, this procedure is performed in adult animals at 11–12 weeks of age and induces cardiac hypertrophy within 48 h [73]. Thus, pressure overload is acute in onset and does not directly mimic the clinical situation of gradually increasing aortic stenosis [74]. Nevertheless, it is a potent stimulus for rapid development of hypertrophy. In these models normally not all animals develop HF with the numbers depending on strain and investigator.

In rats, the procedure is mainly performed in young animals at an age of 4–5 weeks. The aorta is constricted using a clip with a remaining opening of 0.58 mm. As rats grow, pressure overload develops gradually as the outflow tract becomes increasingly constricted due to the impossibility to grow. In this model, LV hypertrophy has been described at 8 weeks and HF associated with systolic dysfunction was found after 18 weeks of pressure overload [75]. This rat model mimics the human situation of a gradual increase in stenosis more closely and is technically easier to accomplish than the mouse model.

TRANSVERSE AORTIC CONSTRICTION

Another common model to induce pressure overload is transverse aortic constriction. In most cases, the transverse aorta is constricted between the brachiocephalic trunk and the left common carotid artery. In mice, transverse aortic constriction is performed at 5–6 weeks of age and a weight of 22–24 g. A 5-0 suture is used to tightly constrict the aorta around a 29 g cannula. This procedure can be performed minimally invasively to reduce surgical trauma and leads to HF in about 50% of the surviving animals [76]. As with ascending aortic constriction this model leads to a sudden increase in pressure and cardiac hypertrophy rapidly develops.

In rats, transverse aortic constriction may be performed in young animals [77]. During growth, stenosis becomes increasingly severe and animals subsequentially develop compensated hypertrophy, HF with preserved ejection fraction, and HF with reduced ejection fraction after 15 weeks of pressure overload [78]. Alternatively, transverse aortic constriction can also be performed in adult rats constricting the aorta with a 3-0 silk suture around a 20 g cannula [79,80]. This induces a rapid hypertrophic response as described in mice, but without impairment of systolic function.

Surgically Induced Hypertension

ABDOMINAL AORTIC CONSTRICTION

Another frequently used model of cardiac hypertrophy and HF is the model of abdominal aortic constriction [81]. Initially, a model of renal occlusion was developed in dogs but it had been shown that abdominal aortic constriction produced the same effect [82]. The model has been subsequently established in rats [83]. Here, suprarenal aortic constriction leads to reduced renal perfusion, increased renin secretion, and an increase in blood pressure. This results in significant cardiac hypertrophy and in the long-range HF [84]. This model is used mainly in rats or dogs, but has been used in mice, baboons, cats, and rabbits as well and mimics the clinical situation of renal hypertension.

RENAL WRAPPING

Wrapping the kidneys in cellophane had been initially described to lead to perinephritis [85], but tubular damage seems to be more important [86]. Both, renal wrapping as well as figure eight ligature of kidney poles leads to renal ischemia. In this model, blood pressure rises and hypertension leads to cardiac hypertrophy and failure. As the model of abdominal aortic constriction, this model is used mainly in rats, dogs, and mice.

Noninvasive Hypertensive Models

DAHL SALT–SENSITIVE RATS

The Dahl salt–sensitive rat has been described as a model to study transition from compensated hypertrophy to HF [45]. The strain develops hypertension receiving a high-salt diet. This leads to LV hypertrophy and HF develops at 15–20 weeks [87].

SPONTANEOUS HYPERTENSION

Animals of the well-established spontaneous hypertensive rat (SHR) model present with genetic hypertension. These animals show normal cardiac function at 12 month and develop HF at 18–24 months. Due to development of HF only after 18–24 months and a rate of about 50% of animals developing HF [88], the model may be very expensive. There are substrains of the SHR, which are more sensitive to a high-salt diet and stroke (e.g., the SHR stroke prone [89,90]), or develop HF much earlier (e.g., the SHHF [91]).

RENIN–ANGIOTENSIN–ALDOSTERONE

Activation of the renin–angiotensin–aldosterone system leads to an increase in blood pressure. Long-term infusion, such as of angiotensin induces hypertension resulting in cardiac hypertrophy and may eventually lead to HF [92]. This model seems to have certain similarities to renal wrapping, however the approach is less invasive.

Right Ventricular Pressure Overload Models

RV pressure overload is achieved by constricting the pulmonary artery. This leads to an increased workload on the right ventricle as pressure has to be higher to achieve the same blood flow. In small animals, RV pressure overload is less frequently performed than LV, which is potentially due to two main reasons: (1) RV hypertrophy and failure have a much lower prevalence in humans than LV HF and (2) surgery for pulmonary artery banding is much more demanding in small animals. The pulmonary artery is much more fragile than the aorta and the right ventricle is not able to withstand stress during manipulation of pulmonary artery. The technique has been described for mice [73] but has also been used in rats [93] and at different ages [94].

Volume Overload-Induced Heart Failure Models

In humans, LV volume overload in a compensated state may be tolerated for a certain time. With a severe and prolonged stimulus, however, it eventually results in ventricular dysfunction and HF. In contrast, RV volume overload as seen in atrial septal defect may be tolerated for extended time. Thus, depending on the type of volume overload, cardiac function is affected more or less.

Valve Insufficiency

One of the most frequent causes for volume overload-induced HF in patients is mitral valve insufficiency with significant mitral regurgitation. Impaired closure of the mitral valve leads to backward flow of blood causing volume overload. In animal models, mitral regurgitation may be induced by rupture of the chordae [95]. Alternatively, shortening of the chordae by attachment to the ventricular wall has the same effect [96]. This model has been used in dogs [97], but may also be applied in other large animals like sheep, pigs, or cats [98–101]. An alternative model of ischemic mitral regurgitation has been established in sheep by occlusion of three specific coronaries, which leads to mitral regurgitation. Some dog breeds spontaneously develop mitral valve regurgitation [102].

Aortic valve insufficiency may be induced in rabbits by perforation of the aortic leaflets. This leads to an increase in end diastolic diameter and subsequently congestive HF [103].

Ventricular Septal Defect

In Yucatan mini swine, a ventricular septal defect occurs naturally. This model appears to mimic human disease with diastolic HF [102].

Aortocaval Shunt

To induce volume overload, an aortocaval shunt may be used. By connecting the aorta and the inferior vena cava distal of the renal arteries, cardiac output dramatically increases. This leads to acute decompensation followed by compensated hypertrophy and hypertrophy with cardiac dysfunction [104]. This model has been primarily used in rats and dogs.

Combination of PO and Ischemia

The most frequently used method to induce HF is permanent ligation of the coronaries. However, the model of permanent coronary ligation does not adequately resemble the human situation. Patients with small-sized infarctions often present with a long-standing history of pressure overload. Thus, new models have been developed combining pressure overload and mild myocardial infarction. In these models, aortic constriction leads to cardiac hypertrophy but not HF. Similarly, isolated myocardial infarction would not cause HF. However, the subsequent application of both, myocardial infarction with or without reperfusion a few weeks after aortic banding reproducibly leads to the development of HF in mice [105] and rats [106].

Rapid-Pacing-Induced Heart Failure

Rapid pacing at high heart rates (e.g., above 200 beats per min in dogs) leads to congestive HF within several weeks with a decrease in ejection fraction. Chronic pacing results in biventricular dilation with changes in LV geometry. The model seems valuable to assess neurohumoral mechanisms and shows similarities to humans with respect to myocardial function and changes in calcium handling and has been used to investigate cardiac metabolism [107]. However, rapid-pacing-induced HF is not stable as it reverses upon stopping of pacing. Rapid pacing models have also been used in sheep, pigs, and rabbits with similar findings [45].

Toxin-Induced Heart Failure

There are many toxins that may induce HF. Ethanol as well as the cytostatic drug doxorubicin may induce cardiomyopathy in rats [108]. Both agents have been found responsible for HF in patients as well. Doxorubicin has been used to induce HF in rabbits with biweekly injections for 6–9 weeks. The model has been suggested to investigate functional consequences of altered ryanodine receptor expression. Toxic cardiomyopathy has been induced in turkeys with furazolidone causing changes in calcium handling and myocardial energetics and in dogs with adriamycin [109] leading to LV dysfunction.

GENETICALLY DETERMINED CARDIOMYOPATHY

The classic model for genetically determined cardiomyopathy has been the syrian hamster, which presents cardiac hypertrophy and HF. The model develops HF in several phases after 7–10 months. It has the advantages of low costs, absence of surgical manipulations, and can be studied in large number of animals. However, attention needs to be given to the time of investigation and there are differences among different strains. A large animal model of genetic cardiomyopathy has been described in the bovine. This model shows similarities to human dilative cardiomyopathy and has the advantage of providing large tissue samples, if necessary [110].

Nowadays, many models of genetic cardiomyopathy have been created using genetically modified mice with null mutations or overexpression of structural proteins. Depending on the modified protein, type of mutation, level of (remaining) function, or intensity of overexpression these models can resemble genetically determined cardiomyopathy in humans to a certain extent.

OBESITY AND DIABETES RELATED MODELS TO ASSESS CARDIAC METABOLISM

Obesity and diabetes (mainly type II) are increasingly prevalent diseases in western societies, partially due to overnutrition and reduced exercise. Furthermore, the onset of disease shifts toward a younger age. Obesity and diabetes are associated with changes in cardiac metabolism, eventually affecting cardiac function. While patients with a BMI above 30 are considered obese, type II diabetes requires increased glucose and insulin levels. Instead, type I diabetes is characterized by lack of insulin due to a defect in pancreatic islet cells. In patients, insulin resistance, a hallmark of type II diabetes, is associated with an increased risk of developing HF. Thus, it is not surprising that obesity and diabetes may affect cardiac metabolism. It is therefore not surprising that a plethora of models of obesity and diabetes are used for research in cardiac metabolism.

Nutritional Models to Assess Cardiac Metabolism

There are nutritional models of obesity and models attempting to induce insulin resistance/type II diabetes by using altered diet. Many different approaches have been described, although some lack well-defined dietary interventions [111].

Diet-Induced Obesity

Induction of obesity in rodents has been mainly performed using a high-fat diet (HFD). Rodents are omnivorous in nature and genetically modified models are available. There are a large number of different dietary approaches using different amounts of fat (e.g., 45, 60, or 80% of calories from fat). Furthermore, the choice of fatty acids (saturated, monounsaturated, or polyunsaturated; long chain, medium chain, or short chain) has been described to be important as well. Saturated fatty acids seem to be more obesogenic than polyunsaturated fatty acids (PUFA) due to a higher degree of satiety resulting from PUFA. However, not only content and type of fatty acids are important, but protein and carbohydrate content as well as carbohydrate composition influence development of diet-induced obesity. There is an excellent review discussing the effects of diet composition on the development of obesity [112].

Dietary Models of Metabolic Syndrome and Diabetes to Assess Cardiac Metabolism

There is no rigid separation between models of obesity, metabolic syndrome, and diabetes. HFD may not only lead to obesity, but to the development of metabolic syndrome and diabetes as well. In C57BL/6J mice, a diet with 40–60% of fat leads to insulin resistance but not to beta-cell failure [113]. It has been suggested that feeding an HFD at an age of 6–8 weeks is most effective to induce obesity and that the degree of hyperglycemia largely depends on type and amount of dietary fat in C57BL/6J mice, which do not develop proper diabetes. In contrast, Sprague–Dawley rats develop insulin resistance in response to HFD. In this model, the pathogenesis of full type II diabetes takes about 10 weeks, which is not the case in some other rat strains [114].

In a different model, fructose and high-sucrose diet induce the development of hypertension and hyperlipidemia, in contrast to starch and glucose, which may be due to its unique metabolism [115]. A combination of HFD (45%) and high fructose/sucrose is considered to reflect dietary composition in the United States and has therefore been named "western diet".

Nondietary Experimentally Induced Models of Metabolic Syndrome and Diabetes

There are several experimental models to induce diabetes in rodents. Application of streptozotocin (STZ) or alloxan may destroy beta cells of the pancreas in adults or neonates and lead to type I diabetes. A model of partial pancreatectomy avoids side effects of the chemicals on other organs, but may be regarded to be between type I and type II diabetes. A side effect of this procedure is the possibility of amylase deficiency, which may lead to digestive problems. Instead, application of several lower doses of STZ seems to lead to a more type II diabetic phenotype [116].

Interestingly, the combination of HFD and low dose of STZ replicates the natural pathogenesis with several characteristics in parallel to human T2DM development [117]. Such protocols have been used with different diets and one or several injections of STZ in both rats and mice [116].

A model of nonobese T2DM has been developed by the use of nicotinamide–streptozotocin injection. This model has been suggested to closely resemble East Asian diabetic patients [116].

A surgical model of T2DM is the intrauterine growth-retardation model. In this model, nutrients available to the fetus during gestation are limited by bilateral uterine artery ligation. Rats in this model become obese, develop increased glucose levels, and reduced insulin secretion with full symptoms of diabetes [116].

Genetic Models for Obesity and Type II Diabetes

Genetic models of obesity and diabetes include db/db mice, ob/ob mice, Zucker diabetic fatty rats, and Otsuka Long–Evans Tokushima fatty (OLETF) rats. In contrast, Goto–Kakizaki (GK) rats are diabetic but nonobese. These models are useful in evaluating mechanisms that may be involved in development of obesity in rodents, but neither diabetes nor the metabolic syndrome in humans are monogenetic disorders. However, these models reflect situations of overeating as satiety signals are disturbed. The RCS 10 mouse instead is a model of polygenetically induced type II diabetes, thus comparable to human diabetes [118].

The ob/ob mouse was an early model for the study of diabetes because of a mutation in the leptin gene leading to dysfunctional leptin signaling. Animals show increased body weights, develop impaired glucose tolerance, and LV hypertrophy but do present with low blood pressure and do not develop dyslipidemia as do patients with the metabolic syndrome. The db/db mice have a mutation in the leptin receptor leading to somehow similar effects as high body weight, reduced glucose tolerance, hyperinsulinemia, and changes in serum lipids. These animals do not show effects on blood pressure. Analogously to the db/db mouse, Zucker Diabetic fatty rats have a mutation in the leptin receptor. ZDF rats are considered a model of early onset diabetes as they become hyperglycemic, hyperinsulinemic, and hypertriglyceridemic at about 14 weeks of age. The rats present with not only moderate increased blood pressure but also increased serum markers of inflammation.

OLETF rats have been used as a model for human diabetes and obesity. Again, this line displays a defect in food-intake control leading to higher food intake and increased meal size, resulting in an increase in body weight. They present with reduced glucose tolerance and increased triglyceride concentrations, LV hypertrophy and diastolic as well as systolic dysfunction.

GK rats are nonobese and spontaneously diabetic. Rats develop hyperglycemia and impaired glucose tolerance quickly and cardiac hypertrophy with reduced contractile function at about 20 weeks of age without changes in blood pressure.

Disrupting insulin signaling in mice, either the insulin receptor in selective tissues (muscle, heart, white or brown fat) or proteins of the insulin-signaling cascade lead frequently to diabetes or features of the metabolic syndrome. Many of these genetic modifications are summarized in an extensive review [119].

INFLAMMATORY CARDIOMYOPATHY

Inflammation of the cardiac muscle is called myocarditis. In humans, myocarditis is most frequently induced by viral infections and, to a lesser extent by nonviral pathogens. Myocarditis is an infection of the heart with inflammatory infiltrate and causing damage to cardiac muscle.

Autoimmune myocarditis can be induced in some mouse strains using immunization with myosin or myosin peptides. Furthermore, immunization with cardiomyosin in Lewis rats may also lead to myocarditis [120]. Additionally, it is possible to induce myocarditis by transfer of cardiac-myosin-stimulated T-cells in SCID mice

and Lewis rats [121]. A third method to induce experimental autoimmune myocarditis consists in the use of bordetella (the bacterium causing pertussis) 3 weeks after two immunizations with cardiac myosin [120]. In humans, virus infections such as adenoviruses, Epstein–Barr virus, enterovirus, or parvovirus may activate T-cells and result in autoimmune myocarditis. Viruses induce continuous low-grade inflammation [122]. Direct virus induction in animals elicits similar response in three phases. However, it is difficult to determine if a virus dose is sufficient for inflammation of the tissue limiting the value of such approach [122].

MODELS TO ASSESS CARDIAC METABOLISM EX VIVO – ISOLATED HEART PERFUSION

Isolated heart perfusion allows for analysis of cardiac metabolism independent of the body. As always this comes with strengths and limits. In the isolated heart, cardiac metabolism may be assessed without interference of other organs and endocrine interference and the composition of the perfusion buffer may be adapted to the research question. The dissociation of cardiac metabolism from the physiologic environment (endocrine regulation) may also be considered a limitation as such situation may be regarded nonphysiologic.

The first isolated heart perfusions were described 1866 by renowned physiologist Carl Ludwig and coworkers from the Department of Physiology at the University of Leipzig, Germany [123]. Although they used frog hearts, it was their pioneering work that led to the development of isolated perfusion of mammalian hearts [124]. In 1895, Oskar Langendorff developed a perfusion system, in which mammalian hearts were "retrogradely" perfused after cannulation of the aorta (i.e., perfusate flows from the aorta toward the aortic valve) [125]. A further development was introduced by James Neely, who in 1967 described the first isolated working heart with anterograde perfusion, where the heart was also ejecting relevant amounts of perfusate [126]. As further modified by Taegtmeyer et al. [127] mainly related to the oxygenator used achieving physiologic work patterns and physiologic levels of cardiac power.

The Langendorff Heart

For the Langendorff heart preparation the ascending aorta is cannulated and connected to a fluid reservoir, leading to retrograde flow of the medium through the aorta with steady pressure on the aortic valve. Due to this pressure, the valve is shut and perfusate flow is led through the coronary arteries, allowing for anterograde coronary perfusion at freely adjustable perfusion pressures. The perfusate drains through the coronary sinus into the right atrium, leaks out of the two cable veins, or is ejected by the right ventricle through the severed pulmonary arteries. It can be easily collected for further analysis. By inserting a balloon into the left ventricle and connecting it to a pressure transducer, a fixed preload set and LV pressures can be measured during systole and diastole [128]. Two modes of perfusion are available: a constant flow mode and a constant pressure mode [129]. Constant flow is achieved by using a pump, which provides a given flow rate to the aorta. The constant pressure model uses a fluid reservoir at a specific height to gain a desired hydrostatic pressure. Roller pumps in combination with a pressure control circuit allow a rapid switch from constant pressure to constant flow mode.

The Isolated Working Heart

The isolated working heart differs from the Langendorff Heart, as perfusate passes orthogradely (i.e., in physiologic direction) through the left atrium into the left ventricle and is ejected through the aortic valve into the ascending aorta. The working heart preparation requires cannulation of the aorta and the left atrium. It offers the advantage of measuring cardiac pump function with freely adjustable pre- and afterloads [129,130] assessing contractile function of the left ventricle under physiologic loading conditions. The first step of setting up a working heart perfusion is equal to the Langendorff heart perfusion, that is, cannulation of the aorta and establishing retrograde perfusion. Then, the left atrium is cannulated via one of the pulmonary veins. Once aorta and atrium are cannulated, retrograde perfusion via the aorta is stopped and anterograde perfusion from an atrial reservoir to the left atrium is started. Preload is determined by the height of the atrial reservoir above the heart. The perfusate is pumped from the left atrium into the left ventricle and out into the cannulated aorta, against a hydrostatic pressure that resembles the afterload.

Recirculating Versus Nonrecirculating Perfusion Mode

Both perfusion models may operate in a recirculating mode, where the coronary effluent (in case of the Langendorff heart) or pumped perfusate (in case of the working heart) is returned into the perfusion fluid reservoir, or a nonrecirculating mode, where any fluid that has passed through the heart (but at least the coronary effluent) is discarded. The working heart experiments are usually performed in the recirculating mode. The same is true for pharmacologica studies in the Langendorff heart, where high costs of used drugs may call for recirculation of the perfusate [130]. However, drug metabolism in the heart or substrate into conversion (i.e., glucose to lactate) may be a confounding factor in the recirculating mode, which can for instance alter the effect of a pharmacologic intervention [131].

Differences Between Various Perfusates

The perfusate needs to be able to deliver sufficient amounts of oxygen to the cardiomyocytes.

The most commonly used perfusate is a modified version of the crystalloid Krebs–Henseleit buffer (KHB). By gassing the buffer with a mixture of 95% oxygen and 5% CO_2, a sufficient arterial pO_2 can be maintained. However, due to the low oxygen-carrying capacity of KHB, investigators have looked out for substitutes with improved oxygen delivery. Perfluorocarbons in perfusates have been used as hemoglobin substitutes, significantly increasing oxygen transport capacity [132]. Compared to crystalloid buffers, perfluorocarbon perfusates have been shown to result in higher LV systolic pressure, cardiac output, and myocardial O_2 consumption [133]. However, the oxygen affinity of perfluorocarbons is so high that oxygen may not be released in the tissue. This limitation has resulted in the abandonment of the perfluorocarbons for heart perfusions.

Several factors, however, suggest that perfusion with KHB is still sufficient to deliver enough oxygen in most experimental settings, including a high venous pO_2 and preserved cardiac function, even when a gassing mixture containing only 70% of oxygen is used. A limitation of the KHB-perfused heart, however, is the artificially high coronary flow that is required to maintain sufficient oxygen delivery. Coronary flow reserve is usually fully exploited in isolated heart perfusions using KHB [134]. Thus, these models are not suitable to assess coronary flow.

Considering the possible confounders with crystalloid perfusates, whole blood has been used as an alternative [135–137]. Although using blood leads to stable experimental conditions, the set-up is more complex [130]: blood is often gained from a donor animal of a different species. Major disadvantages include hemolysis as well as the possibility of immunologic interactions due to contact of the donors humoral immune system with the exogenous, isolated heart. Despite the challenging methodology, blood perfused heart preparations exhibit less-impaired coronary flow reserves than preparations using KHB and are often used to assess the

effects of ischemia and reperfusion on the heart specifically if coronary flow reserve is the target of investigation [128,138,139].

To improve oxygen-carrying capacity, erythrocytes can be added to crystalloid perfusion buffers. Compared to pure KHB perfusion experiments, more physiologic conditions regarding coronary flow rate can be achieved. During the purification procedure, erythrocytes are cleared off leucocytes and proteins, minimizing confounding immunogenic effects [128].

CELL CULTURE MODELS

Cell culture experiments are frequently used to assess molecular mechanisms and to influence protein expression and modification. However, analysis of disease mechanisms in cardiac cell culture is restricted by several limitations: (1) Cardiomyocytes *in vivo* as well as *ex vivo* are only able to divide for a limited time, *in vivo* this process ends a few days after birth. (2) In cell culture, cardiomyocytes tend to dedifferentiate which means that they lose their cardiac characteristics. (3) There are no immortalized ventricular cell lines, thus, experiments have to be carried out on primary ventricular myocytes, skeletal muscle cells, or other cell lines. Targeting cardiomyocytes is a challenge; stable genetic manipulation of the cells may only be achieved by retroviral infection.

Neonatal Cardiomyocytes

Neonatal rat cardiomyocytes are the most frequently used cell culture models for research in cardiomyocytes. They are recognized as a valid cell culture model. The cells are relatively easy to obtain from 1-day to 3-day old pups by enzymatic digestion with collagenase of cardiac tissue. The process is very sensitive to changes in enzyme activity and experimental conditions. Neonatal cardiomyocytes may perform a few divisions. The isolation process not only delivers cardiomyocytes but also endothelial cells and fibroblasts, the latter may easily overgrow a cardiomyocyte culture. Thus, it is necessary to selectively enrich cardiomyocytes for example, by differential adhesion. Neonatal cardiomyocytes may also be obtained from mice allowing for genetically modified cells from the appropriate animals. However, neonatal cardiomyoces may not be obtained from humans.

Adult Cardiomyocytes

The advantage of using adult cells is that they may be isolated from animals with cardiac hypertrophy or HF as well as from patients with heart disease. Thus, especially human adult cardiomyocytes may better represent disease states in patients. However, the problem of fibroblasts overgrowing cultures may be even more pronounced as in neonatal cells, mostly because adult cells do not divide.

HL-1

HL-1 cells are the only commercially available isolated cardiac myocyte cell line. HL-1 cells have been developed from a mouse atrial tumor cell line and the cells maintain a differentiated phenotype and contract under appropriate conditions [140]. These cells are more suited for genetic manipulation as they divide, can be cultivated for extended periods and selected. Genetic manipulation has to be carried out using retroviruses. Their atrial nature has to be taken into account when interpreting results.

SUMMARY

There are many experimental models that function as surrogates for human diseases (Table 8.1). This chapter was intended as a general overview without the expectation of completeness. With present day and future genetic tools it is easy to envision that many more

TABLE 8.1 Advantages and Disadvantages of Using Animal Models as Surrogates for Human Diseases

Advantages	Disadvantages
Experimentally controlled conditions	Missing human variability
Possibility to assess single stimuli	Do not take comorbidities into account
Combination of different interventions possible	Disease models do not represent complete spectrum of disease
Rapid onset of disease	Contrast to often slow development in humans (taking years or tens of years)
Surgical models allow for fast induction	Influence of surgery often ignored
Models mostly available as needed	Disease of the elderly are frequently assessed in young animals
Influence of genetic background may be assessed	Different genetics compared to humans
Ex vivo investigations possible	Differences in physiology between models and humans already in health
Biopsy/sample availability as needed	Assessment of severity of disease not always established and may be different from humans
Mechanistic assessment in cell culture or transgenic animals	

models will be developed. We have tried to illustrate the currently used models for analysis of disease pathomechanisms.

References

[1] Endoh M. Force-frequency relationship in intact mammalian ventricular myocardium: physiological and pathophysiological relevance. Eur J Pharmacol 2004;500:73–86.

[2] Barnabei MS, Palpant NJ, Metzger JM. Influence of genetic background on *ex vivo* and *in vivo* cardiac function

in several commonly used inbred mouse strains. Physiol Genomics 2010;42A:103–13.

[3] Hartman JL IV, Garvik B, Hartwell L. Principles for the buffering of genetic variation. Science 2001;291:1001–4.

[4] Hongo M, Ryoke T, Ross J Jr. Animal models of heart failure recent developments and perspectives. Trends Cardiovasc Med 1997;7:161–7.

[5] Schoepe M, Schrepper A, Schwarzer M, Osterholt M, Doenst T. Exercise can induce temporary mitochondrial and contractile dysfunction linked to impaired respiratory chain complex activity. Metabolism 2012;61: 117–26.

[6] Haram PM, Kemi OJ, Lee SJ, Bendheim M.Ø., Al-Share QY, Waldum HL, et al. Aerobic interval training vs. continuous moderate exercise in the metabolic syndrome of rats artificially selected for low aerobic capacity. Cardiovasc Res 2009;81:723–32.

[7] Konhilas JP, Widegren U, Allen DL, Paul AC, Cleary A, Leinwand LA. Loaded wheel running and muscle adaptation in the mouse. Am J Physiol 2005;289:H455–65.

[8] Natali AJ, Turner DL, Harrison SM, White E. Regional effects of voluntary exercise on cell size and contraction-frequency responses in rat cardiac myocytes. J Exp Biol 2001;204:1191–9.

[9] Novak CM, Burghardt PR, Levine JA. The use of a running wheel to measure activity in rodents: relationship to energy balance, general activity, and reward. Neurosci Biobehav Rev 2012;36:1001–14.

[10] Meijer JH, Robbers Y. Wheel running in the wild. Proc Biol Sci 2014;281(1786):DOI: 10.1098/rspb.2014.0210. Published 21 May 2014.

[11] Konhilas JP, Maass AH, Luckey SW, Stauffer BL, Olson EN, Leinwand LA. Sex modifies exercise and cardiac adaptation in mice. Am J Physiol 2004;287:H2768–76.

[12] Konhilas JP, Chen H, Luczak E, McKee LA, Regan J, Watson PA, et al. Diet and sex modify exercise and cardiac adaptation in the mouse. Am J Physiol 2015;308:H135–45.

[13] Hoydal MA, Wisloff U, Kemi OJ, Ellingsen O. Running speed and maximal oxygen uptake in rats and mice: practical implications for exercise training. Eur J Cardiovasc Prev Rehabil 2007;14:753–60.

[14] Wang Y, Wisloff U, Kemi OJ. Animal models in the study of exercise-induced cardiac hypertrophy. Physiol Res 2010;59:633–44.

[15] Rosa EF, Silva AC, Ihara SS, Mora OA, Aboulafia J, Nouailhetas VL. Habitual exercise program protects murine intestinal, skeletal, and cardiac muscles against aging. J Appl Physiol 2005;99:1569–75.

[16] Fewell JG, Osinska H, Klevitsky R, Ng W, Sfyris G, Bahrehmand F, et al. A treadmill exercise regimen for identifying cardiovascular phenotypes in transgenic mice. Am J Physiol 1997;273:H1595–605.

[17] Moore RL, Musch TI, Yelamarty RV, Scaduto RC Jr, Semanchick AM, Elensky M, et al. Chronic exercise alters contractility and morphology of isolated rat cardiac myocytes. Am J Physiol 1993;264:C1180–9.

[18] Moran M, Saborido A, Megias A. Ca2+ regulatory systems in rat myocardium are altered by 24 weeks treadmill training. Pflugers Arch 2003;446:161–8.

[19] Kemi OJ, Haram PM, Loennechen JP, Osnes JB, Skomedal T, Wisloff U, et al. Moderate vs. high exercise intensity: differential effects on aerobic fitness, cardiomyocyte contractility, and endothelial function. Cardiovasc Res 2005;67:161–72.

[20] Iemitsu M, Maeda S, Otsuki T, Goto K, Miyauchi T. Time course alterations of myocardial endothelin-1 production during the formation of exercise training-induced cardiac hypertrophy. Exp Biol Med (Maywood) 2006;231:871–5.

[21] Zhang XQ, Song J, Carl LL, Shi W, Qureshi A, Tian Q, et al. Effects of sprint training on contractility and [Ca(2+)](i) transients in adult rat myocytes. J Appl Physiol 2002;93:1310–7.

[22] Musch TI. Effects of sprint training on maximal stroke volume of rats with a chronic myocardial infarction. J Appl Physiol 1992;72:1437–43.

[23] Laughlin MH, Hale CC, Novela L, Gute D, Hamilton N, Ianuzzo CD. Biochemical characterization of exercise-trained porcine myocardium. J Appl Physiol 1991;71:229–35.

[24] White FC, McKirnan MD, Breisch EA, Guth BD, Liu YM, Bloor CM. Adaptation of the left ventricle to exercise-induced hypertrophy. J Appl Physiol 1987;62:1097–110.

[25] Stepien RL, Hinchcliff KW, Constable PD, Olson J. Effect of endurance training on cardiac morphology in Alaskan sled dogs. J Appl Physiol 1998;85:1368–75.

[26] Carroll JF, Kyser CK. Exercise training in obesity lowers blood pressure independent of weight change. Med Sci Sports Exerc 2002;34:596–601.

[27] Edwards JG. Swim training increases ventricular atrial natriuretic factor (ANF) gene expression as an early adaptation to chronic exercise. Life Sci 2002;70:2753–68.

[28] Evangelista FS, Brum PC, Krieger JE. Duration-controlled swimming exercise training induces cardiac hypertrophy in mice. Braz J Med Biol Res 2003;36:1751–9.

[29] Harri M, Kuusela P. Is swimming exercise or cold exposure for rats? Acta Physiol Scand 1986;126:189–97.

[30] Abel EL. Behavioral and physiological effects of different water depths in the forced swim test. Physiol Behav 1994;56:411–4.

[31] Iemitsu M, Miyauchi T, Maeda S, Sakai S, Fujii N, Miyazaki H, et al. Cardiac hypertrophy by hypertension and exercise training exhibits different gene expression of enzymes in energy metabolism. Hypertens Res 2003;26:829–37.

[32] Notomi T, Okimoto N, Okazaki Y, Tanaka Y, Nakamura T, Suzuki M. Effects of tower climbing exercise on bone mass, strength, and turnover in growing rats. J Bone Miner Res 2001;16:166–74.

[33] Tamaki T, Uchiyama S, Nakano S. A weight-lifting exercise model for inducing hypertrophy in the hindlimb muscles of rats. Med Sci Sports Exerc 1992;24:881–6.

[34] Leite RD, Durigan Rde C, de Souza Lino AD, de Souza Campos MV, Souza M, Selistre-de-Araujo HS, et al. Resistance training may concomitantly benefit body composition, blood pressure and muscle MMP-2 activity on the left ventricle of high-fat fed diet rats. Metabolism 2013;62:1477–84.

[35] Lanier GM, Vaishnava P, Kosmas CE, Wagman G, Hiensch R, Vittorio TJ. An update on diastolic dysfunction. Cardiol Rev 2012;20:230–6.

[36] Liem AL, van 't Hof AW, Hoorntje JC, de Boer MJ, Suryapranata H, Zijlstra F. Influence of treatment delay on infarct size and clinical outcome in patients with acute myocardial infarction treated with primary angioplasty. J Am Coll Cardiol 1998;32:629–33.

[37] Maxwell MP, Hearse DJ, Yellon DM. Species variation in the coronary collateral circulation during regional myocardial ischaemia: a critical determinant of the rate of evolution and extent of myocardial infarction. Cardiovasc Res 1987;21:737–46.

[38] Hood WB Jr, McCarthy B, Lown B. Myocardial infarction following coronary ligation in dogs. Hemodynamic effects of isoproterenol and acetylstrophanthidin. Circ Res 1967;21:191–9.

[39] Iwanaga K, Takano H, Ohtsuka M, Hasegawa H, Zou Y, Qin Y, et al. Effects of G-CSF on cardiac remodeling after acute myocardial infarction in swine. Biochem Biophys Res Commun 2004;325:1353–9.

[40] Guo Y, Flaherty MP, Wu WJ, Tan W, Zhu X, Li Q, et al. Genetic background, gender, age, body temperature, and arterial blood pH have a major impact on myocardial infarct size in the mouse and need to be carefully measured and/or taken into account: results of a comprehensive analysis of determinants of infarct size in 1,074 mice. Basic Res Cardiol 2012;107:288.

[41] Garcia-Menendez L, Karamanlidis G, Kolwicz S, Tian R. Substrain specific response to cardiac pressure overload in C57BL/6 mice. Am J Physiol 2013;305:H397–402.

[42] Ahn D, Cheng L, Moon C, Spurgeon H, Lakatta EG, Talan MI. Induction of myocardial infarcts of a predictable size and location by branch pattern probability-assisted coronary ligation in C57BL/6 mice. Am J Physiol 2004;286:H1201–7.

[43] Pfeffer MA, Pfeffer JM, Fishbein MC, Fletcher PJ, Spadaro J, Kloner RA, et al. Myocardial infarct size and ventricular function in rats. Circ Res 1979;44:503–12.

[44] Liu YH, Yang XP, Nass O, Sabbah HN, Peterson E, Carretero OA. Chronic heart failure induced by coronary artery ligation in Lewis inbred rats. Am J Physiol 1997;272:H722–7.

[45] Hasenfuss G. Animal models of human cardiovascular disease, heart failure and hypertrophy. Cardiovasc Res 1998;39:60–76.

[46] Gehrmann J, Frantz S, Maguire CT, Vargas M, Ducharme A, Wakimoto H, et al. Electrophysiological characterization of murine myocardial ischemia and infarction. Basic Res Cardiol 2001;96:237–50.

[47] Dixon JA, Spinale FG. Large animal models of heart failure: a critical link in the translation of basic science to clinical practice. Circ Heart Fail 2009;2:262–71.

[48] Reimer KA, Jennings RB. The "wavefront phenomenon" of myocardial ischemic cell death. II. Transmural progression of necrosis within the framework of ischemic bed size (myocardium at risk) and collateral flow. Lab Invest 1979;40:633–44.

[49] Geens JH, Trenson S, Rega FR, Verbeken EK, Meyns BP. Ovine models for chronic heart failure. Int J Artif Organs 2009;32:496–506.

[50] Weaver ME, Pantely GA, Bristow JD, Ladley HD. A quantitative study of the anatomy and distribution of coronary arteries in swine in comparison with other animals and man. Cardiovasc Res 1986;20:907–17.

[51] Zeng L, Hu Q, Wang X, Mansoor A, Lee J, Feygin J, et al. Bioenergetic and functional consequences of bone marrow-derived multipotent progenitor cell transplantation in hearts with postinfarction left ventricular remodeling. Circulation 2007;115:1866–75.

[52] Gorman JH III, Gorman RC, Plappert T, Jackson BM, Hiramatsu Y, St John-Sutton MG, et al. Infarct size and location determine development of mitral regurgitation in the sheep model. J Thorac Cardiovasc Surg 1998;115:615–22.

[53] Kim WG, Cho SR, Sung SH, Park HJ. A chronic heart failure model by coronary artery ligation in the goat. Int J Artif Organs 2003;26:929–34.

[54] Harada K, Grossman W, Friedman M, Edelman ER, Prasad PV, Keighley CS, et al. Basic fibroblast growth factor improves myocardial function in chronically ischemic porcine hearts. J Clin Invest 1994;94:623–30.

[55] Roth DM, White FC, Mathieu-Costello O, Guth BD, Heusch G, Bloor CM, et al. Effects of left circumflex Ameroid constrictor placement on adrenergic innervation of myocardium. Am J Physiol 1987;253:H1425–34.

[56] Klocke R, Tian W, Kuhlmann MT, Nikol S. Surgical animal models of heart failure related to coronary heart disease. Cardiovasc Res 2007;74:29–38.

[57] Heusch G, Schulz R, Haude M, Erbel R. Coronary microembolization. J Mol Cell Cardiol 2004;37:23–31.

[58] Sherwood LC, Sobieski MA, Koenig SC, Giridharan GA, Slaughter MS. Benefits of aggressive medical management in a bovine model of chronic ischemic heart failure. ASAIO J 2013;59:221–9.

[59] Vandervelde S, van Amerongen MJ, Tio RA, Petersen AH, van Luyn MJ, Harmsen MC. Increased inflammatory

response and neovascularization in reperfused vs. non-reperfused murine myocardial infarction. Cardiovasc Pathol 2006;15:83–90.

[60] Perez de Prado A, Cuellas-Ramon C, Regueiro-Purrinos M, Gonzalo-Orden JM, Perez-Martinez C, Altonaga JR, et al. Closed-chest experimental porcine model of acute myocardial infarction-reperfusion. J Pharmacol Toxicol Methods 2009;60:301–6.

[61] Sjaastad I, Grund F, Ilebekk A. Effects on infarct size and on arrhythmias by controlling reflow after myocardial ischaemia in pigs. Acta Physiol Scand 2000;169:195–201.

[62] Nossuli TO, Lakshminarayanan V, Baumgarten G, Taffet GE, Ballantyne CM, Michael LH, et al. A chronic mouse model of myocardial ischemia-reperfusion: essential in cytokine studies. Am J Physiol 2000;278:H1049–55.

[63] Kudej RK, Ghaleh B, Sato N, Shen YT, Bishop SP, Vatner SF. Ineffective perfusion-contraction matching in conscious, chronically instrumented pigs with an extended period of coronary stenosis. Circ Res 1998;82:1199–205.

[64] McFalls EO, Murad B, Haspel HC, Marx D, Sikora J, Ward HB. Myocardial glucose uptake after dobutamine stress in chronic hibernating swine myocardium. J Nucl Cardiol 2003;10:385–94.

[65] Domkowski PW, Hughes GC, Lowe JE. Ameroid constrictor versus hydraulic occluder: creation of hibernating myocardium. Ann Thorac Surg 2000;69:1984.

[66] St Louis JD, Hughes GC, Kypson AP, DeGrado TR, Donovan CL, Coleman RE, et al. An experimental model of chronic myocardial hibernation. Ann Thorac Surg 2000;69:1351–7.

[67] Capasso JM, Jeanty MW, Palackal T, Olivetti G, Anversa P. Ventricular remodeling induced by acute nonocclusive constriction of coronary artery in rats. Am J Physiol 1989;257:H1983–93.

[68] Lee SH, Yang DK, Choi BY, Lee YH, Kim SY, Jeong D, et al. The transcription factor Eya2 prevents pressure overload-induced adverse cardiac remodeling. J Mol Cell Cardiol 2009;46:596–605.

[69] Miyamoto MI, del Monte F, Schmidt U, DiSalvo TS, Kang ZB, Matsui T, et al. Adenoviral gene transfer of SERCA2a improves left-ventricular function in aortic-banded rats in transition to heart failure. Proc Natl Acad Sci USA 2000;97:793–8.

[70] Norton GR, Woodiwiss AJ, Gaasch WH, Mela T, Chung ES, Aurigemma GP, et al. Heart failure in pressure overload hypertrophy. The relative roles of ventricular remodeling and myocardial dysfunction. J Am Coll Cardiol 2002;39:664–71.

[71] Rockman HA, Ross RS, Harris AN, Knowlton KU, Steinhelper ME, Field LJ, et al. Segregation of atrial-specific and inducible expression of an atrial natriuretic factor transgene in an *in vivo* murine model of cardiac hypertrophy. Proc Natl Acad Sci USA 1991;88:8277–81.

[72] Szymanska G, Stromer H, Kim DH, Lorell BH, Morgan JP. Dynamic changes in sarcoplasmic reticulum function in cardiac hypertrophy and failure. Pflugers Arch 2000;439:339–48.

[73] Tarnavski O, McMullen JR, Schinke M, Nie Q, Kong S, Izumo S. Mouse cardiac surgery: comprehensive techniques for the generation of mouse models of human diseases and their application for genomic studies. Physiol Genomics 2004;16:349–60.

[74] Patten RD, Hall-Porter MR. Small animal models of heart failure: development of novel therapies, past and present. Circ Heart Fail 2009;2:138–44.

[75] Litwin SE, Katz SE, Weinberg EO, Lorell BH, Aurigemma GP, Douglas PS. Serial echocardiographic-Doppler assessment of left ventricular geometry and function in rats with pressure-overload hypertrophy. Chronic angiotensin-converting enzyme inhibition attenuates the transition to heart failure. Circulation 1995;91:2642–54.

[76] Faerber G, Barreto-Perreia F, Schoepe M, Gilsbach R, Schrepper A, Schwarzer M, et al. Induction of heart failure by minimally invasive aortic constriction in mice: reduced peroxisome proliferator-activated receptor gamma coactivator levels and mitochondrial dysfunction. J Thorac Cardiovasc Surg 2011;141:492–500. 00 e1.

[77] Zaha V, Grohmann J, Gobel H, Geibel A, Beyersdorf F, Doenst T. Experimental model for heart failure in rats – induction and diagnosis. Thorac Cardiovasc Surg 2003;51:211–5.

[78] Shingu Y, Amorim PA, Nguyen TD, Osterholt M, Schwarzer M, Doenst T. Echocardiography alone allows the determination of heart failure stages in rats with pressure overload. Thorac Cardiovasc Surg 2013;61:718–25.

[79] Schwarzer M, Faerber G, Rueckauer T, Blum D, Pytel G, Mohr FW, et al. The metabolic modulators, Etomoxir and NVP-LAB121, fail to reverse pressure overload induced heart failure *in vivo*. Basic Res Cardiol 2009;104:547–57.

[80] Doenst T, Goodwin GW, Cedars AM, Wang M, Stepkowski S, Taegtmeyer H. Load-induced changes *in vivo* alter substrate fluxes and insulin responsiveness of rat heart *in vitro*. Metabolism 2001;50:1083–90.

[81] Cantor EJ, Babick AP, Vasanji Z, Dhalla NS, Netticadan T. A comparative serial echocardiographic analysis of cardiac structure and function in rats subjected to pressure or volume overload. J Mol Cell Cardiol 2005;38:777–86.

[82] Goldblatt H, Lynch J, Hanzal RF, Summerville WW. Studies on experimental hypertension: I. The production of persistent elevation of systolic blood pressure by means of renal ischemia. J Exp Med 1934;59:347–79.

[83] Fernandes M, Onesti G, Weder A, Dykyj R, Gould AB, Kim KE, et al. Experimental model of severe renal hypertension. J Lab Clin Med 1976;87:561–7.

[84] Pawlush DG, Moore RL, Musch TI, Davidson WR Jr. Echocardiographic evaluation of size, function, and mass of normal and hypertrophied rat ventricles. J Appl Physiol 1993;74:2598–605.

[85] Page IH. The production of persistent arterial hypertension by cellophane perinephritis. J Am Med Assoc 1939;113:2046–8.

[86] Vanegas V, Ferrebuz A, Quiroz Y, Rodriguez-Iturbe B. Hypertension in Page (cellophane-wrapped) kidney is due to interstitial nephritis. Kidney Int 2005;68:1161–70.

[87] Inoko M, Kihara Y, Morii I, Fujiwara H, Sasayama S. Transition from compensatory hypertrophy to dilated, failing left ventricles in Dahl salt-sensitive rats. Am J Physiol 1994;267:H2471–82.

[88] Conrad CH, Brooks WW, Hayes JA, Sen S, Robinson KG, Bing OH. Myocardial fibrosis and stiffness with hypertrophy and heart failure in the spontaneously hypertensive rat. Circulation 1995;91:161–70.

[89] Nagaoka A, Iwatsuka H, Suzuoki Z, Okamoto K. Genetic predisposition to stroke in spontaneously hypertensive rats. Am J Physiol 1976;230:1354–9.

[90] Griffin KA, Churchill PC, Picken M, Webb RC, Kurtz TW, Bidani AK. Differential salt-sensitivity in the pathogenesis of renal damage in SHR and stroke prone SHR. Am J Hypertens 2001;14:311–20.

[91] Holycross BJ, Summers BM, Dunn RB, McCune SA. Plasma renin activity in heart failure-prone SHHF/Mccfacp rats. Am J Physiol 1997;273:H228–33.

[92] Tsukamoto Y, Mano T, Sakata Y, Ohtani T, Takeda Y, Tamaki S, et al. A novel heart failure mice model of hypertensive heart disease by angiotensin II infusion, nephrectomy, and salt loading. Am J Physiol 2013;305:H1658–67.

[93] Takeyama D, Kagaya Y, Yamane Y, Shiba N, Chida M, Takahashi T, et al. Effects of chronic right ventricular pressure overload on myocardial glucose and free fatty acid metabolism in the conscious rat. Cardiovasc Res 1995;29:763–7.

[94] Kuroha M, Isoyama S, Ito N, Takishima T. Effects of age on right ventricular hypertrophic response to pressure-overload in rats. J Mol Cell Cardiol 1991;23:1177–90.

[95] Leroux AA, Moonen ML, Pierard LA, Kolh P, Amory H. Animal models of mitral regurgitation induced by mitral valve chordae tendineae rupture. J Heart Valve Dis 2012;21:416–23.

[96] Cremer SE, Zois NE, Moesgaard SG, Ravn N, Cirera S, Honge JL, et al. Strong association between activated valvular interstitial cells and histopathological lesions in porcine model of induced mitral regurgitation. Int J Cardiol 2014;174:443–6.

[97] Kleaveland JP, Kussmaul WG, Vinciguerra T, Diters R, Carabello BA. Volume overload hypertrophy in a closed-chest model of mitral regurgitation. Am J Physiol 1988;254:H1034–41.

[98] Cremer SE, Zois NE, Moesgaard SG, Ravn N, Cirera S, Honge JL, et al. Serotonin markers show altered transcription levels in an experimental pig model of mitral regurgitation. Vet J 2015;203:192–8.

[99] Goel R, Witzel T, Dickens D, Takeda PA, Heuser RR. The QuantumCor device for treating mitral regurgitation: an animal study. Catheter Cardiovasc Interv 2009;74:43–8.

[100] Minakawa M, Robb JD, Morital M, Koomalsinghl KJ, Vergnat M, Gillespie MJ, et al. A model of ischemic mitral regurgitation in pigs with three-dimensional echocardiographic assessment. J Heart Valve Dis 2014;23:713–20.

[101] McCullagh WH, Covell JW, Ross J Jr. Left ventricular dilatation and diastolic compliance changes during chronic volume overloading. Circulation 1972;45:943–51.

[102] Carabello BA. Models of volume overload hypertrophy. J Card Fail 1996;2:55–64.

[103] Magid NM, Wallerson DC, Borer JS, Mukherjee A, Young MS, Devereux RB, et al. Left ventricular diastolic and systolic performance during chronic experimental aortic regurgitation. Am J Physiol 1992;263:H226–33.

[104] Newman WH, Webb JG, Privitera PJ. Persistence of myocardial failure following removal of chronic volume overload. Am J Physiol 1982;243:H876–83.

[105] Weinheimer CJ, Lai L, Kelly DP, Kovacs A. Novel mouse model of left ventricular pressure overload and infarction causing predictable ventricular remodelling and progression to heart failure. Clin Exp Pharmacol Physiol 2015;42:33–40.

[106] Chen J, Chemaly ER, Liang LF, LaRocca TJ, Yaniz-Galende E, Hajjar RJ. A new model of congestive heart failure in rats. Am J Physiol 2011;301:H994–H1003.

[107] Osorio JC, Stanley WC, Linke A, Castellari M, Diep QN, Panchal AR, et al. Impaired myocardial fatty acid oxidation and reduced protein expression of retinoid X receptor-alpha in pacing-induced heart failure. Circulation 2002;106:606–12.

[108] Breckenridge RA. Animal models of myocardial disease. In: Conn PM, editor. Animal Models for the Study of Human Disease. Boston: Academic Press; 2013. p. 145–71 [chapter 7].

[109] Monnet E, Orton EC. A canine model of heart failure by intracoronary adriamycin injection: hemodynamic and energetic results. J Card Fail 1999;5:255–64.

[110] Eschenhagen T, Diederich M, Kluge SH, Magnussen O, Mene U, Muller F, et al. Bovine hereditary cardiomyopathy: an animal model of human dilated cardiomyopathy. J Mol Cell Cardiol 1995;27:357–70.

[111] Warden CH, Fisler JS. Comparisons of diets used in animal models of high-fat feeding. Cell Metab 2008;7:277.

[112] Hariri N, Thibault L. High-fat diet-induced obesity in animal models. Nutr Res Rev 2010;23:270–99.

[113] Surwit RS, Kuhn CM, Cochrane C, McCubbin JA, Feinglos MN. Diet-induced type II diabetes in C57BL/6J mice. Diabetes 1988;37:1163–7.

[114] Chang S, Graham B, Yakubu F, Lin D, Peters JC, Hill JO. Metabolic differences between obesity-prone and obesity-resistant rats. Am J Physiol 1990;259: R1103–10.

[115] Sharma N, Okere IC, Barrows BR, Lei B, Duda MK, Yuan CL, et al. High-sugar diets increase cardiac dysfunction and mortality in hypertension compared to low-carbohydrate or high-starch diets. J Hypertens 2008;26:1402–10.

[116] Islam MS, Wilson RD. Experimentally induced rodent models of type 2 diabetes. Methods Mol Biol 2012;933:161–74.

[117] Chen D, Wang MW. Development and application of rodent models for type 2 diabetes. Diabetes Obes Metab 2005;7:307–17.

[118] Reifsnyder PC, Leiter EH. Deconstructing and reconstructing obesity-induced diabetes (diabesity) in mice. Diabetes 2002;51:825–32.

[119] Nandi A, Kitamura Y, Kahn CR, Accili D. Mouse models of insulin resistance. Physiol Rev 2004;84:623–47.

[120] Kodama M, Matsumoto Y, Fujiwara M, Masani F, Izumi T, Shibata A. A novel experimental model of giant cell myocarditis induced in rats by immunization with cardiac myosin fraction. Clin Immunol Immunopathol 1990;57:250–62.

[121] Wegmann KW, Zhao W, Griffin AC, Hickey WF. Identification of myocarditogenic peptides derived from cardiac myosin capable of inducing experimental allergic myocarditis in the Lewis rat. The utility of a class II binding motif in selecting self-reactive peptides. J Immunol 1994;153:892–900.

[122] Ou L, Li W, Liu Y, Zhang Y, Jie S, Kong D, et al. Animal models of cardiac disease and stem cell therapy. Open Cardiovasc Med J 2010;4:231–9.

[123] Cyon E. Über den Einfluss der Temperaturveränderungen auf Zahl, Dauer und Stärke der Herzschläge. Arbeiten aus der Physiol Anstalt zu Leipzig 1866;77–127, http://vlpneu.mpiwg-berlin.mpg.de/references?id=lit1341.

[124] Zimmer H-G. The isolated perfused heart and its pioneers. News Physiol Sci 1998;13:203–10.

[125] Langendorff O. Untersuchungen am überlebenden Säugethierherzen. Pflügers Arch 1895;61:291–332.

[126] Neely JR, Liebermeister H, Battersby EJ, Morgan HE. Effect of pressure development on oxygen consumption by isolated rat heart. Am J Physiol 1967;212:804–14.

[127] Taegtmeyer H, Hems R, Krebs HA. Utilization of energy-providing substrates in the isolated working rat heart. Biochem J 1980;186:701–11.

[128] Bell RM, Mocanu MM, Yellon DM. Retrograde heart perfusion: the Langendorff technique of isolated heart perfusion. J Mol Cell Cardiol 2011;50:940–50.

[129] Liao R, Podesser BK, Lim CC. The continuing evolution of the Langendorff and ejecting murine heart: new advances in cardiac phenotyping. Am J Physiol Heart Circ Physiol 2012;303:H156–67.

[130] Sutherland FJ, Hearse DJ. The isolated blood and perfusion fluid perfused heart. Pharmacol Res 2000;41:613–27.

[131] Webster I, Smith A, Lochner A, Huisamen B. Sanguinarine non-versus re-circulation during isolated heart perfusion – a Jekyll and Hyde effect? Cardiovasc Drugs Ther 2014;28:489–91.

[132] Lowe KC. Perfluorinated blood substitutes and artificial oxygen carriers. Blood Rev 1999;13:171–84.

[133] Chemnitius JM, Burger W, Bing RJ. Crystalloid and perfluorochemical perfusates in an isolated working rabbit heart preparation. Am J Physiol 1985;249:H285–92.

[134] Bratkovsky S, Aasum E, Birkeland CH, Riemersma RA, Myhre ES, Larsen TS. Measurement of coronary flow reserve in isolated hearts from mice. Acta Physiol Scand 2004;181:167–72.

[135] Bergmann SR, Clark RE, Sobel BE. An improved isolated heart preparation for external assessment of myocardial metabolism. Am J Physiol 1979;236:H644–61.

[136] Gamble WJ, Conn PA, Kumar AE, Plenge R, Monroe RG. Myocardial oxygen consumption of blood-perfused, isolated, supported, rat heart. Am J Physiol 1970;219:604–12.

[137] Qiu Y, Manche A, Hearse DJ. Contractile and vascular consequences of blood versus crystalloid cardioplegia in the isolated blood-perfused rat heart. Eur J Cardiothorac Surg 1993;7:137–45.

[138] Walters HL III, Digerness SB, Naftel DC, Waggoner JR III, Blackstone EH, Kirklin JW. The response to ischemia in blood perfused vs. crystalloid perfused isolated rat heart preparations. J Mol Cell Cardiol 1992;24:1063–77.

[139] Deng Q, Scicli AG, Lawton C, Silverman NA. Coronary flow reserve after ischemia and reperfusion of the isolated heart. Divergent results with crystalloid versus blood perfusion. J Thorac Cardiovasc Surg 1995;109:466–72.

[140] Claycomb WC, Lanson NA Jr, Stallworth BS, Egeland DB, Delcarpio JB, Bahinski A, et al. HL-1 cells: a cardiac muscle cell line that contracts and retains phenotypic characteristics of the adult cardiomyocyte. Proc Natl Acad Sci USA 1998;95:2979–84.

Diurnal Variation in Cardiac Metabolism

Martin E. Young

Division of Cardiovascular Diseases, Department of Medicine,
University of Alabama at Birmingham, Birmingham, AL, USA

INTRODUCTION

Classically, homeostasis is defined as maintenance of a constant internal environment. However, in reality, the internal (and external) environment for an organism is far from constant. Marked fluctuations occur throughout the lifespan of an organism, on multiple time scales, ranging from seconds/minutes (e.g., metabolic fluxes) to months/years (e.g., developmental processes). Indeed, the dynamic nature with which cells/organs fluctuate biological processes is essential for life, facilitating adaptation to different environmental stimuli/stresses. One could argue that a form of rigidity or inflexibility is more conducive to a pathologic, as opposed to physiologic state; this concept is exemplified by cardiac metabolism, wherein the heart exhibits metabolic inflexibility during several disease states [1,2]. At the whole body level, multiple feedforward and feedback mechanisms are employed to maintain nutrients/metabolites within physiologic boundaries. For example, following ingestion of a mixed meal, glucose levels initially increase in the circulation, which subsequently decrease upon activation of various neurohumoral axes designed to activate pathways of glucose utilization [3]. Noteworthy, for a discrete period of time, blood glucose levels are elevated; it is only when this momentary hyperglycemia persists for prolonged periods of time that pathology ensues [4]. Accordingly, a more accurate definition of homeostasis might be maintenance of biological processes within a physiologic range; when these processes persist at the extremes of this physiologic range, inflexibility is associated with pathology (Fig. 9.1).

An important component of the homeostatic model described earlier involves time. More specifically, biological processes fluctuate/oscillate in a time-dependent manner, which are essential for normal physiologic functions. The time scale that is the focus of this chapter is the 24 h day. Over the course of a normal day, organisms on earth are subjected to dramatic changes in their environment. These range from alterations in lighting, temperature, and humidity, to availability of food and the presence of predators. At

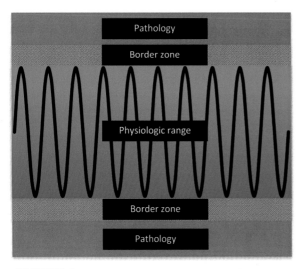

FIGURE 9.1 **Homeostatic model.** Biological processes oscillate within a physiologic range, with respect to time. The time span may be in the range of milliseconds (e.g., ion homeostasis), minutes (e.g., pulsatile insulin secretion), hours (e.g., pulsatile growth hormone secretion), a day (e.g., behavior) and/or month/year (e.g., reproductive potential). Persistence at extremities of oscillations (i.e., inflexibility) is associated with pathology.

a biological level, organisms exhibit daily oscillations in functions such as behavior, neurohumoral factors, cell cycle progression, signaling, transcription, translation, and metabolism [5,6]. It is essential that these processes are orchestrated in a temporally appropriate manner. Cell cycle progression is an excellent example of this concept, wherein DNA synthesis and DNA repair occur at distinct times of the day (DNA repair occurs during the light phase, while DNA replication is restricted to the dark phase, thereby minimizing transmission of UV-induced mutations to daughter cells) [7]. The same concept applies to metabolism, wherein temporal organization of opposing pathways (e.g., anabolic versus catabolic pathways) often parallel daily alterations in behaviors. Returning to the example of postprandial glucose utilization, increased circulating insulin levels following consumption of a mixed meal during the awake period is a critical component of glucose homeostasis. However, it

is important to note that the level of a stimulus is only one half of stimulus-response coupling, the second half involving sensitivity to the stimulus. Not only do stimuli vary over the course of the day, but so too does sensitivity of a cell/organ/ organism to the stimulus. For example, insulin sensitivity increases in humans in the early hours of the morning, perhaps in anticipation of food consumption upon waking, thus synchronizing the level of the stimulus (i.e., insulin) with responsiveness (i.e., insulin sensitivity) [8,9]. The importance of appropriate synchronization between temporal organization of biological processes with anticipated stimuli/stresses is illustrated beautifully by DNA synthesis/repair; inappropriate exposure of cells/organisms to UV irradiation during the period of DNA synthesis increases risk for tumorogenesis [7].

The purpose of the current chapter is to discuss temporal organization of cardiac metabolism on the 24 h timescale. This discussion will include a detailed overview of known oscillations in cardiac metabolism, information regarding mediators of these oscillations, and the importance of maintaining metabolic synchrony for normal cardiac function. In addition, the experimental implications of these oscillations, in terms of assessment of cardiac metabolism, will be highlighted.

TIME-OF-DAY-DEPENDENT OSCILLATIONS IN CARDIAC METABOLISM

The heart is subjected to dramatic time-of-day-dependent oscillations in stimuli and stresses. On a daily basis, dramatic fluctuations in energetic demand and supply occur, in association with sleep/wake and fasting/feeding cycles. It is therefore not surprising that cardiac metabolism is subject to dramatic swings each and every day, of which the peaks and troughs approach levels often observed in pathologic states (e.g., insulin resistance is an underlying feature of type 2 diabetes mellitus, yet a relative decrease in insulin sensitivity occurs on a daily basis). In terms of metabolism, the heart is undoubtedly a different organ during the day versus night. Day-night differences are observed at the transcriptional (e.g., expression of genes encoding for metabolically-relevant enzymes), translational (e.g., expression of proteins that play key roles in metabolism), posttranslational (e.g., regulatory modifications of metabolically-active proteins, such as phosphorylation), metabolite (e.g., steady state levels of metabolic intermediates, cofactors, and substrate storage forms), and metabolic flux (e.g., carbon flux through metabolic pathways) levels. Within this subsection, current knowledge regarding time-of-day-dependent oscillations in cardiac metabolism will be discussed. It is important to note that the majority of these studies have been performed in laboratory rats and mice housed in a 12 h:12 h light:dark cycle. Accordingly, diurnal variations, as opposed to circadian rhythms (which are revealed only under constant conditions; usually constant darkness), have primarily been investigated. Laboratory rodents are nocturnal, meaning that they are more active during the dark phase, and are less active during the light phase. Thus, energetic demands on the heart of laboratory rodents are greatest during the dark phase (which is antiphase relative to humans).

Diurnal Rhythms in Metabolic Fluxes

Glucose Metabolism

In quantitative terms, primary energy sources for the heart are lipid (triglyceride and fatty acids) and carbohydrate (glucose and lactate). It has been estimated that up to 70% of the ATP required for contraction of the heart is derived from the oxidation of lipid/fatty acids, whereas carbohydrate utilization (e.g., glucose and lactate oxidation) accounts for the vast majority of the remaining energetic demands [10]. It is noteworthy that recent studies suggest a more

balanced energetic reliance between lipid, carbohydrate, and ketone bodies in the mouse heart [11]. Importantly, increased workload (e.g., during periods of physical activity) selectively increases myocardial glucose and lactate utilization as a way to meet elevated energetic demand [12]. In addition, myocardial glucose utilization is extremely responsive to insulin, an endocrine factor whose circulating levels are dependent on feeding status [13]. Given daily fluctuations in behaviors such as fasting/feeding and sleep/wake cycles, it is not surprising that myocardial glucose utilization exhibits a marked diurnal variation at multiple levels (as described in subsequent sections).

Initial studies investigated day-night differences in glucose oxidation in *ex vivo* perfused rat hearts [14]. Rates of glucose oxidation were approximately twofold higher for hearts isolated in the middle of the dark phase, versus hearts investigated in the middle of the light phase. An important point to note here was that glucose oxidation rates were interrogated in an *ex vivo* setting, where perfusion conditions (i.e., buffer composition, workload, etc.) were fixed at constant levels, and hearts were allowed to reach a steady state for 30 min following isolation. In doing so, it was reasoned that acute neurohumoral influences were eliminated, thereby revealing diurnal rhythms in the intrinsic properties of the heart. Subsequent studies in both rat and mouse hearts have confirmed these observations (i.e., increased glucose utilization in the rodent heart when perfused *ex vivo* in the working mode during the dark, versus light, phase) [15,16]. One important caveat of the original rat heart study was that cardiac function was also higher for *ex vivo* perfused hearts when isolated during the dark phase [14]. As mentioned earlier, the heart increases reliance on glucose (and lactate) during periods of increased contractility, raising the possibility that day-night differences in glucose oxidation were secondary to energetic demand. However, subsequent studies in *ex vivo* perfused mouse hearts revealed increased glucose utilization during the dark phase even in the absence of diurnal variations in contractility/function [16].

Glucose oxidation is not the sole index of glucose utilization that exhibits a diurnal variation in the rodent heart. Consistent with rhythms in glucose oxidation, rates of both glycolysis and glycogen synthesis are elevated in the middle of the dark phase in *ex vivo* working mouse hearts [16]. A coordinated increase in glycolysis, glucose oxidation, and glycogen synthesis during the dark phase has led to speculation that glucose uptake is increased in the mouse/rat heart at this time. However, time-of-day-dependent oscillations in myocardial glucose uptake have not been reported.

The previously described studies have been performed in an *ex vivo* setting. Less is known regarding diurnal variations in myocardial glucose metabolism *in vivo*. Intuitively, one would expect that fluctuations in energetic demand and feeding status would augment day-night differences in myocardial glucose utilization in intact animals. In mice, synchronization between increased workload, circulating insulin levels, and the intrinsic drive to promote glucose uptake during the dark phase would be anticipated to result in augmented rates of glucose utilization at this time. Indirect evidence in support of this concept includes accumulation of glycogen and protein O-GlcNAcetylation (requiring glucosyl unit flux through the hexosamine biosynthetic pathway, a branching pathway from glycolysis) during the dark phase in the rodent heart (as well as a reciprocal decline in these parameters during the light phase) [15,16]. Collectively, these observations have revealed marked diurnal variations in both oxidative and nonoxidative glucose metabolism in rodent hearts.

Fatty Acid Metabolism

Unlike myocardial glucose utilization, the oxidative metabolism of fatty acids by the heart does not appear to exhibit a marked diurnal variation (at least in an *ex vivo* setting). Both

initial studies in isolated perfused rat hearts, as well as subsequent studies in mouse hearts, were unable to detect significant day-night differences in myocardial fatty acid oxidation rates [14,15,17]. It is noteworthy that these studies were performed using the nonesterified fatty acid oleate, raising the question whether diurnal variations exist in the oxidative metabolism of distinct, quantitatively significant lipid/fatty acid species, such as palmitate or lipoproteins (e.g., chylomicrones). Previous studies have suggested a differential channeling of oleate and palmitate between oxidative and nonoxidative metabolic fates [18,19]. Interestingly, rates of net triglyceride synthesis exhibits a marked diurnal variation in *ex vivo* perfused mouse hearts, peaking in the middle of the dark phase [17]. Subsequent pulse-chase tracer studies indicated that rates of lipolysis significantly contribute toward day-night differences in net triglyceride synthesis, with increased lipolysis rates reported during the light phase [17].

Similar to the situation for myocardial glucose utilization, currently no studies have investigated diurnal rhythms in myocardial fatty acid metabolism in the intact animal. However, consistent with *ex vivo* flux measurements, triglyceride levels oscillate over the course of the day in the rodent heart, with decreased levels observed in the middle of the light phase [15,17]. Collectively, these observations suggest that the rodent heart exhibits increased lipolysis during the less active/sleep phase both *in vivo* and *ex vivo*. It is tempting to speculate that decreased insulin levels during the less active/sleep phase, coupled with the intrinsic properties of the heart, synergize to increase lipolysis at this time.

Amino Acid Metabolism

The heart is considered a metabolic omnivore, which is capable of utilizing essentially all energy substrates [20]. Although quantitatively less important, in terms of being a source of energy for contraction, the heart can readily utilize amino acids in both catabolic (e.g., oxidation) and anabolic (e.g., protein synthesis) fashions. With regards to time-of-day-dependent oscillations in amino acid metabolism, relatively little is known for the heart. Following injection of rats with a bolus of ^{3}H-labeled leucine, Rau and Meyer reported that net incorporation of this radio labeled tracer into myocardial protein was greatest at the end of the light phase [21]. However, it is important to note that this study did not take into account, potential differences in circulating levels of endogenous leucine, or differential rates of leucine oxidation. We have previously reported that several amino acids are elevated in the heart during the sleep phase [22]. In addition, recent unpublished studies suggest increased rates of protein synthesis in the isolated perfused mouse heart during the light phase (M.E. Young, unpublished observations). Collectively, these observations suggest that protein synthesis may be increased during the light phase in the rodent heart, both *in vivo* and *ex vivo*. Consistent with this concept, isoproterenol-induced growth of the heart is greatest during the light, versus the dark, phase [23]. These observations appear to be somewhat counterintuitive with the anabolic properties of insulin; increased circulating insulin levels during the dark phase for the rodent would be anticipated to promote protein synthesis. Additional studies are required to elucidate fully time-of-day-dependent oscillations in myocardial amino acid metabolism, and the mechanisms responsible.

Diurnal Variations in Metabolically Relevant Transcription, Translation, and Posttranslational Modifications

Hypothesis generating (e.g., transcriptome, proteome) and hypothesis testing (e.g., RT-PCR, Western blotting) approaches are often taken as indirect indices of cardiac metabolism. With this in mind, it is important to note that at the transcriptional, translational, and posttranslational levels, the heart is a distinct organ during the day versus night. Within this subsection, current

knowledge regarding diurnal variations in these metabolically relevant parameters will be reviewed.

Transcription

Through the use of gene expression microarray approaches, it has been estimated that approximately 9–13% of the cardiac trancriptome exhibits a time-of-day-dependent oscillation [24,25]. Both diurnal and circadian rhythms have been investigated. Martino and coworkers housed mice in a standard 12 h:12 h light:dark cycle, isolated hearts at 3 h intervals, after which diurnal variations in the transcriptome were investigated [24]. Computational analysis revealed that rhythmic genes in the heart could be subdivided based on the characteristics of the oscillation (e.g., timing of peak expression, waveform (single vs. multiple peaks/troughs) and function of encoded protein). Gene ontology analysis suggested that genes known to play important roles in processes such as growth and renewal, metabolism, transcription/translation, and molecular signal pathways were regulated in a time-of-day-dependent manner. In an attempt to determine, the relative importance of the light/dark cycle on rhythmic gene expression in the heart, Storch and coworkers housed mice in constant darkness for 2 days prior to interrogation of time-of-day-dependent rhythms in the cardiac transcriptome; in doing so, the study revealed circadian rhythms in cardiac gene expression [25]. Comparison of the diurnal and circadian studies reveals that only a relatively small fraction of rhythmic genes in the heart appear to be dependent on the light/dark cycle. These data suggest that mechanisms intrinsic of the mouse are important mediators of 24 h rhythms in cardiac gene expression.

In addition to transcriptome analyses, candidate gene approaches have been taken in an attempt to delineate the mediators of time-of-day-dependent oscillations in cardiac metabolism. The first class of genes investigated in this manner included fatty acid responsive genes.

Briefly, fatty acids are known to signal in a feedforward manner at a transcriptional level, through induction of enzymes that promote their utilization [26]. In doing so, these transcriptionally based mechanisms facilitate maintenance of intracellular fatty acid species within a physiologic range (i.e., homeostasis). Failure to adequately synchronize fatty acid utilization with availability results in accumulation of multiple intramyocellular lipid species, aberrant signaling, impaired cellular functions (e.g., contraction), and even cell death (i.e., lipotoxicity) [27]. The peroxisome proliferator activated receptors (PPARs) are critical components for this feedforward mechanism. For example, genetic disruption of PPARα in mice impairs transcriptional responsiveness of the heart to fatty acids, leading to attenuated metabolic adaptation of the heart to conditions associated with elevated fatty acid availability, such as fasting [28]. Known PPARα target genes that play essential regulatory roles in cardiac metabolism include pyruvate dehydrogenase kinase 4 (PDK4; augments reliance on fatty acid oxidation by attenuating oxidative metabolism of glucose via phosphorylation and inhibition of pyruvate dehydrogenase), uncoupling proteins (UCPs; promotes fatty acid oxidation through effects on the proton motive force and/or CoA cycling), malonyl-CoA decarboxylase (MCD; lowers malonyl-CoA levels, a potent inhibitor of long chain fatty acid entry into the mitochondrial matrix) and core β-oxidation enzymes (e.g., acyl-CoA dehydrogenases) [29–32]. In both the rat and mouse heart, these classic PPARα target genes oscillate in a time-of-day-dependent manner, with increased expression during the dark phase. For example, in the rat heart, *pdk4* and *ucp3* transcripts are 2.5-fold and 4.2-fold higher in the middle of the dark phase versus the middle of the light phase [33,34]. In the mouse heart, these oscillations are phase shifted slightly, such that peak expression occurs more toward the end of the dark phase [35]. It is has been suggested that rhythms in fatty acid responsive genes in the heart are secondary

to fluctuations in the sensitivity of the PPARα system. Evidence in support of this concept includes the observations that challenging mice with fatty acids and/or a synthetic PPARα agonist at different times of the day results in greater induction of genes such as *pdk4* and *ucp3* during the dark phase [33]. It may seem somewhat counterintuitive for the rodent heart to increase sensitivity to fatty acids during the dark phase, due to apparent dyssynchrony between circulating fatty acid levels and myocardial responsiveness; in the laboratory rat/mouse fed a normal chow in an *ad libitum* fashion, circulating fatty acids are higher during the light/sleep phase. However, consideration of the animal in the wild and evolutionary pressures may provide a possible explanation. The animal in the wild must be active upon awakening, as it forages for food and avoids predation. Even if the animal is not successful in its forage for food, it must maintain physical activity, and therefore requires a fuel source to meet energetic demands. During this scenario, an important fuel source would be circulating fatty acids; continued lipolysis during the onset of the awake period would result in a rapid induction of enzymes promoting β-oxidation, until the animal is successful in foraging for food. As such, there is synchrony between fatty acid availability and responsiveness of the heart to fatty acids during prolongation of the sleep phase fast.

Candidate gene approaches have been taken in attempts to determine the mechanisms contributing toward diurnal variations in myocardial glucose utilization and triglyceride turnover. In the latter case, expression of several enzymes known to play critical roles in triglyceride turnover has been reported to exhibit diurnal variations in the heart. These include diacylglyceride acyltransferase 2 (DGAT2; promotes triglyceride synthesis), adiponutrin (ADPN; promotes lipolysis), and hormone sensitive lipase (HSL; promotes lipolysis) [17]. Time-of-day-dependent fluctuations in expression of genes encoding known regulators of glucose uptake and utilization as also been reported in the rodent heart. In addition to *pdk4* (discussed earlier), glucose transporter 4 (GLUT4; facilitates glucose uptake) exhibits a diurnal variation in both the rat and mouse heart, peaking during the dark phase (when glucose utilization is elevated) [14,35,36]. Similarly, glutamine:fructose 6-phosphate aminotransferase 2 (GFAT2; catalyzes the committed step for the hexosamine biosynthetic pathway) and O-GlcNAc transferase (OGT; catalyzes the addition of O-GlcNAc to target proteins) both exhibit peak expression in the heart during the dark phase, consistent with increased protein O-GlcNAcylation at this time [16]. In contrast, surprisingly little is known regarding diurnal variations in the expression of genes involved in glycolysis and glycogen turnover in the heart.

Translation and Posttranslational Modifications

Although alterations in gene expression have the potential of providing mechanistic insight, it is important to recall that the activity of the encoded protein holds a greater potential to directly impact metabolic fluxes. The activity of a protein is influenced not only by its level of expression (which is dependent on rates of protein synthesis and degradation), but also through modulation of substrate availability, allosteric effector levels, and covalent modifications (both irreversible (e.g., proteolysis of an apoprotein to mature protein) and reversible (e.g., phosphorylation)). The purpose of this subsection is to review current knowledge regarding diurnal variations in protein levels and reversible posttranslational modifications (PTMs).

Comparable to transcription-based studies, diurnal variations in protein expression have been investigated through the use of both hypothesis generating and testing approaches. For example, proteomic studies have recently revealed that approximately 8% of cardiac proteins exhibit a time-of-day-dependent oscillation in the murine heart [37]. Direct comparison of proteomic and microarray data suggest that

only approximately 50% of the oscillating proteins can be readily explained by reciprocal oscillations in corresponding mRNA transcripts; comparable observations have been reported for the liver [38]. Interestingly, proteomic analysis of the murine heart revealed that a large proportion of proteins exhibiting diurnal variations are directly involved in metabolism. Identified proteins included a surprisingly large number of nuclear-encoded mitochondrial enzymes, which play important roles in oxidative metabolism. These include enzymes involved in alcohol metabolism, fatty acid β-oxidation, and oxidative phosphorylation [37]. Previous studies suggest that the turnover of mitochondria in the heart is relatively long, with an estimated half-life of 30 days [39]. The recent proteomic studies identifying a subset of mitochondrial proteins exhibiting a diurnal variation suggest that even though cardiac mitochondria have a half-life well in excess of 24 h, mitochondrial function and oxidative metabolism potentially exhibit time-of-day-dependent oscillations. To date, no studies have directly investigated day-night differences in cardiac mitochondrial function.

Western blotting analysis has been performed to investigate diurnal variations in the expression of candidate proteins and PTMs. For example, total GLUT4 protein levels are elevated in rat hearts during the dark (vs. light) phase (although subcellular localization was not investigated) [14]. OGT exhibits increased protein expression during the dark phase in the murine heart, consistent with increased protein O-GlcNAcylation at that time [16]. Phosphorylation status of components of signaling axes known to directly influence myocardial metabolism have also been investigated. This includes the insulin signaling cascade. For example, phosphorylation of AKT and GSK3β both increase in the murine heart during the dark phase; although these oscillations in insulin signaling likely reflect increased circulating insulin levels during the dark/active phase, the possibility remains that the heart exhibits increased insulin sensitivity at this time [40]. Indeed, whole body insulin sensitivity peaks in both rodents and humans at the beginning of the active period [8,9]. Another signaling protein that directly regulates myocardial metabolism is AMPK. Consistent with increased glucose utilization during the dark phase, both the phosphorylation and activity of AMPK is elevated in the heart at this time [17]. It is worthy to note that additional PTMs have been shown to oscillate at both the whole proteome and individual protein level in extra-cardiac tissues, such as the liver. These PTMs include acetylation, ADP-ribosylation, and ubiquitination [41–48]. In the latter case, it is important to note that protein degradation is essential for protein turnover, particularly in the 24 h timeframe. It is likely that rhythms in these PTMs, as well as others, play an important role in modulating cardiac metabolism over the course of the day.

EXTRINSIC VERSUS INTRINSIC INFLUENCES

Cardiac metabolism undoubtedly fluctuates over the course of the day. As highlighted earlier, evidence exists in support of the concept that the intrinsic metabolic properties of the heart fluctuate in a time-of-day-dependent manner. This includes persistence of day-night differences in: (1) metabolic fluxes of *ex vivo* perfused hearts; (2) responsiveness of the heart to stimuli/stresses (e.g., fatty acid availability); and (3) the expression and posttranslational modification of proteins known to play critical roles in metabolism. Observations such as these have led to questions focused on identification of the mechanism(s) responsible for diurnal rhythms in these metabolism-related parameters. Two main possibilities exist. Rhythms are driven by extracardiac (e.g., neurohumoral factors) or intracardiac influences. It is likely that behavior-associated differences in neurohumoral factors (e.g., cortisol, insulin, norepinephrine, growth hormone, etc.) influence the heart both acutely

(e.g., immediate stimulation/repression of a metabolic process) and chronically (e.g., alterations in gene/protein expression) [49]. For example, cortisol is a steroid hormone that alters cellular function through transcriptional mechanisms, and exhibits a marked diurnal variation in both rodents and humans, peaking at the beginning of the awake phase [50,51]. However, recent studies have highlighted a mechanism that is intrinsic to the cardiomyocyte as a critical mediator of diurnal variations in cardiac metabolism. This mechanism is the circadian clock.

Circadian clocks are cell autonomous molecular mechanisms that enable the cell/organ to temporally partition biological processes [5,6]. In doing so, circadian clocks confer the selective advantage of anticipation, allowing preparation prior to the onset of a predicted stimulus/stress. In other words, circadian clocks orchestrate biological processes, such that temporal synchronization between the stimulus and responsiveness is achieved in an appropriate manner. The mammalian circadian clock is comprised of at least 15 transcription factors, whose levels/activities oscillate over the course of the day, with a periodicity of approximately 24 h [6]. At the heart of this mechanism are two transcription factors, CLOCK and BMAL1. Upon heterodimerization, these transcriptional regulators bind to E-boxes in target genes, resulting in induction [52,53]. The protein products of several CLOCK/BMAL1 target genes are components of the circadian clock mechanism, and act in a feedback manner, by influencing the activity of CLOCK/BMAL1. These include the period and cryptochrome genes (PER1/PER2/PER3 and CRY1/CRY2, respectively); PER/CRY heterodimers translocate into the nucleus, and inhibit CLOCK/BMAL1 [54,55]. An additional negative feedback loop involves REV-ERBα, which binds to the Bmal1 promoter, thereby inhibiting ROR-dependent BMAL1 induction [56]. It is important to note that CLOCK/BMAL1 induce a large number of additional genes, which are not core circadian clock components. These genes,

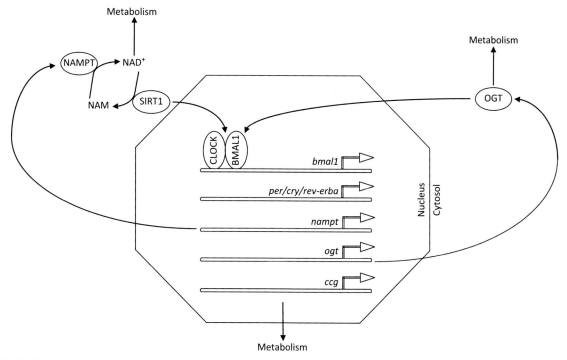

FIGURE 9.2 **Metabolism as an integral clock component.** The CLOCK/BMAL1 heterodimer binds to E-boxes within the promoter of target genes, including core circadian clock components (e.g., bmal1, per, cry, rev-erbα genes) and clock controlled genes (ccg). Two ccg's encode for NAMPT and OGT, which form feedback loops by affecting BMAL1 acetylation and O-GlcNAcylation, respectively. The circadian clock influences cellular metabolism through modulation of NAD levels, protein O-GlcNAcylation, and other ccg's.

termed clock controlled or clock output genes, influence cellular function over the course of the day. One cellular process that appears to be under direct circadian clock control is metabolism. Indeed, metabolism has emerged as an integral component of the mammalian circadian clock (Fig. 9.2).

The Cardiomyocyte Circadian Clock

Circadian clocks are ubiquitous in nature, having been identified in all mammalian cells investigated to date. Circadian clocks can be divided into two main categories, namely central versus peripheral circadian clocks. The central circadian clock is located within the suprachiasmatic nucleus (SCN), a collection of specialized

neurons within the hypothalamus [6]. The SCN perceives light signals via melanopsin-containing ganglion cells within the retina, which are subsequently transmitted to the SCN via the retinohypothalamic tract [57]. A hierarchy is believed to exist, wherein the SCN orchestrates synchrony via neurohumoral signals to reset peripheral clocks (i.e., clocks located in other regions of the body, including the heart). However, it is important to note that this synchrony is easily disrupted through behaviors such as eating, which can dissociate central and peripheral oscillators.

Consistent with its ubiquitous nature, genetic disruption of the circadian clock in murine models in a germline fashion impacts a plethora of processes, ranging from behaviors (e.g., sleep/wake and fasting/feeding cycles) to neuroendocrine functions, cellular signaling, and metabolism. For example, BMAL1 null mice exhibit arrhythmicity in locomotor activity, are lean on control diets, are susceptible to diet-induced obesity, exhibit glucose intolerance, develop cardiomyopathy, and have a reduced lifespan [58–60]. Such a phenotype makes use of germline models of clock disruption difficult for investigation of the role of a cell autonomous clock on a specific biological process. This includes determining of the influence of the cardiomyocyte circadian clock on cardiac metabolism. Accordingly, two novel models of cardiomyocyte-specific circadian clock disruption have been developed to investigate the impact that this molecular mechanism has on cardiac metabolism. These are cardiomyocyte-specific CLOCK mutant (CCM) and cardiomyocyte-specific BMAL1 knockout (CBK) mice [61,62].

Given its transcriptional nature, an initial tactic for identification of those metabolic processes that are regulated by the cardiomyocyte circadian clock included gene expression microarray studies. Comparison of time-of-day-dependent oscillations in the transcriptome between hearts isolated from CCM/CBK mice versus littermate controls revealed: (1) up to 8% of the cardiac transcriptome is regulated by the cardiomyocyte circadian clock; (2) at a transcriptional level, CCM and CBK hearts are temporally suspended at the beginning of the light phase; and (3) the cardiomyocyte circadian clock regulates a large number of biological processes, including transcription, protein turnover, signaling cascades, ion homeostasis, and metabolism [61,62]. An in-depth computational analysis of 128 microarrays, involving *in silico* comparisons with previously published promotor sequence and liver ChIPseq information, identified multiple candidate direct CLOCK/BMAL1 target genes in the heart. These included genes with known functions in lipid metabolism (e.g., dgat2), NAD salvage (e.g., nampt), and insulin signaling (e.g., pik3r1); validation studies at the gene and/or

protein level have confirmed regulation by the cardiomyocyte circadian clock [62].

Both CCM and CBK mice exhibit striking cardiac metabolism phenotypes. For example, diurnal variations in glucose metabolism normally observed in wild-type hearts are completely absent in CCM hearts [16]. Consistent with temporal suspension during the light/sleep phase, CCM hearts exhibit chronically low rates of glucose oxidation. Similar observations have been reported for CBK hearts (i.e., depressed rates of glucose oxidation), although only one time of day was investigated (middle of the dark/active phase, when glucose oxidation normally peaks) [62]. Oscillations in additional glucose utilization parameters (e.g., glycolysis, glycogen synthesis, protein O-GlcNAcylation) are also diminished in CCM hearts, suggesting that the cardiomyocyte circadian clock coordinates temporal regulation of multiple enzymes/proteins and/or modulates glucose transport [16]. Interestingly, the increase in AMPK phosphorylation and activity observed in the murine heart during the dark phase is abolished in CCM hearts [17]. Given that AMPK promotes glucose transport, cardiomyocyte clock dependent increased activity during the dark/active phase may drive increased glucose utilization at this time. Indeed, glucose transport is decreased in CCM hearts during the dark phase (relative to wild-type littermates) [16].

As discussed earlier, net triglyceride synthesis exhibits a marked diurnal variation in wild-type hearts. Interestingly, mRNA levels of the CLOCK/BMAL1 target gene *dgat2* peak approximately 6 h before the peak in myocardial triglyceride synthesis [17]. Both oscillations in *dgat2* mRNA and triglyceride synthesis are absent in CCM hearts, consistent with the hypothesis that the cardiomyocyte circadian clock drives time-of-day-dependent rhythms in non-oxidative fatty acid metabolism [17]. It is noteworthy that although oxidative metabolism of fatty acids does not exhibit a diurnal variation in wild-type rodent hearts, both CCM and CBK hearts exhibit increased rates of oleate oxidation, in a time-of-day-independent manner [61,62]. These observations are counterintuitive to reports of mitochondrial dysfunction in both CCM and CBK hearts, as well as decreased β-oxidation in liver mitochondria from BMAL1 null mice [36,61]. The mechanism(s) responsible for increased myocardial fatty acid oxidation in CCM/CBK hearts remain unknown, but may be secondary to decreased glucose oxidation (i.e., reverse Randle cycle) and/or regulation of fatty acyl group transport (i.e., cellular or mitochondrial uptake). Conversely, ketone body oxidation is severely depressed in both CCM and CBK hearts in a time-of-day-independent manner, due to CLOCK/BMAL1 regulation of β-hydroxybutyrate dehydrogenase [62]. In summary, it is clear that the cardiomyocyte circadian clock regulates cardiac metabolism, and indeed, metabolism appears to be an integral circadian clock mechanism (Fig. 9.2).

PRACTICAL IMPLICATIONS

The impact of diurnal variations in cardiac processes on experimental strategies and outcomes cannot be overestimated. It is apparent that cardiac metabolism is a "moving target," and as such, choice of the time-of-day at which an experiment is performed markedly impacts the results observed. Failure to appreciate/account for these oscillations during experimental design can lead to erroneous conclusions. Within this subsection, experimental considerations in light of diurnal variations in metabolic processes will be discussed.

It is important to note that commonly utilized rodent models (e.g., rats, mice) are nocturnal. Accordingly, laboratory rodents tend to be less active during the light phase, and are more active during the dark phase. This is an important, yet often overlooked, consideration. The majority of animal model-based research is performed with the goal of improving knowledge

regarding underlying mechanisms responsible for human health and disease. Yet, for the sake of convenience, many rodent-based studies are performed during the light phase, which is equivalent to observing human subjects during the middle of the night. This can become increasingly problematic when interventions are performed in laboratory rodents and/or data are correlated to findings in humans. Simple interventions, such as dietary manipulations (including fasting), exercise, and pharmacological interventions are often performed, without consideration of time-of-day. For example, mice are routinely fasted overnight prior to investigation, an intervention that is equivalent to fasting a human subject throughout a day. The effects of distinct diets on metabolic parameters are commonly investigated during the light phase, a time at which food intake reaches a natural trough in rodents, thereby minimizing identification of mechanisms contributing to perturbations during food consumption. Many rodent-based studies exercise mice during the light phase, which is equivalent to forcing a human subject to be physically active in the middle of the night. Finally, both human and rodent studies indicate that sensitivity to many pharmacological agents exhibits a time-of-day-dependence (termed chronopharmacology). Collectively, failure to consider the nocturnal nature of rodent models will likely limit translational capacity.

When assessing cardiac metabolism, it is important to appreciate that many of these metabolic processes oscillate in a time-of-day-dependent manner. Many investigations attempt to circumvent this issue, by performing all studies at the same time-of-day. However, choice of the time-of-day at which metabolic studies are performed markedly impacts the results obtained and conclusions drawn. Consider a hypothetical study designed to investigate the mechanisms responsible for high fat diet-induced perturbations in cardiac metabolism in rodents. The majority of outcomes would be assessed during the light phase. However, diet consumption occurs primarily during the dark phase, resulting in an elevation of "the stimulus" at that time. One such stimulus includes fatty acids. The latter induce changes in gene and protein expression, which persist when cardiac metabolism is assessed during the light phase. However, circulating fatty acid levels return to control levels during the light phase [33]. Such observations could lead to the erroneous conclusion that circulating fatty acids are not important for perturbations in cardiac metabolism following high fat feeding. Similarly, if the impact of putative glucose oxidation inhibitor was assessed during the light phase, it is possible that an erroneous null result could be obtained, as cardiac glucose oxidation is already depressed at this time of day.

Aberrant oscillations in cardiac metabolism are likely to be more common than currently appreciated. During various disease states (e.g., hypertrophy, diabetes mellitus), the heart has been described as exhibiting metabolic inflexibility [1]. Such a term has been utilized to highlight an inability of the myocardium to respond to both acute and chronic interventions. It is likely that the heart exhibits metabolic inflexibility over the course of the day during these disease states, although studies directly interrogating this issue have not been reported to date. Consistent with this hypothesis, the circadian clock in the heart is altered in various animal models of cardiac disease, including pressure overload and streptozotocin-induced diabetes mellitus [63,64]. It is also noteworthy that many genetically modified mouse models are likely models of disrupted metabolic rhythms. Overexpression or knockout of a gene that normally exhibits a diurnal variation in the heart essentially locks expression at the peak or trough. If such a gene plays an important role in metabolism, this in turn has the potential to temporally suspend a metabolic process. One example may include PGC1α, a transcriptional coactivator influencing mitochondrial biogenesis and distinct metabolic pathways. PGC1α expression oscillates in both the heart and extracardiac tissues, suggesting that genetic manipulation of PGC1α likely impacts the temporal partitioning of pathways influenced by this coactivator [65].

SUMMARY

In summary, the heart exhibits marked diurnal variations in glucose, fatty acid/lipid, and amino acid/protein metabolism. These rhythms appear to be orchestrated through interactions of extrinsic (e.g., neurohumoral factors) and intrinsic (e.g., circadian clock) influences. These rhythms likely play a critical role in synchronization of metabolic pathways with time-of-day-dependent fluctuations in energetic demand, nutrient availability, and presence of pro-oxidants. Importance of this synchrony is revealed following genetic- and/or environment- induced perturbations in normal diurnal variations, which results in cardiac dysfunction.

Acknowledgment

This work was supported by the National Heart, Lung, and Blood Institute (HL123574 and HL122975).

References

[1] Taegtmeyer H, Golfman L, Sharma S, Razeghi P, van Arsdall M. Linking gene expression to function: metabolic flexibility in the normal and diseased heart. Ann NY Acad Sci 2004;1015:202–13.
[2] Riera CE, Dillin A. Tipping the metabolic scales towards increased longevity in mammals. Nat Cell Biol 2015;17:196–203.
[3] Plum L, Belgardt BF, Bruning JC. Central insulin action in energy and glucose homeostasis. J Clin Invest 2006;116:1761–6.
[4] American Diabetes Association. Diagnosis and classification of diabetes mellitus. Diabetes Care 2013;36(Suppl. 1): S67–74.
[5] Edery I. Circadian rhythms in a nutshell. Physiol Genomics 2000;3:59–74.
[6] Takahashi JS, Hong HK, Ko CH, McDearmon EL. The genetics of mammalian circadian order and disorder: implications for physiology and disease. Nat Rev Genet 2008;9:764–75.

[7] Mitra S. Does evening sun increase the risk of skin cancer? Proc Nat Acad Sci USA 2011;108:18857–8.

[8] Lee A, Ader M, Bray GA, Bergman RN. Diurnal variation in glucose tolerance. Cyclic suppression of insulin action and insulin secretion in normal-weight, but not obese, subjects. Diabetes 1992;41:750–9.

[9] Whichelow MJ, Sturge RA, Keen H, Jarrett RJ, Stimmler L, Grainger S. Diurnal variation in response to intravenous glucose. Brit Med J 1974;1:488–91.

[10] Lopaschuk GD, Ussher JR, Folmes CD, Jaswal JS, Stanley WC. Myocardial fatty acid metabolism in health and disease. Physiol Rev 2010;90:207–58.

[11] Stowe KA, Burgess SC, Merritt M, Sherry AD, Malloy CR. Storage and oxidation of long-chain fatty acids in the c57/bl6 mouse heart as measured by nmr spectroscopy. FEBS Lett 2006;580:4282–7.

[12] Allard M, Schonekess B, Henning S, English D, Lopaschuk G. Contribution of oxidative metabolism and glycolysis to atp production in hypertrophied hearts. Am J Physiol 1994;267:H742–50.

[13] Russell RR III, Cline GW, Guthrie PH, Goodwin GW, Shulman GI, Taegtmeyer H. Regulation of exogenous and endogenous glucose metabolism by insulin and acetoacetate in the isolated working rat heart. A three tracer study of glycolysis, glycogen metabolism, and glucose oxidation. J Clin Invest 1997;100:2892–9.

[14] Young M, Razeghi P, Cedars A, Guthrie P, Taegtmeyer H. Intrinsic diurnal variations in cardiac metabolism and contractile function. Circ Res 2001;89:1199–208.

[15] Durgan D, Moore M, Ha N, Egbejimi O, Fields A, Mbawuike U, Egbejimi A, Shaw C, Bray M, Nannegari V, Hickson-Bick D, Heird W, Dyck J, Chandler M, Young M. Circadian rhythms in myocardial metabolism and contractile function: influence of workload and oleate. Am J Physiol Heart Circ Physiol 2007;293: H2385–93.

[16] Durgan DJ, Pat BM, Laczy B, Bradley JA, Tsai JY, Grenett MH, Ratcliffe WF, Brewer RA, Nagendran J, Villegas-Montoya C, Zou C, Zou L, Johnson RL Jr, Dyck JR, Bray MS, Gamble KL, Chatham JC, Young ME. O-glcnacylation, novel post-translational modification linking myocardial metabolism and cardiomyocyte circadian clock. J Biol Chem 2011;286:44606–19.

[17] Tsai JY, Kienesberger PC, Pulinilkunnil T, Sailors MH, Durgan DJ, Villegas-Montoya C, Jahoor A, Gonzalez R, Garvey ME, Boland B, Blasier Z, McElfresh TA, Nannegari V, Chow CW, Heird WC, Chandler MP, Dyck JR, Bray MS, Young ME. Direct regulation of myocardial triglyceride metabolism by the cardiomyocyte circadian clock. J Biol Chem 2010;285:2918–29.

[18] Gaster M, Rustan AC, Beck-Nielsen H. Differential utilization of saturated palmitate and unsaturated oleate: evidence from cultured myotubes. Diabetes 2005;54:648–56.

[19] Kolwicz SC, Tian R. Abstract 15951: palmitate, not oleate, is the preferred fatty acid for cardiac triacylglycerol synthesis. Circulation 2014;130:A15951.

[20] Taegtmeyer H. Metabolism–the lost child of cardiology. J Am Coll Cardiol 2000;36:1386–8.

[21] Rau E, Meyer DK. A diurnal rhythm of incorporation of l-[3h] leucine in myocardium of the rat. Recent Adv Stud Card Struct Metab 1975;7:105–10.

[22] Tsai J, Young M. Diurnal variations in myocardial metabolism. Heart Metab 2009;44:5–9.

[23] Durgan DJ, Tsai JY, Grenett MH, Pat BM, Ratcliffe WF, Villegas-Montoya C, Garvey ME, Nagendran J, Dyck JR, Bray MS, Gamble KL, Gimble JM, Young ME. Evidence suggesting that the cardiomyocyte circadian clock modulates responsiveness of the heart to hypertrophic stimuli in mice. Chronobiol Int 2011;28:187–203.

[24] Martino T, Arab S, Straume M, Belsham DD, Tata N, Cai F, Liu P, Trivieri M, Ralph M, Sole MJ. Day/night rhythms in gene expression of the normal murine heart. J Mol Med 2004;82:256–64.

[25] Storch KF, Lipan O, Leykin I, Viswanathan N, Davis FC, Wong WH, Weitz CJ. Extensive and divergent circadian gene expression in liver and heart. Nature 2002;417: 78–83.

[26] Huss JM, Kelly DP. Nuclear receptor signaling and cardiac energetics. Circ Res 2004;95:568–78.

[27] Goldberg IJ, Trent CM, Schulze PC. Lipid metabolism and toxicity in the heart. Cell Metab 2012;15:805–12.

[28] Gelinas R, Labarthe F, Bouchard B, Mc Duff J, Charron G, Young ME, Des Rosiers C. Alterations in carbohydrate metabolism and its regulation in pparalpha null mouse hearts. Am J Physiol Heart Circ Physiol 2008;294:H1571–80.

[29] Wu P, Peters JM, Harris RA. Adaptive increase in pyruvate dehydrogenase kinase 4 during starvation is mediated by peroxisome proliferator-activated receptor alpha. Biochem Biophys Res Comm 2001;287:391–6.

[30] Young ME, Patil S, Ying J, Depre C, Ahuja HS, Shipley GL, Stepkowski SM, Davies PJ, Taegtmeyer H. Uncoupling protein 3 transcription is regulated by peroxisome proliferator-activated receptor (alpha) in the adult rodent heart. FASEB J 2001;15:833–45.

[31] Young ME, Goodwin GW, Ying J, Guthrie P, Wilson CR, Laws FA, Taegtmeyer H. Regulation of cardiac and skeletal muscle malonyl-coa decarboxylase by fatty acids. Am J Physiol Endocrinol Metab 2001;280:E471–9.

[32] Gulick T, Cresci S, Caira T, Moore DD, Kelly DP. The peroxisome proliferator-activated receptor regulates mitochondrial fatty acid oxidative enzyme gene expression. Proc Nat Acad Sci USA 1994;91:11012–6.

[33] Stavinoha M, RaySpellicy J, Hart-Sailors M, Mersmann H, Bray M, Young M. Diurnal variations in the responsiveness of cardiac and skeletal muscle to fatty acids. Am J Physiol 2004;287:E878–87.

[34] Stavinoha MA, RaySpellicy JW, Essop MF, Graveleau C, Abel ED, Hart-Sailors ML, Mersmann HJ, Bray MS, Young ME. Evidence for mitochondrial thioesterase 1 as a peroxisome proliferator-activated receptor-alpha-regulated gene in cardiac and skeletal muscle. Am J Physiol Endocrinol Metab 2004;287:E888–95.

[35] Durgan D, Trexler N, Egbejimi O, McElfresh T, Suk H, Petterson L, Shaw C, Hardin P, Bray M, Chandler M, Chow C, Young M. The circadian clock within the cardiomyocyte is essential for responsiveness of the heart to fatty acids. J Biol Chem 2006;281:24254–69.

[36] Kohsaka A, Das P, Hashimoto I, Nakao T, Deguchi Y, Gouraud SS, Waki H, Muragaki Y, Maeda M. The circadian clock maintains cardiac function by regulating mitochondrial metabolism in mice. PloS ONE 2014;9:e112811.

[37] Podobed P, Pyle WG, Ackloo S, Alibhai FJ, Tsimakouridze EV, Ratcliffe WF, Mackay A, Simpson J, Wright DC, Kirby GM, Young ME, Martino TA. The day/night proteome in the murine heart. Am J Physiol Regul Integr Comp Physiol 2014;307:R121–37.

[38] Reddy AB, Karp NA, Maywood ES, Sage EA, Deery M, O'Neill JS, Wong GK, Chesham J, Odell M, Lilley KS, Kyriacou CP, Hastings MH. Circadian orchestration of the hepatic proteome. Curr Biol 2006;16:1107–15.

[39] Kasumov T, Dabkowski ER, Shekar KC, Li L, Ribeiro RF, Walsh K, Previs SF, Sadygov RG, Willard B, Stanley WC. Assessment of cardiac proteome dynamics with heavy water: slower protein synthesis rates in interfibrillar than subsarcolemmal mitochondria. Am J Physiol Heart Circ Physiol 2013;304:H1201–14.

[40] Durgan DJ, Pulinilkunnil T, Villegas-Montoya C, Garvey ME, Frangogiannis NG, Michael LH, Chow CW, Dyck JR, Young ME. Short communication: ischemia/reperfusion tolerance is time-of-day-dependent: mediation by the cardiomyocyte circadian clock. Circ Res 2010;106:546–50.

[41] Chatham J, Laczy B, Durgan D, Young M. Direct interrelationship between protein o-glcnacation and the cardiomyocyte circadian clock. Cir Res 2009;105:e50.

[42] Hardin PE, Yu W. Circadian transcription: passing the hat to clock. Cell 2006;125:424–6.

[43] Cardone L, Hirayama J, Giordano F, Tamaru T, Palvimo JJ, Sassone-Corsi P. Circadian clock control by sumoylation of bmal1. Science 2005;309:1390–4.

[44] Katada S, Sassone-Corsi P. The histone methyltransferase mll1 permits the oscillation of circadian gene expression. Nat Struct Mol Biol 2010;17:1414–21.

[45] Dardente H, Cermakian N. Molecular circadian rhythms in central and peripheral clocks in mammals. Chronobiol Int. 2007;24:195–213.

[46] Bray M, Shaw C, Moore M, Garcia R, Zanquetta M, Durgan D, Jeong W, Tsai J, Bugger H, Zhang D, Rohrwasser A, Rennison J, Dyck J, Litwin S, Hardin P, Chow C, Chandler M, Abel E, Young M. Disruption of the circadian clock within the cardiomyocyte influences myocardial contractile function; metabolism; and gene expression. Am J Physiol Heart Circ Physiol 2008;294:H1036–47.

[47] Durgan DJ, Young ME. The cardiomyocyte circadian clock: emerging roles in health and disease. Circ Res 2010;106:647–58.

[48] Tamaru T, Hirayama J, Isojima Y, Nagai K, Norioka S, Takamatsu K, Sassone-Corsi P. Ck2alpha phosphorylates bmal1 to regulate the mammalian clock. Nat Struct Mol Biol 2009;16:446–8.

[49] Gamble KL, Berry R, Frank SJ, Young ME. Circadian clock control of endocrine factors. Nat Rev Endocrinol 2014;10:466–75.

[50] Kalsbeek A, van Heerikhuize JJ, Wortel J, Buijs RM. A diurnal rhythm of stimulatory input to the hypothalamo-pituitary-adrenal system as revealed by timed intrahypothalamic administration of the vasopressin v1 antagonist. J Neurosci 1996;16:5555–65.

[51] Dickmeis T. Glucocorticoids and the circadian clock. J Endocrinol 2009;200:3–22.

[52] Gekakis N, Staknis D, Nguyen H, Davis F, Wilsbacher L, King D, Takahashi J, Weitz C. Role of the clock protein in the mammalian circadian mechanism. Science 1998;280:1564–9.

[53] Hogenesch J, Gu Y, Jain S, Bradfield C. The basic-helix-loop-helix-pas orphan mop3 forms transcriptionally active complexes with circadian and hypoxia factors. Proc Natl Acad Sci USA 1998;95:5474–9.

[54] Zylka MJ, Shearman LP, Weaver DR, Reppert SM. Three period homologs in mammals: differential light responses in the suprachiasmatic circadian clock and oscillating transcripts outside of brain. Neuron 1998;20:1103–10.

[55] Miyamoto Y, Sancar A. Vitamin B2-based blue-light photoreceptors in the retinohypothalamic tract as the photoactive pigments for setting the circadian clock in mammals. Proc Nat Acad Sci USA 1998;95:6097–102.

[56] Preitner N, Damiola F, Lopez-Molina L, Zakany J, Duboule D, Albrecht U, Schibler U. The orphan nuclear receptor rev-erbalpha controls circadian transcription within the positive limb of the mammalian circadian oscillator. Cell 2002;110:251–60.

[57] Berson DM, Dunn FA, Takao M. Phototransduction by retinal ganglion cells that set the circadian clock. Science 2002;295:1070–3.

[58] Bunger MK, Walisser JA, Sullivan R, Manley PA, Moran SM, Kalscheur VL, Colman RJ, Bradfield CA. Progressive arthropathy in mice with a targeted disruption of the mop3/bmal-1 locus. Genesis 2005;41:122–32.

[59] Shi SQ, Ansari TS, McGuinness OP, Wasserman DH, Johnson CH. Circadian disruption leads to insulin resistance and obesity. Curr Biol 2013;23:372–81.

[60] Lefta M, Campbell KS, Feng HZ, Jin JP, Esser KA. Development of dilated cardiomyopathy in bmal1-deficient mice. Am J Physiol Heart Circ Physiol 2012;303: H475–485.

[61] Bray MS, Shaw CA, Moore MW, Garcia RA, Zanquetta MM, Durgan DJ, Jeong WJ, Tsai JY, Bugger H, Zhang D, Rohrwasser A, Rennison JH, Dyck JR, Litwin SE, Hardin PE, Chow CW, Chandler MP, Abel ED, Young ME. Disruption of the circadian clock within the cardiomyocyte influences myocardial contractile function, metabolism, and gene expression. Am J Physiol Heart Circ Physiol 2008;294:H1036–47.

[62] Young ME, Brewer RA, Peliciari-Garcia RA, Collins HE, He L, Birky TL, Peden BW, Thompson EG, Ammons BJ, Bray MS, Chatham JC, Wende AR, Yang Q, Chow CW, Martino TA, Gamble KL. Cardiomyocyte-specific bmal1 plays critical roles in metabolism, signaling, and maintenance of contractile function of the heart. J Biol Rhythms 2014;29:257–76.

[63] Young M, Razeghi P, Taegtmeyer H. Clock genes in the heart: characterization and attenuation with hypertrophy. Circ Res 2001;88:1142–50.

[64] Young M, Wilson C, Razeghi P, Guthrie P, Taegtmeyer H. Alterations of the circadian clock in the heart by streptozotocin-induced diabetes. J Mol Cell Cardiol 2002;34:223–31.

[65] Duez H, Staels B, Rev-erb-alpha. An integrator of circadian rhythms and metabolism. J Appl Physiol 2009;107:1972–80.

10

Nutritional and Environmental Influences on Cardiac Metabolism and Performance

Marc van Bilsen

Departments of Physiology and Cardiology, Cardiovascular Research Institute
Maastricht, Maastricht University, Maastricht, The Netherlands

INTRODUCTION

The heart is an omnivorous organ, in the sense that it is able to use many kinds of substrates to generate the chemical energy required to sustain its pumping activity. Although fatty acids and glucose are considered the main energy supplying substrates, their oxidation covering approximately 60–70% and 20–30% of the cardiac energy needs at rest, respectively, the healthy heart will easily and quickly switch to alternative substrates in case plasma levels of these substrates rise [1,2]. This might be the case, for instance, during extended periods of fasting when the liver starts to produce ketone bodies, or after a bout of intense exercise when skeletal muscle releases lactate into the circulation. In case the period of fasting is not overly long, or the physical exercise is less exhaustive, the changes in circulating substrate levels are subtle and predominantly driven by a drop in the insulin/glucagon ratio.

Under these catabolic conditions adipose tissue becomes the main source of circulating fatty acids and cardiac metabolism shifts toward the oxidation of even more fatty acids, while glucose oxidation becomes less.

Next to these relatively acute changes in circulating substrate levels, sometimes the heart is also faced with more chronic changes in its nutritional environment for instance under hyperlipidemic or diabetic conditions. Under these conditions there appears to be ample time for the heart to adapt to these longstanding alterations in nutritional environment. However, it seems that this happens at the expense of its omnivorous capacity, that is, the heart loses some of its flexibility to readily switch to other substrates [2].

In addition to changes at the supply side (nutritional environment), from the demand side the need for substrates of the cardiac tissue is largely dependent on the workload of the heart.

The healthy heart is subject to acute changes in workload that may be relatively subtle such as changes during the sleep/wake cycle, or very robust as during exercise whereby cardiac output may increase up to fivefold. Chronic increases in cardiac workload also occur in case of, for instance, hypertension, when the heart is forced to generate a higher pressure to maintain cardiac output. Moreover, longstanding hypertension also evokes remodeling of the heart (hypertrophy, dilation). These alterations in cardiac geometry by itself have an effect on cardiac wall stress (dependent on ventricular pressure, radius or ventricular cavity, and wall thickness as dictated by Laplace's law) and, thereby, on the energy needs of the myocardium. As with chronic changes in the nutritional milieu, cardiac remodeling due to chronic changes in cardiac workload has been shown to lead to adaptations in the substrate preference of the heart and limit the capacity of the heart to use alternative substrates.

In the present chapter the focus will first be on the effect of relatively short-term variations in nutritional environment on cardiac substrate handling and on cardiac function and phenotype. Thereafter the more long-term effects of metabolic derangements, associated with obesity and diabetes, and the effects of dietary interventions on the heart are being discussed.

DIURNAL VARIATIONS

From day-to-day our physiologic and behavioral rhythms are largely set by the circadian clock localized in the suprachiasmatic nucleus in the hypothalamus, which also determines our overall energy intake and energy expenditure. Various hormones controlling plasma nutrient levels show a circadian rhythm, the most well-known being cortisol, the plasma levels of which peak in the late night/early morning hours, thereby priming us for our daytime activities. A detailed description of this circadian rhythm can be found in Chapter 7. Importantly, recent studies indicate that in addition to the master clock in the brain, almost all tissues, including liver, adipose tissue, skeletal muscle and heart, possess an autonomous circadian rhythm. By transcriptomic analysis it was demonstrated that the expression of about 13% of the genes in the murine heart, including many metabolic genes, is subject to a circadian rhythm [3]. Under in vivo conditions, however, it is impossible to discern if these changes are secondary to changes in neurohumoral factors that are under control of the central clock, or if they are intrinsic to the heart itself. By using isolated adult cardiomyocytes, deprived of potential changes in levels of neurohumoral factors, it was demonstrated that gene expression still follows a circadian rhythm, thereby providing unequivocal evidence for the existence of a functional, intrinsic circadian clock within cardiomyocytes [4].

It is commonly acknowledged that the evolutionary advantage of the existence of central and peripheral clocks is that it allows the organism and its tissues, to anticipate the upcoming changes. For the heart this mainly refers to anticipation of the daily increase in physical activity during the active time of the day (light phase for man, dark phase for rodents) and the associated feeding/fasting cycles. Studies in rats indicate that increases in cardiac energy demand during the active phase are largely met by increases in carbohydrate oxidation, with only modest effects on fatty acid oxidation [5]. These changes are consistent with the circadian changes in the expression of pyruvate dehydrogenase kinase (PDK4), an inhibitor of the pyruvate dehydrogenase (PDH) complex, and the activity of the PDH complex, a major regulatory step for carbohydrate oxidation. Paradoxically, at the end of the sleep-phase plasma fatty acid levels peak [6,7], while cardiac fatty acid oxidation capacity remains relatively constant during the active phase [5]. The heart seems to cope with this situation by increasing triglyceride synthesis [8]. In this respect it is noteworthy that feeding a high-fat diet to dogs primarily increased plasma fatty acid levels during the sleep phase [7]. This rise in nocturnal fatty acids has been directly linked to the development of insulin resistance. It would be interesting to test if the rise in plasma fatty acid levels during the sleep phase also leads to an enhanced accumulation of triglycerides in the myocardium, possibly predisposing to lipotoxicity. It is important to realize when studying the role of circadian rhythms on cardiac metabolism and function that the feeding pattern (*ad libitum* or time-restricted) of laboratory animals is clearly important. Notwithstanding these limitations it has been convincingly shown that in various cardiac disease conditions the rhythmic expression of genes across the day, also those involved in cardiac metabolism, becomes dampened [5], thereby limiting the ability of the heart to anticipate changes in nutritional environment.

At the same time, several elegant studies in which genes that are components of the circadian clock have been knocked out at the whole body level or in a tissue specific manner, demonstrated the importance of a functional clock in controlling overall energy homeostasis and, consequently, the risk of becoming obese or diabetic (reviewed in [9]). Recently, it was shown that cardiac-specific disruption of the circadian clock by knockout of Bmal1, leads to changes in cardiac metabolism that were manifested during fasting in particular [10,11]. In addition, disruption of the cardiac circadian clock was associated with the development of dilated cardiomyopathy, thereby revealing the importance of the intrinsic clock in maintaining proper function of the heart.

FASTING AND CIRCULATING FATTY ACID LEVELS

Like for the sleep/wake cycle, the heart is also responsive to changes in its nutritional environment associated with feeding/fasting cycles. It has even been argued that entrainment of peripheral clocks is the result of feeding and fasting primarily. Generally speaking, the early postprandial phase is characterized by high glucose (and insulin) levels, while during the fasting-phase plasma fatty acid (and glucagon) levels rise. Cardiac metabolism largely follows these changes being somewhat more dependent on glucose oxidation in the fed state, and more dependent on fatty acid oxidation in the fasted state. It should be noted that more prolonged fasting also leads to a rise in the levels of circulating ketone bodies, produced by the liver, which act as a preferred substrate for the heart [12]. The role of ketogenesis is often ignored, but may be significant in mice, possibly owing to their high metabolic rate.

We and others have shown that fasting activates a transcriptional program, involving certain nuclear hormone receptors, the so-called perosixome proliferator-activated receptors (PPARs), that act as lipid sensors [13,14]. In this way, the rise in plasma fatty acids following fasting leads to activation of PPARs resulting in an enhanced expression of genes involved in cardiac fatty acid uptake and conversion (Fig. 10.1). In the same way the expression of PDK4 is also induced in a PPAR-responsive manner [15], thereby limiting the ability of the heart to oxidize glucose. Accordingly, this response allows the heart to match its substrate preference to the prevailing nutritional environment as imposed by fasting. The central role of dietary lipids and the PPARs in initiating this cardiac transcriptional program has also been demonstrated using transcriptomic analysis by giving wildtype and PPARα knockout mice a single oral dose of synthetic triglycerides [16]. By challenging isolated neonatal rat cardiomyocytes with fatty acids or synthetic PPAR ligands, it was demonstrated that the transcriptional response is mediated by PPARs indeed [17,18]. Subsequent studies with synthetic PPAR ligands, among others, showed that in cardiomyocytes the fatty acids acted as ligands for PPARα as well as PPARβ [19].

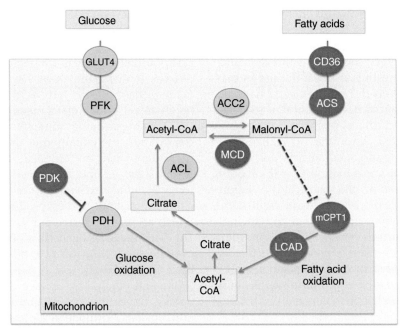

FIGURE 10.1   **Increased supply of fatty acids to the heart, commonly seen during fasting and exercise and in chronic diseases such as diabetes, leads to ligand-mediated activation of PPARs.** This results in enhanced transcription of various genes (marked red) involved in sarcolemmal fatty acid uptake (CD36), activation (ACS), mitochondrial import (mCPT1) and β-oxidation (LCAD, HADHA). At the same time PPAR increases expression of PDK4. Phosphorylation of the PDH complex by PDK inhibits PDH activity, thereby reducing glucose oxidation. The combined effect is a further increase in cardiac fatty acid oxidation at the expense of glucose oxidation.

It is well established that changes in nutritional environment directly influence cardiac function and affect the susceptibility of the heart for challenges like (transient) ischemia. With respect to the circadian rhythm it was already observed that hearts, extirpated from rats during the sleep phase and perfused *ex vivo*, generated less force than hearts isolated during the active phase. Subjecting healthy persons to a low-calorie diet for only three days, led to mobilization of fat stores, as evidenced by the increase in plasma fatty acids, and increased myocardial triglyceride content as assessed by ^{1}H-MRS [20]. Notably, this intervention did not affect cardiac systolic function, but diastolic function was significantly reduced in these persons. In other studies the effect of prior fasting on ischemia tolerance of the

heart has been investigated. However, the outcome is inconclusive, with some reporting reduced ischemia tolerance [21] and others reporting enhanced ischemia tolerance in response to prior fasting [22,23]. All studies employed *ex vivo* perfused isolated hearts to study the effect of prior fasting on ischemia–reperfusion damage and the outcome seems to depend largely on the substrate conditions chosen during isolated heart perfusion. Notwithstanding this discrepancy, the majority of studies clearly indicate that stimulation of glucose oxidation is beneficial for the heart during ischemia–reperfusion [24,25]. When extrapolating these findings to the *in vivo* situation, the findings suggest that the heart *in situ* is better protected against acute ischemic insults when glucose is high, that is, in the fed state. It is tempting to speculate, that the observation, that the frequency of ischemic attacks in man is higher in the early morning hours is somehow related to the nutritional environment prevailing at the end of the sleep phase.

EXERCISE

Regular physical exercise is known to reduce the risk for both cardiovascular disease and diabetes and to improve prognosis and quality of life of cardiac patients to a significant extent. As far as the heart is concerned many of the health benefits are secondary to improved skeletal muscle performance, decreased adiposity, and increased insulin sensitivity. At the same time the physical exercise also asks for adaptation of the heart itself, due to changes in circulating hormone levels, nutrient levels, and cardiac workload. As far as hormones are concerned it is important to realize that next to changes in well-known humoral factors, like the rise in adrenaline, more recent findings indicate that exercising skeletal muscle also releases factors acting as hormones, so-called myokines, that may influence cardiac metabolism and function in the short term and cardiac remodeling in the long term. Exercise dramatically increases the workload and, hence, the energy demand of the heart, which must be met by increasing mitochondrial oxidative metabolism. As exercise is associated with rapid mobilization of fatty acids from adipose tissue, resulting in an up to fourfold rise in circulating fatty acid levels with moderate exercise [26], it is likely that the heart *in situ* will respond to this by increasing fatty acid oxidation as well. The more acute changes in cardiac substrate metabolism during exercise involve multiple steps, some of which are related to the change in nutritional environment (higher circulating fatty acid levels), while others are driven by the increase in cardiac high-energy phosphate demand. From a molecular/biochemical perspective the effects are largely mediated by

the phosphorylation of enzymes and the allosteric modulation of enzyme activity. The drop in ATP/ADP ratio associated with increased work demand, and the consequent rise in AMP, leads to activation of AMP-activated kinase (AMPK), which stimulates both fatty acid uptake and glucose uptake via translocation of, respectively, the trans-sarcolemmal fatty acid transporter CD36 and GLUT4 to the membrane. Furthermore, AMPK activation stimulates fatty acid oxidation by phosphorylating, and thereby inhibiting, acetyl-CoA carboxylase (ACC). This results in a drop in malonyl-CoA levels and deinhibition of carnitine palmitoyl-transferase-1 (CPT1), the enzyme responsible for the entry of long-chain acyl-CoA into the mitochondria (Fig. 10.2). Intriguingly, studies with *ex vivo* perfused hearts of normal, untrained rats have shown that a sudden increase in workload as induced by administration of adrenaline is largely met by first increasing the oxidation from glucose derived from intracellular glycogen stores, followed by a more gradual increase in the oxidation of exogenous glucose. The oxidation of fatty acids remains fairly constant under these *ex vivo* conditions, whereby nutrient levels remain constant [27]. The increase in glucose metabolism as seen with sudden increases in workload appears to be AMPK independent and might involve the PI3K/Akt pathway [28]. It is still unclear if these *ex vivo* observations can be translated to the *in vivo* situation.

Repeated exercise results in remodeling of the heart (physiologic hypertrophy), which also extends to cardiac metabolism. Transcriptomic analysis and biochemical analysis showed that exercise training enhanced the capacity for both cardiac glucose and fatty acid oxidation in rats and dogs [29–31]. The mechanisms underlying this metabolic adaptation most likely also involve activation of AMPK, the kinase that is able to sense the energy status of the cell, and of the transcriptional coactivator PPARγ-coactivator-1 (PGC1). PGC1 not only activates PPARs, and thereby fatty acid oxidation, but also activates

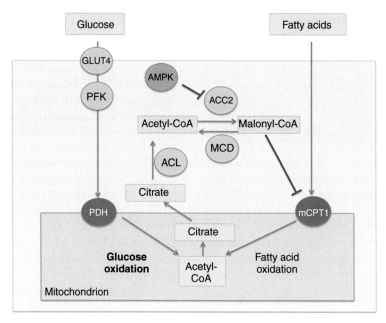

FIGURE 10.2   **The relative rates of cardiac glucose and fatty acid oxidation are primarily controlled at the level of the mitochondrial membrane by means of PDH and mCPT1, respectively.** Both pathways regulate each other's activity reciprocally. An enhanced glucose uptake and metabolism results in increased mitochondrial acetyl-CoA and citrate levels. In case of excess, mitochondrial citrate will be transported into the cytosol, eventually leading (via the combined activity of ACL and ACC) to an increase in malonyl-CoA levels. Malonyl-CoA is a potent allosteric inhibitor of mCPT1, in this way reducing the entry of fatty acids into the mitochondria and, thus fatty acid oxidation. In case of increased energy demand the AMP/ATP ratio will rise, leading to the activation of AMPK. ACC2 is one of the targets of AMPK and ACC2 phosphorylation inhibits its activity. As a consequence, malonyl-CoA levels will drop, allowing increased entry of fatty acids into the mitochondria for subsequent β-oxidation.

transcription factors involved in mitochondrial biogenesis (Fig. 10.3). More recent findings also suggest involvement of the Sirtuins, the activity of which is dependent on the $NAD^+/NADH$ redox status, and of hypoxia-inducible factor-1a (HIF-1α). The latter may not only activate transcription of genes involved in carbohydrate utilization, but also promote angiogenesis, via induction of vascular endothelial growth factor (VEGF), and thus oxygen supply to the myocardium. The factors mentioned earlier closely interact to increase oxidative capacity, on the one hand, and nutrient and oxygen supply, on the other (for recent reviews see [32,33]).

Exercise training has been consistently shown to protect the heart against subsequent ischemic insults [30,34]. Moreover, various studies demonstrated that exercise training after myocardial infarction has also been shown to be of therapeutic value [35]. Whether the same therapeutic effect is achieved in patients with nonischemic heart failure is still a matter of debate and is possibly dependent on heart failure etiology [36]. In patients with dilated cardiomyopathy it was shown that exercise training increased cardiac efficiency, defined as oxygen consumption per gram of tissue normalized to external work, possibly because of the use of more energy-efficient substrates such as glucose [37]. As far as the underlying mechanism is concerned it remains to be established if the exercise-mediated effects on the diseased heart can be attributed to improved endothelial function, improved cardiomyocyte calcium homeostasis and/or improved cardiac energy metabolism.

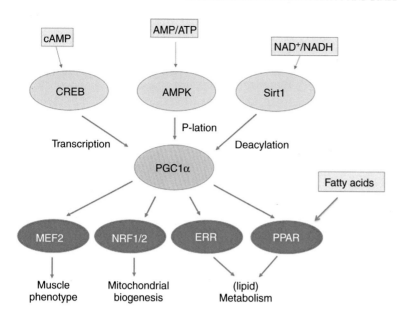

FIGURE 10.3 **PGC1 acts as a nodal point in the regulation of cardiac energy homeostasis.** The metabolic status of the cardiac myocyte (AMP/ATP and NAD$^+$/NADH ratios) and the levels of second messengers like cAMP are sensed by different proteins. Proteins like CREB, AMPK, and Sirt1 all convey changes in the metabolic status to PGC1α via different mechanisms. Activation of PGC1α has many metabolic effects, involving the stimulation of (primarily, but not exclusively) fatty acid metabolism by activating the estrogen-related receptor ERR and the PPARs, and by increasing the total mitochondrial oxidative capacity by stimulating mitochondrial biogenesis via the NRF transcription factors.

CARDIAC METABOLISM IN OBESITY AND DIABETES

Obesity and diabetes have a profound impact on cardiac metabolism. When trying to understand the changes in cardiac metabolism during obesity and diabetes it is important to discriminate between extrinsic and intrinsic factors. The extrinsic factors refer to the changes at the systemic level to which the heart is exposed during these metabolic diseases. This encompasses changes in nutritional environment, hormone levels, and adipokine levels. Strictly speaking this also includes changes in workload, as hypertension is a prevalent comorbid factor in patients with diabetes.

Conversely, the intrinsic changes refer to changes in cardiac metabolism that are associated with the structural and functional remodeling of the heart. This metabolic remodeling has been studied most extensively in relation to cardiac hypertrophy and failure whereby the relative contribution of glucose utilization increases at the expense of fatty acids [38–40]. It is commonly acknowledged that the shift to-

ward the use of glucose reflects the substrate preference of the fetal/neonatal heart and, as such, is part of the reactivation of the fetal gene program, a hallmark of cardiac hypertrophy. Conceivably, the failing heart benefits from this shift in substrate preference, because the oxidation of glucose is more energy efficient than that of fatty acids in terms of ATP generated per oxygen atom used (ATP/O ratio). Notably, in obesity and diabetes the opposite happens, that is, the heart becomes even more dependent on fatty acids in response to changes in both external and internal factors. By inference this would imply that the heart becomes less energy efficient. In clinical studies employing positron-emission tomography (PET) to measure cardiac glucose and fatty acid metabolism and oxygen consumption, it was shown first that obesity was associated with increased cardiac fatty acid oxidation, oxygen consumption, and blood flow. Body mass index (BMI) was found to correlate positively with cardiac oxygen consumption and correlated inversely with cardiac efficiency (oxygen consumption per unit of cardiac work) [41]. As plasma glucose, fatty acid, and triglyceride levels were comparable between obese and nonobese persons, differences in nutrient supply cannot account for the increased oxidation of fatty acids. However, the obese persons had become insulin resistant already (insulin levels almost twice as high). Subsequent studies of the same group showed that cardiac reliance on fatty acids was further elevated in obese women with established diabetes [42]. In this group, plasma glucose and fatty acid levels were elevated, and insulin levels were higher than in obese nondiabetic women. It is noteworthy that blood pressure was also significantly higher in the diabetes group, explaining part of the increase in myocardial oxygen consumption in this group. Finally, in other studies it was shown that moderate weight loss (diet) and, in particular, severe weight loss (gastric bypass surgery) partly reversed these metabolic changes as reflected by a reduction in cardiac fatty acid uptake and oxidation and concurrent decline in cardiac oxygen consumption [43,44].

When studying cardiac metabolism under *in vivo* conditions it is often difficult to discriminate between the role of extrinsic and intrinsic factors. To some extent the intrinsic changes can be studied using isolated, *ex vivo* perfused heart preparations, where substrate levels, hormone levels (e.g., insulin), and workload can be kept constant for hearts from control and diseased animals. Preclinical studies with isolated hearts have indeed shown that cardiac oxygen consumption is increased and cardiac efficiency is reduced when nondiabetic hearts are exposed to increased fatty acid levels and in hearts form diabetic mice [45,46].

Nutrient supply: In obese persons, plasma substrate levels (glucose, fatty acids, triglycerides), as routinely measured in the fasting state, do not differ substantially, if any, from those in nonobese persons. However, as discussed earlier, during the day there may be differences in substrate levels associated with sleep/wake and feeding/fasting cycles, such as the more pro-

nounced nocturnal surge in plasma fatty acid levels in the obese state, that may impact cardiac metabolism. The more or less chronic changes in nutritional environment (e.g., hyperglycemia, hyperlipidemia) associated with diabetes either force or enable the heart to alter its substrate preference. Under acute, severely hyperglycemic conditions the contribution of glucose may account for up to 70% of total energy production of the normal, nondiabetic heart [47]. However, in insulin resistant and diabetic conditions the cardiac uptake of glucose becomes less, despite elevated circulating glucose levels (see later in the chapter), while fatty acid uptake increases.

At this point a word of caution is in place as in the majority of rodent studies the animals are fed *ad libitum*. Recent studies documented the implications of this common practice, especially in relation to studies concerning aspects of metabolic disease. In a simple but elegant study Hatori et al. [48] demonstrated that mice were protected against the development of obesity and hyperinsulinemia when access to the high-fat diet was restricted to the active phase of the day, despite the fact that total caloric intake was similar as in mice fed *ad libitum*. This important finding allows us to rethink the design of studies aimed at evaluating the effect of dietary interventions on disease outcome.

Hormone levels: Disease-related changes in hormone levels, reflected for instance by the high-circulating insulin levels in the insulin resistant, prediabetic state and the low insulin levels after prolonged type II diabetes, likely affect cardiac metabolism directly. Under normal conditions the direct action of insulin on the heart is to enhance trans-sarcolemmal glucose uptake by promoting the translocation of the insulin-dependent glucose transporter GLUT4 from an inactive, cytosolic pool to the sarcolemma. This process is affected in case of insulin resistance. It is important to note however, that compared to skeletal muscle, the stimulatory effect of insulin on glucose uptake in the isolated working heart is modest, amounting to 50% approximately [49].

It is well feasible that this is related to the fact that, unlike skeletal muscle, the heart is continuously active and that cellular activity by itself already leads to recruitment of GLUT4 to the sarcolemma. Furthermore, next to diminished translocation, in diabetes protein expression of GLUT4 also becomes downregulated, although this may be compensated to some extent by an increased expression of the insulin-independent glucose transporter GLUT1.

Adipokines: More recently the role of adipokines, that is hormones and cytokines secreted from adipose tissue, is increasingly gaining attention. The type and amount of adipokines that are being secreted depend on adipose tissue mass. It is generally believed that obesity leads to a dysbalance in adipokine secretion, with a shift toward secretion of more proinflammatory cytokines, among which TNFα [50]. The resulting chronic systemic low-grade inflammation is likely to influence the metabolism of various organs, including the heart. Serum levels of TNFα correlate with the degree of obesity, increasing with weight gain and declining following weight loss [51]. TNFα has been linked directly to the development of insulin resistance as it interferes with signal transduction directly downstream of the insulin receptor [52]. From the perspective of cardiac intermediate metabolism proinflammatory cytokines like TNFα are known to activate NFκB, the p65 subunit of which has been shown to interact with both the transcriptional coactivator PGC1 and the transcription factor PPAR [53,54], thereby inhibiting transcription of genes involved in fatty acid oxidation and mitochondrial function. Remarkably, the satiety hormone leptin, the levels of which are elevated in obese persons, was found to directly stimulate cardiac fatty acid oxidation in isolated perfused hearts, through an as yet unknown mechanism [55]. Accordingly, the various adipokines may influence cardiac metabolism in distinct ways, making it hard to predict what the net effect of the "obesity adipokine signature" on cardiac metabolism will be.

In diabetes there appears to be an increased risk for myocardial lipid accumulation, and hence lipotoxicity, as the increase in myocardial fatty acid uptake is higher than their oxidation under these conditions. It is well conceivable that mitochondrial dysfunction, as observed in morbid obese, insulin-resistant mice, may be responsible for the mismatch between the uptake and oxidation of fatty acids under these conditions [56]. The decrease in cardiac efficiency in diabetes, may therefore be related to the increase in fatty acid oxidation (requiring more oxygen per ATP produced than glucose oxidation) and partial uncoupling of the dysfunctional mitochondria. Respiratory uncoupling in mitochondria in obesity and diabetes may be caused by intrinsic changes in the respiratory chain complexes, but may also be caused by increased expression and activity of uncoupling proteins UCP2 and UCP3, dissipating the electrochemical proton gradient across the inner mitochondrial membrane, thereby generating heat instead of ATP [57]. Finally, there is evidence of increased futile cycling under these conditions, related to the increased expression and activity of fatty acyl-CoA thioesterases that liberate fatty acids once transported into the mitochondria [58]. A subject less studied is whether the increased accumulation of triglycerides in the diabetic myocardium also stimulates futile cycling of fatty acids through the enlarged triglyceride pool. Futile cycling results in ATP loss, as the activation of one long-chain fatty acid by acyl-CoA synthase goes at the expense of two high-energy phosphate bonds (ATP being converted into AMP directly).

Metabolic remodelling: Importantly, the expression of many of the enzymes involved in fatty acid oxidation, and also that of the uncoupling proteins and thioesterases are under transcriptional control of the PPARs. It is commonly acknowledged that the increased availability of fatty acids leads to the activation of the PPARs under these conditions, particularly in the early stages of diabetes [59]. By stimulating the transcription of genes involved in fatty acid uptake (CD36), metabolism (ACS, CPT1), and oxidation (LCAD, HADHA), the diabetic heart adapts to the "lipid-rich environment" and becomes even more dependent on fatty acids as the source of energy. Importantly, the kinase PDK4, responsible for the inactivation of the PDH complex and thereby for the inhibition of glucose oxidation, is also under the transcriptional control of PPAR. Accordingly, the gene program activated by the PPARs can be considered the transcriptional equivalent of the Randle cycle. Furthermore, PPAR also promotes transcription of its own coactivator protein PGC1α, thereby enforcing its own function and, at the same time activating the mitochondrial biogenesis program via the PCG1/NRF axis [60]. Notwithstanding, this concerted action at the level of mitochondrial metabolism, the occurrence of lipotoxicity in the diabetic heart indicates that the stimulation of fatty acid uptake exceeds fatty acid oxidation rate. It has also been argued that with prolonged, untreated diabetes the activity of the PGC1/PPAR axis is eventually reduced. It remains to be established whether this reduction in PPAR activity, and the accompanying metabolic changes, are causally linked to the development of diabetic cardiomyopathy, or whether this is merely a consequence of the cardiac remodeling associated with diabetic cardiomyopathy.

An alternative (but not mutually exclusive) theory is that the mitochondria become dysfunctional due to increased production of reactive oxygen species (ROS) under obese and diabetic conditions. Increased cardiac ROS production can originate from cytosolic sources, like NADPH oxidase or xanthine oxidase, or from the mitochondria themselves, secondary to nutrient oversupply. The increased activity of NADPH oxidase is being attributed to the activation of the renin–angiotensin–aldosterone system under these conditions [61]. Excess mitochondrial ROS production, along with proton leak and reverse electron flow, might be the result of a higher

mitochondrial membrane potential ($\Delta\Psi$) that is commonly associated with increased fatty acid oxidation. Excessive ROS generation has been implicated in both the enhancement of insulin resistance by interfering with insulin receptor substrate-1 (IRS-1) signaling [62] and in the aggravation of mitochondrial function itself. Recent observations also highlight the importance of proper insulin signaling in mitochondrial dynamics (fission and fusion) and mitophagy [63]. The end result of all this is a self-perpetuating negative spiral culminating in an ever-increasing mitochondrial damage and reduced ATP generating capacity in the diabetic heart.

THE WESTERN DIET AND DIETARY INTERVENTIONS

The relation between the western diet, obesity, and the risk for cardiovascular disease (CVD) in man is well established (for reviews see [64,65]. First, this has led to dietary guidelines aimed at a reduction of total caloric intake or fat intake. More recently it has been recognized that a high intake of carbohydrates, especially those with a high-glycemic index, is at least as bad as high fat intake [66,67]. Furthermore, the detrimental effect of high fat intake is highly dependent on the source of fat (meat, plant, or fish), as revealed by epidemiologic studies showing beneficial effects of Mediterranean-type of diets (rich in olive oil) and diets rich in fish oils [64] on CVD morbidity and mortality. For instance, recently it was reported that a diet enriched for Mediterranean-type of fatty acids (olive oil, nuts) reduced the incidence of CVD when compared to a low-fat diet [68]. Another intriguing observation, referred to as the "obesity paradox," is that obese individuals have a higher risk of developing CVD, on the one hand, but that obese individuals with established CVD have a better prognosis than nonobese individuals, on the other [69]. These findings have instigated a renewed interest in the effect of dietary components, fatty acids in particular, on cardiovascular disease. Initially this research was merely confined to the relation between diet and the risk of developing atherosclerosis, but more recently the effect of diet on cardiac function and phenotype is receiving much attention.

It should be noted that there is still no consensus regarding the effect of a high-fat diet, rich in saturated fatty acids, on the heart. On the one hand, it is still commonly believed that obesity predisposes to cardiac dysfunction and remodeling, independent of coronary atherosclerosis and hypertension, a condition referred to as diabetic cardiomyopathy. Many studies demonstrated indeed that feeding otherwise healthy rats and mice a western-type diet (>45% of calories from fat) for a couple of months sufficed to induce cardiac hypertrophy and cardiac dysfunction [70–73]. On the other hand, there are also various studies that do not support this contention and show that high-fat feeding has limited effects on the rodent heart and may even induce an increase in cardiac function, despite inducing obesity and insulin resistance [74,75]. Based on the findings on feeding-related cardiac growth in Burmese pythons [76,77] Riquelme et al. explored the growth-promoting effect of certain fatty acid species *in vitro* in isolated cardiomyocytes and *in vivo* in mice and found that a mixture of (mono)unsaturated fatty acids (C14:0, C16:0, and C16:1) acted as a humoral trigger for a cardiac hypertrophic response that was physiologic, rather than pathologic, in nature. Even more mind boggling are the observations that in the setting of pressure-overload-induced hypertrophy, feeding rats a high-fat diet was able to reduce the increase in cardiac mass and to attenuate cardiac dysfunction (for review see [78]). At first sight the latter observations show some analogy to the "obesity paradox" observed in epidemiologic studies, which also point to the fact that a high fat-diet is detrimental for the nondiseased heart, but may afford protection to the diseased heart. The mechanism underlying this putative protective effect of a high-fat diet remains to be elucidated still.

Another important issue is whether the fatty acid composition of the diet is crucial for the car-diac effects. Generally speaking, diets enriched in poly-unsaturated fatty acids, particularly with the fish-oil derived ω-3 fatty acids eicosa-pentaenoic acid (C20:5) and docosahexaenoic acid (C22:6), are considered to improve cardiovascular outcome in various ways. In mice subjected to transverse aortic constriction, a diet enriched in fish-oils was found to reduce cardiac hypertrophy and fibrosis, and to improve cardiac function [79]. In a transgenic rat model of hypertensive heart disease, dietary ω-3 fatty acids did not reduce hypertrophy, but were shown to have antifibrotic, anti-inflammatory, and antiarrhythmic effects [80]. In a rabbit model of pressure–volume overload-induced heart failure, supplementation of the diet with relatively small amounts (only 1.25% w/w) of sunflower oil (enriched for monounsaturated ω-9 fatty acids) or fish oil was shown to attenuate the development of cardiac hypertrophy and failure in both cases [81]. On top of that, fish-oil supplementation, but not sunflower oil supplementation, was found to have additional antiarrhythmic effects.

Although not the topic of this chapter, it is important to realize that, next to being energy-rich substrates, certain fatty acid species also serve important functions in cellular signal transduction (for recent review see [82]), by acting as ligands for a variety of membrane-bound and nuclear receptors and by modulating the activity of proteins involved in various signaling pathways. In this way, changes in diet (total amount and composition of the lipid fraction of the diet) and alterations in cardiac fatty acid uptake and metabolism, are likely to lead to the activation of specific signaling pathways, in addition to direct consequences for cardiac energy metabolism.

Recently, there has also been lots of interest in long-term caloric restriction as a means to increase lifespan and to reduce cardiovascular disease. Caloric restriction has been shown to reduce atherosclerosis [83] and to be associated with a better diastolic function in nondiabetic aged humans and in diabetic rats [84,85]. However, at this stage it is difficult to discern whether the effects of a low calorie diet are due solely to its favorable effects on many of the risk factors for cardiac disease (obesity, serum lipid profile, hypertension) or whether there are also more direct effects on the heart involved.

To date the literature regarding the effects of nutritional factors on the heart has often led to controversial findings. Future studies should be aimed at acquiring a better understanding of all the regulatory factors involved and their mutual interactions. It is only then that it will become feasible to match cardiac substrate preference and cardiac energy needs with systemic substrate supply in order to try to reduce the prevalence of cardiac disease in obesity and diabetes.

References

[1] Van der Vusse GJ, Glatz JFC, Stam HCG, Reneman RS. Fatty acid homeostasis in the normoxic and ischemic heart. Physiol Rev 1992;72:881–940.

[2] Lopaschuk GD, Ussher JR, Folmes CDL, Jaswal JS, Stanley WC. Myocardial fatty acid metabolism in health and disease. Physiol Rev 2010;90(1):207–58.

[3] Martino T, Arab S, Straume M, Belsham DD, Tata N, Cai F, et al. Day/night rhythms in gene expression of the normal murine heart. J Mol Med 2004;82(4):256–64.

[4] Durgan DJ, Hotze MA, Tomlin TM, Egbejimi O, Graveleau C, Abel ED, et al. The intrinsic circadian clock within the cardiomyocyte. Am J Physiol Heart Circ Physiol 2005;289(4):H1530–41.

[5] Young ME, Razeghi P, Cedars AM, Guthrie PH, Taegtmeyer H. Intrinsic diurnal variations in cardiac metabolism and contractile function. Circ Res 2001;89(12):1199–208.

[6] Stavinoha MA, RaySpellicy JW, Hart-Sailors ML, Mersmann HJ, Bray MS, Young ME. Diurnal variations in the responsiveness of cardiac and skeletal muscle to fatty acids. Am J Physiol Endocrinol Metab 2004;287(5):E878–87.

[7] Kim SP, Catalano KJ, Hsu IR, Chiu JD, Richey JM, Bergman RN. Nocturnal free fatty acids are uniquely elevated in the longitudinal development of diet-induced insulin resistance and hyperinsulinemia. Am J Physiol Endocrinol Metab 2007;292(6):E1590–8.

[8] Tsai J-Y, Kienesberger PC, Pulinilkunnil T, Sailors MH, Durgan DJ, Villegas-Montoya C, et al. Direct regulation of myocardial triglyceride metabolism by the cardiomyocyte circadian clock. J Biol Chem 2010;285(5):2918–29.

[9] Gooley JJ, Chua EC-P. Diurnal regulation of lipid metabolism and applications of circadian lipidomics. J Genet Genomics 2014;41(5):231–50.

[10] Young ME, Brewer RA, Peliciari-Garcia RA, Collins HE, He L, Birky TL, et al. Cardiomyocyte-specific BMAL1 plays critical roles in metabolism, signaling, and maintenance of contractile function of the heart. J Biol Rhythms 2014;29(4):257–76.

[11] Kohsaka A, Das P, Hashimoto I, Nakao T, Deguchi Y, Gouraud SS, et al. The circadian clock maintains cardiac function by regulating mitochondrial metabolism in mice. PLoS ONE 2014;9(11):e112811.

[12] Russell RR, Cline GW, Guthrie PH, Goodwin GW, Shulman GI, Taegtmeyer H. Regulation of exogenous and endogenous glucose metabolism by insulin and acetoacetate in the isolated working rat heart. A three tracer study of glycolysis, glycogen metabolism, and glucose oxidation. J Clin Invest 1997;100(11):2892–9.

[13] Leone TC, Weinheimer CJ, Kelly DP. A critical role for the peroxisome proliferator-activated receptor α (PPARα) in the cellular fasting response: the PPARα-null mouse as a model of fatty acid oxidation disorders. Proc Natl Acad Sci USA 1999;96:7473–8.

[14] Gilde AJ, Van Bilsen M. Peroxisome proliferator-activated receptors (PPARs): regulators of gene expression in heart and skeletal muscle. Acta Physiol 2003;178(4):425–34.

[15] Wu P, Peters JM, Harris RA. Adaptive increase in pyruvate dehydrogenase kinase 4 during starvation is mediated by peroxisome proliferator-activated receptor alpha. Biochem Biophys Res Commun 2001;287(2):391–6.

[16] Georgiadi A, Boekschoten MV, Müller M, Kersten S. Detailed transcriptomics analysis of the effect of dietary fatty acids on gene expression in the heart. Physiol Genomics 2012;44(6):352–61.

[17] Van der Lee KAJM, Vork MM, De Vries JE, Willemsen PHM, Glatz JFC, Reneman RS, et al. Long-chain fatty acid-induced changes in gene expression in neonatal cardiac myocytes. J Lipid Res 2000;41(1):41–7.

[18] Lockridge J, Sailors M, Durgan D, Egbejimi O, Jeong W, Bray M, et al. Bioinformatic profiling of the transcriptional response of adult rat cardiomyocytes to distinct fatty acids. J Lipid Res 2008;49(7):1395–408.

[19] Gilde A, van der Lee K, Willemsen P, Chinetti G, van der Leij F, van der Vusse G, et al. Peroxisome proliferator-activated receptor (PPAR) alpha and PPAR-beta/delta, but not PPARgamma, modulate the expression of genes involved in cardiac lipid metabolism. Circ Res 2003;92(5):518–24.

[20] Van der Meer RW, Hammer S, Smit JWA, Frölich M, Bax JJ, Diamant M, et al. Short-term caloric restriction induces accumulation of myocardial triglycerides and decreases left ventricular diastolic function in healthy subjects. Diabetes 2007;56(12):2849–53.

[21] Liepinsh E, Makrecka M, Kuka J, Makarova E, Vilskersts R, Cirule H, et al. The heart is better protected against myocardial infarction in the fed state compared to the fasted state. Metabolism 2014;63(1):127–36.

[22] Schneider CA, Taegtmeyer H. Fasting *in vivo* delays myocardial cell damage after brief periods of ischemia in the isolated working rat heart. Circ Res 1991;68(4):1045–50.

[23] Montessuit C, Papageorgiou I, Tardy I, Lerch R. Effect of nutritional state on substrate metabolism and contractile function in postischemic rat myocardium. Am J Physiol 1996;271(5 Pt 2):H2060–70.

[24] Liu B, Clanachan AS, Schulz R, Lopaschuk GD. Cardiac efficiency is improved after ischemia by altering both the source and fate of protons. Circ Res 1996;79(5):940–8.

[25] Ussher JR, Wang W, Gandhi M, Keung W, Samokhvalov V, Oka T, et al. Stimulation of glucose oxidation protects against acute myocardial infarction and reperfusion injury. Cardiovasc Res 2012;94(2):359–69.

[26] Jensen MD. Fate of fatty acids at rest and during exercise: regulatory mechanisms. Acta Physiol Scand 2003;178(4):385–90.

[27] Goodwin GW, Taylor CS, Taegtmeyer H. Regulation of energy metabolism of the heart during acute increase in heart work. J Biol Chem 1998;273(45):29530–9.

[28] Beauloye C, Marsin A-S, Bertrand L, Vanoverschelde J-L, Rider MH, Hue L. The stimulation of heart glycolysis by increased workload does not require AMP-activated protein kinase but a wortmannin-sensitive mechanism. FEBS Lett 2002;531(2):324–8.

[29] Stuewe SR, Gwirtz PA, Agarwal N, Mallet RT. Exercise training enhances glycolytic and oxidative enzymes in canine ventricular myocardium. J Mol Cell Cardiol 2000;32(6):903–13.

[30] Burelle Y, Wambolt RB, Grist M, Parsons HL, Chow JCF, Antler C, et al. Regular exercise is associated with a protective metabolic phenotype in the rat heart. Am J Physiol Heart Circ Physiol 2004;287(3):H1055–63.

[31] Strøm CC, Aplin M, Ploug T, Christoffersen TEH, Langfort J, Viese M, et al. Expression profiling reveals differences in metabolic gene expression between exercise-induced cardiac effects and maladaptive cardiac hypertrophy. FEBS J 2005;272(11):2684–95.

[32] Mann N, Rosenzweig A. Can exercise teach us how to treat heart disease? Circulation 2012;126(22):2625–35.

[33] Rowe GC, Safdar A, Arany Z. Running forward: new frontiers in endurance exercise biology. Circulation 2014;129(7):798–810.

[34] Freimann S, Scheinowitz M, Yekutieli D, Feinberg MS, Eldar M, Kessler-Icekson G. Prior exercise training improves the outcome of acute myocardial infarction in the rat: heart structure, function, and gene expression. J Am Coll Cardiol 2005;45(6):931–8.

[35] De Waard MC, van Haperen R, Soullié T, Tempel D, de Crom R, Duncker DJ. Beneficial effects of exercise training after myocardial infarction require full eNOS expression. J Mol Cell Cardiol 2010;48(6):1041–9.

[36] Duncker D, van Deel E, de Waard M, de Boer M, Merkus D, van der Velden J. Exercise training in adverse cardiac remodeling. Pflugers Arch Eur J Physiol 2014;466(6):1079–91.

[37] Stolen KQ, Kemppainen J, Ukkonen H, Kalliokoski KK, Luotolahti M, Lehikoinen P, et al. Exercise training improves biventricular oxidative metabolism and left ventricular efficiency in patients with dilated cardiomyopathy. J Am Coll Cardiol 2003;41(3):460–7.

[38] Van Bilsen M, van Nieuwenhoven F, van der Vusse G. Metabolic remodelling of the failing heart: beneficial or detrimental? Cardiovasc Res 2009;81(3):420–8.

[39] Neubauer S. The failing heart – an engine out of fuel. New Engl J Med 2007;356(11):1140–51.

[40] Doenst T, Nguyen T, Abel E. Cardiac metabolism in heart failure: implications beyond ATP production. Circ Res 2013;113(6):709–24.

[41] Peterson LR, Herrero P, Schechtman KB, Racette SB, Waggoner AD, Kisrieva-Ware Z, et al. Effect of obesity and insulin resistance on myocardial substrate metabolism and efficiency in young women. Circulation 2004;109(18):2191–6.

[42] McGill J, Peterson L, Herrero P, Saeed I, Recklein C, Coggan A, et al. Potentiation of abnormalities in myocardial metabolism with the development of diabetes in women with obesity and insulin resistance. J Nucl Cardiol 2011;18(3):421–9.

[43] Viljanen AP, Karmi A, Borra R, Parkka JP, Lepomaki V, Parkkola R, et al. Effect of caloric restriction on myocardial fatty acid uptake, left ventricular mass, and cardiac work in obese adults. Am J Cardiol 2009;103(12):1721–6.

[44] Lin CH, Kurup S, Herrero P, Schechtman KB, Eagon JC, Klein S, et al. Myocardial oxygen consumption change predicts left ventricular relaxation improvement in obese humans after weight loss. Obesity 2011;19(9):1804–12.

[45] How O-J, Aasum E, Kunnathu S, Severson DL, Myhre ESP, Larsen TS. Influence of substrate supply on cardiac efficiency, as measured by pressure-volume analysis in *ex vivo* mouse hearts. Am J Physiol Heart Circ Physiol 2005;288(6):H2979–85.

[46] How O-J, Aasum E, Severson DL, Chan WYA, Essop MF, Larsen TS. Increased myocardial oxygen consumption reduces cardiac efficiency in diabetic mice. Diabetes 2006;55(2):466–73.

[47] Bertrand L, Horman S, Beauloye C, Vanoverschelde J-L. Insulin signalling in the heart. Cardiovasc Res 2008;79(2):238–48.

[48] Hatori M, Vollmers C, Zarrinpar A, DiTacchio L, Bushong E, Gill S, et al. Time-restricted feeding without

reducing caloric intake prevents metabolic diseases in mice fed a high-fat diet. Cell Metab 2012;15(6):848–60.

[49] Belke DD, Betuing S, Tuttle MJ, Graveleau C, Young ME, Pham M, et al. Insulin signaling coordinately regulates cardiac size, metabolism, and contractile protein isoform expression. J Clin Invest 2002;109(5):629–39.

[50] Nakamura K, Fuster JJ, Walsh K. Adipokines: a link between obesity and cardiovascular disease. J Cardiol 2014;63(4):250–9.

[51] Ziccardi P, Nappo F, Giugliano G, Esposito K, Marfella R, Cioffi M, D'Andrea F, Molinari AM, Giugliano D. Reduction of inflammatory cytokine concentrations and improvement of endothelial functions in obese women after weight loss over one year. Circulation 2002;105(7):804–9.

[52] Hotamisligil GS, Shargill NS, Spiegelman BM. Adipose expression of tumor necrosis factor-alpha: direct role in obesity-linked insulin resistance. Science 1993;259(5091):87–91.

[53] Planavila A, Rodríguez-Calvo R, Jové M, Michalik L, Wahli W, Laguna J, et al. Peroxisome proliferator-activated receptor beta/delta activation inhibits hypertrophy in neonatal rat cardiomyocytes. Cardiovasc Res 2005;65(4):832–41.

[54] Smeets P, Teunissen B, Planavila A, de Vogel-van den Bosch H, Willemsen P, van der Vusse G, et al. Inflammatory pathways are activated during cardiomyocyte hypertrophy and attenuated by peroxisome proliferator-activated receptors PPARα and PPARβ. J Biol Chem 2008;283(43):29109–18.

[55] Atkinson LL, Fischer MA, Lopaschuk GD. Leptin activates cardiac fatty acid oxidation independent of changes in the AMP-activated protein kinase-acetyl-CoA carboxylase-malonyl-CoA axis. J Biol Chem 2002;277(33):29424–30.

[56] Boudina S, Sena S, O'Neill BT, Tathireddy P, Young ME, Abel ED. Reduced mitochondrial oxidative capacity and increased mitochondrial uncoupling impair myocardial energetics in obesity. Circulation 2005;112(17):2686–95.

[57] Murray AJ, Panagia M, Hauton D, Gibbons GF, Clarke K. Plasma free fatty acids and peroxisome proliferator–activated receptorα in the control of myocardial uncoupling protein levels. Diabetes 2005;54(12):3496–502.

[58] Wilson CR, Tran MK, Salazar KL, Young ME, Taegtmeyer H. Western diet, but not high fat diet, causes derangements of fatty acid metabolism and contractile dysfunction in the heart of Wistar rats. Biochem J 2007;406(3):457–67.

[59] Finck BN, Lehman JJ, Leone TC, Welch MJ, Bennett MJ, Kovacs A, et al. The cardiac phenotype induced by PPARalpha overexpression mimics that caused by diabetes mellitus. J Clin Invest 2002;109(1):121–30.

[60] Mitra R, Nogee DP, Zechner JF, Yea K, Gierasch CM, Kovacs A, et al. The transcriptional coactivators, PGC-1α and β, cooperate to maintain cardiac mitochondrial function during the early stages of insulin resistance. J Mol Cell Cardiol 2012;52(3):701–10.

[61] Cooper SA, Whaley-Connell A, Habibi J, Wei Y, Lastra G, Manrique C, et al. Renin-angiotensin-aldosterone system and oxidative stress in cardiovascular insulin resistance. Am J Physiol Heart Circ Physiol 2007;293(4):H2009–23.

[62] Kim J-A, Wei Y, Sowers JR. Role of mitochondrial dysfunction in insulin resistance. Circ Res 2008;102(4):401–14.

[63] Westermeier F, Navarro-Marquez M, Lopez-Crisosto C, Bravo-Sagua R, Quiroga C, Bustamante M, et al. Defective insulin signaling and mitochondrial dynamics in diabetic cardiomyopathy. Biochim Biophys Acta 2015;1853(5):1113–8.

[64] Erkkilä A, de Mello V, Risérus U, Laaksonen D. Dietary fatty acids and cardiovascular disease: an epidemiological approach. Prog Lipid Res 2008;47(3):172–87.

[65] Baum S, Kris-Etherton P, Willett W, Lichtenstein A, Rudel L, Maki K, et al. Fatty acids in cardiovascular health and disease: a comprehensive update. J Clin Lipidol 2012;6(3):216–34.

[66] Jakobsen MU, Dethlefsen C, Joensen AM, Stegger J, Tjonneland A, Schmidt EB, et al. Intake of carbohydrates compared with intake of saturated fatty acids and risk of myocardial infarction: importance of the glycemic index. Am J Clin Nutr 2010;91:1764–8.

[67] Chahoud G, Aude Y, Mehta J. Dietary recommendations in the prevention and treatment of coronary heart disease: do we have the ideal diet yet? Am J Cardiol 2004;94(10):1260–7.

[68] Estruch R, Ros E, Salas-Salvadó J, Covas M-I, Corella D, Arós F, et al. Primary prevention of cardiovascular disease with a Mediterranean diet. New Engl J Med 2013;368(14):1279–90.

[69] Fonarow G, Srikanthan P, Costanzo M, Cintron G, Lopatin M, Committee ASA, et al. An obesity paradox in acute heart failure: analysis of body mass index and inhospital mortality for 108,927 patients in the Acute Decompensated Heart Failure National Registry. Am Heart J 2007;153(1):74–81.

[70] Ouwens D, Boer C, Fodor M, de Galan P, Heine R, Maassen J, et al. Cardiac dysfunction induced by high-fat diet is associated with altered myocardial insulin signalling in rats. Diabetologia 2005;48(6):1229–37.

[71] Ceylan-Isik AF, Kandadi MR, Xu X, Hua Y, Chicco AJ, Ren J, Nair S. Apelin administration ameliorates high fat diet-induced cardiac hypertrophy and contractile dysfunction. J Mol Cell Cardiol 2013;63:4–13.

[72] Fang C, Dong F, Thomas D, Ma H, He L, Ren J. Hypertrophic cardiomyopathy in high-fat diet-induced

obesity: role of suppression of forkhead transcription factor and atrophy gene transcription. Am J Physiol Heart Circ Physiol 2008;295(3):H1206–15.

[73] Ge F, Hu C, Hyodo E, Arai K, Zhou S, Lobdell H IV, et al. Cardiomyocyte triglyceride accumulation and reduced ventricular function in mice with obesity reflect increased long chain fatty acid uptake and *de novo* fatty acid synthesis. J Obes 2012;205648.

[74] Brainard RE, Watson LJ, DeMartino AM, Brittian KR, Readnower RD, Boakye AA, et al. High fat feeding in mice is insufficient to induce cardiac dysfunction and does not exacerbate heart failure. PLoS ONE 2013;8(12):e83174.

[75] Harmancey R, Lam TN, Lubrano GM, Guthrie PH, Vela D, Taegtmeyer H. Insulin resistance improves metabolic and contractile efficiency in stressed rat heart. FASEB J 2012;26(8):3118–26.

[76] Andersen J, Rourke B, Caiozzo V, Bennett A, Hicks J. Physiology: postprandial cardiac hypertrophy in pythons. Nature 2005;434(7029):37–8.

[77] Riquelme C, Magida J, Harrison B, Wall C, Marr T, Secor S, et al. Fatty acids identified in the Burmese python promote beneficial cardiac growth. Science 2011;334(6055):528–31.

[78] Stanley W, Dabkowski E, Ribeiro R, O'Connell K. Dietary fat and heart failure: moving from lipotoxicity to lipoprotection. Circ Res 2012;110(5):764–76.

[79] Chen J, Shearer G, Chen Q, Healy C, Beyer A, Nareddy V, et al. Omega-3 fatty acids prevent pressure overload-induced cardiac fibrosis through activation of cyclic GMP/protein kinase G signaling in cardiac fibroblasts. Circulation 2011;123(6):584–93.

[80] Fischer R, Dechend R, Qadri F, Markovic M, Feldt S, Herse F, et al. Dietary n-3 polyunsaturated fatty acids and direct renin inhibition improve electrical remodeling in a model of high human renin hypertension. Hypertension 2008;51(2):540–6.

[81] Den Ruijter H, Verkerk A, Schumacher C, Houten S, Belterman C, Baartscheer A, et al. A diet rich in unsaturated fatty acids prevents progression toward heart failure in a rabbit model of pressure and volume overload. Circ Heart Fail 2012;5(3):376–84.

[82] Van Bilsen M, Planavila A. Fatty acids and cardiac disease: fuel carrying a message. Acta Physiol 2014;211(3):476–90.

[83] Fontana L, Meyer TE, Klein S, Holloszy JO. Long-term calorie restriction is highly effective in reducing the risk for atherosclerosis in humans. Proc Natl Acad Sci USA 2004;101(17):6659–63.

[84] Meyer TE, Kovacs SJ, Ehsani AA, Klein S, Holloszy JO, Fontana L. Long-term caloric restriction ameliorates the decline in diastolic function in humans. J Am Coll Cardiol 2006;47(2):398–402.

[85] Takatsu M, Nakashima C, Takahashi K, Murase T, Hattori T, Ito H, et al. Calorie restriction attenuates cardiac remodeling and diastolic dysfunction in a rat model of metabolic syndrome. Hypertension 2013;62(5): 957–65.

Influence of Ischemia-Reperfusion Injury on Cardiac Metabolism

David I. Brown, Monte S. Willis*,†,
Jessica M. Berthiaume***

*McAllister Heart Institute, University of North Carolina at Chapel Hill, Chapel
Hill, NC, USA; †Department of Pathology & Laboratory Medicine, University
of North Carolina Medicine, Chapel Hill, NC, USA; **Department of Physiology
& Biophysics, School of Medicine, Case Western Reserve University,
Cleveland, OH, USA

FROM BEDSIDE TO BENCH: ISCHEMIA–REPERFUSION INJURY IN THE HEART

Ischemia refers to the reduction of blood flow in a given tissue or organ. In the heart, this commonly occurs as a result of an obstruction of the vessels primarily involved in myocardial perfusion – the coronary arteries. Ischemia abolishes the delivery of substrates for energy metabolism, namely glucose and fatty acids (FAs), and diminishes the supply of oxygen (O_2) to create hypoxic conditions (in addition to impaired cellular waste removal) within the myocardium. As the heart perfuses itself in addition to the body, any interruption in pump function may also reduce blood flow to the myocardium. A number of clinical conditions feature cardiac ischemia including; myocardial infarction, hypertension,

cardiac arrest, and the various clinical outcomes of coronary artery disease such as angina, acute coronary syndromes, and silent ischemia (which is asymptomatic). Many of these pathologic etiologies progress to heart failure that also frequently involve a reduction in coronary flow to further promote ischemia in the heart. Myocardial infarction (MI) is a prominent clinical concern and it also constitutes one of the more easily recapitulated and controllable injury models to study I/R injury. Interestingly, MI was not conclusively understood as a condition of ischemia until the 1980s [1], spurring investigation into the cellular basis for I/R injury. The underlying effects of this altered nutrient and oxygen delivery most certainly impact cardiac metabolism and the following serves to explore the consequences of ischemic insult and the reperfusion phase in the myocardium.

VULNERABILITY OF THE HEART TO ISCHEMIA–REPERFUSION INJURY

As discussed in previous chapters, the heart is a unique organ requiring a constant and substantial supply of ATP to support the function of ion channels that allow for polarization and depolarization of the membrane, the intracellular control of ions (primarily calcium to regulate contraction), and to drive the work done by the contractile apparatus. The majority of this energetic demand is met through oxidative metabolism by the large population of mitochondria in the cell (~30% cell volume) to continually regenerate a small pool of ATP; the robust turnover rate of this pool is often illustrated by the fact that it would only fuel several heart beats. The bulk of ATP generated in cardiomyocytes comes from oxidizing FA, which are more energetically dense than carbohydrates such as glucose. However, generating myocardial ATP from FA comes at a cost; a greater input of O_2 per mole of FA. A higher relative oxygen use for FA in the normal heart is inconsequential as even under maximal work conditions the myocardium can meet energetic demand. However, the ischemic heart requires an alternative strategy for generating ATP and the resulting changes in metabolism ironically put the myocardium at risk for further injury during reperfusion. The effect of ischemia and reperfusion at the level of the cardiomyocyte includes numerous detrimental changes that directly and indirectly alter energy metabolism.

METABOLIC CHANGES DURING ISCHEMIA

A number of pathologic events induce ischemia in the heart and a great deal of research has offered insight on the cellular processes underlying I/R injury. It is worth noting that several caveats exist in our understanding of I/R injury that, in part, are related to models employed that often do not fully mimic the conditions of the beating heart within the organism and may not fully recapitulate the clinical setting. Cultured cells typically possess high rates of glycolysis (especially immortalized cell lines) and exist in a very different gaseous environment as they reside as a monolayer in a bulk solution. *Ex vivo* perfused hearts are easily manipulated to model injury and substrate changes, but are prone to nonphysiologic oxygen exchange at the epicardium. In light of the limitations in experimental design, *ex vivo* heart perfusion models employing low-flow ischemia have been valuable in gaining insight on the time course of I/R injury, however most of these studies use global ischemia making correlation to the clinical syndromes with heterogeneous tissue changes complicated. *In vivo* studies often focus on long-term effects at the exclusion of examining acute biochemical and molecular changes, are typically heterogeneous in sampling, and are confounded by surgical procedures that invoke systemic responses. The reader should be mindful that depending on the type of injury, the metabolic profile of the heart as a whole may be quite distinct as the locale and extent of ischemia may vary. Regardless, data from various models demonstrate that, an immediate consequence of ischemia is a change in cardiac metabolism; a loss of oxygen (and nutrient) delivery forces the high energy requiring cardiomyocyte to respond in seconds to the need to generate ATP (summarized in Fig. 11.1) [2]. The loss of blood flow creates a fixed pool of oxygen and nutrients to induce both direct and indirect effects on metabolism and leads to the disruption of other cellular processes, which may reinforce defects in energy metabolism.

Accelerated anaerobic glycolysis increases lactate and proton (H^+) production for several minutes, then gradually diminishes while intracellular pH drops >½ unit [3]. ATP demand

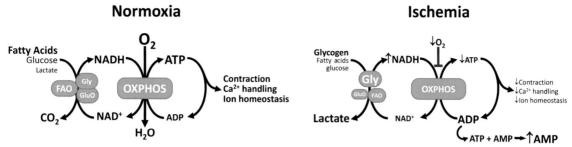

FIGURE 11.1 **Summary of biochemical effects on energy metabolism with onset of ischemia.** Under normal (normoxic) conditions with adequate myocardial perfusion, the heart receives adequate oxygen and circulating substrates to generate the ATP input needed to support cardiac function (contraction, calcium handling, and ion-channel function controlling depolarization). The primary substrate is FAs with the remainder supplied by glucose (and lactate) and oxidation of substrates occurs to maximize reducing equivalent production (NADH, and also FADH$_2$, *not shown*) that allows ATP to be made via oxidative phosphorylation (OXPHOS). With the onset of ischemia, the loss of oxygen stimulates anaerobic glycolysis to utilize glycogen. Glucose oxidation (GluO) becomes "uncoupled" from glycolysis, FAO decreases and the heart becomes a net lactate producer.

rapidly exceeds production and intracellular [ADP] rises. The salvage reaction of adenylate kinase converts ADP to ATP and AMP, but this energy production is finite. An increase in [AMP] activates AMP-activated protein kinase (AMPK, discussed later in the chapter), which activates a number of metabolic targets, including glucose transport and glycolysis, along with cell survival pathways and fatty acid oxidation (FAO), which can impede metabolic adaptation to ischemia and recovery during reperfusion [4,5]. The growing ATP deficit resulting from limited oxygen and substrates is exacerbated by nonoxidative reactions that yield far less ATP per molecule of substrate (glucose). Oxygen is the final electron acceptor for the mitochondrial electron transport chain (ETC) in the process of oxidative phosphorylation, thus ETC flux falls, but the complexes become reduced and cellular [NADH] increases, which has consequences in the reperfusion phase. ATP production through anaerobic glycolysis in the cytosol nets only 2 ATP per molecule and further oxidation in the mitochondrion does not occur (to generate an additional 34 ATP), effectively "uncoupling" glycolysis from glucose oxidation [6]. As the

heart generally oxidizes FAs under normal conditions, the gap in metabolic production of ATP is even greater as 1 molecule of FA can generate >100 ATP. The lack of robust energy metabolism and nutrient delivery also dysregulates extracellular ion concentrations. Numerous studies have also shown a large efflux of potassium during acute ischemia and intracellular calcium overload. This dysregulation of ion gradients results in arrhythmia and eventual cardiac arrest leaving the heart in a noncontractile state. The ensuing cellular dysfunction must be rectified by reperfusion, however, as will be discussed this paradoxically induces further damage to the myocardium.

ISCHEMIA–REPERFUSION INJURY

Although direct revascularization allowing reperfusion in the affected myocardial tissue is key to attenuating myocardial dysfunction, the localized loss then restoration of blood flow causes a myriad of damaging intracellular effects that can ultimately precipitate cardiomyocyte apoptosis or necrosis during reperfusion.

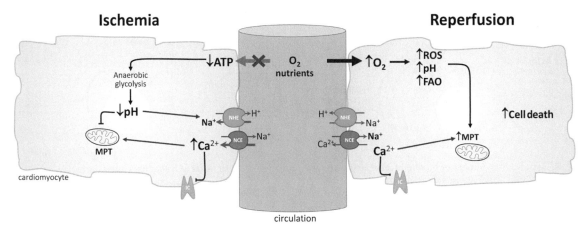

FIGURE 11.2 **Intracellular changes associated with I/R injury.** I/R injury predominantly drives cellular dysfunction by damaging cells directly and/or inducing cell death through multiple mechanisms. During ischemia, the loss of oxygen (O_2) delivery greatly diminishes energy production, requiring enhanced rates of anaerobic glycolysis that produces lactate and protons (H^+). The pH decrease associated with H^+ accumulation alters ion transport by the sodium–hydrogen exchanger (NHE) and the NCE to overload and dyregulate intracellular calcium (Ca^+). The increased Ca+ causes mitochondrial swelling, however the acidic pH inhibits the MPT.

Thus, despite the fact that the detrimental phase of ischemia is ended by reperfusion, reperfusion itself may inflict further damage namely the reperfusion injury.

While a comprehensive picture of mechanisms of I/R-mediated damage remains forthcoming, to date the key contributors of injury include alterations in intracellular calcium handling, mitochondrial dysfunction, and the generation of reactive oxygen species (ROS) – all of which impinge on the cell's ability to generate ATP (see Fig. 11.2). Myocardial ischemia creates a condition in which the metabolic demands of cardiomyocytes cannot be met and the switch to anaerobic glycolysis lowers intracellular pH. This acidic environment, resulting from the combination of increased Na^+ entering the cell via Na^+–H^+ exchange and Ca^{2+} entering the cell via Na^+–Ca^{2+} exchange, increases intracellular Ca^{2+} levels. Normally, rapid increases in Ca^{2+} induce the mitochondrial permeability transition (MPT) pore, however the low intracellular pH is inhibitory. Without eventual restoration of blood flow, the lack of ATP and high Ca^{2+} levels trigger myocyte atrophy and eventually

apoptosis and necrosis. The time-dependent evolution of myocardial cell death following coronary artery occlusion dictates the window in which reperfusion can successfully attenuate the amount of cardiomyocyte death. The size of an infarction is often well-correlated with ischemic time as cells at the focal point of injury die and cell death propagates out from the center over the duration of ischemia. Short durations of ischemia generally yield cellular dysfunction (up to ~15 min in animal models), though prolonged ischemia (60 min or greater) will incur a great deal of irreversible cell damage and frank necrosis. Generally, the longer the ischemic episode, the greater the irreversible damage upon reperfusion.

With the return of oxygen to hypoxic cells during reperfusion, glucose and FA oxidation resume in addition to oxidative phosphorylation and the resulting increase in ATP supports active transport of lactate out of the cell and intracellular pH begins to recover. Reperfusion is necessary to salvage the myocardium and to promote homeostasis, but a lag in recovery occurs that involves energy metabolism. Early in reperfusion, anaerobic glycolysis rates remain elevated and FA oxidation recovers quickly to further suppress glucose oxidation through the Randle cycle as FAO produces acetyl-CoA and NADH to feedback inhibit pyruvate dehydrogenase to reinforce uncoupling of glycolysis and glucose oxidation. Higher oxygen and ATP levels promote the recovery of intracellular pH due to the metabolism and export of lactic acid, however continued elevation of intracellular Ca^{2+} levels leads to the activation of the MPT. Apoptosis and particularly necrosis occurring at reperfusion releases inflammatory cytokines, which activate leukocytes, further exacerbating inflammation and cell death of cells surrounding the ischemic area leading to the formation of an infarct. At the whole-organ level, infarction reduces cardiac contractility through permanent loss of functional myocardium, in addition to progressive remodeling and fibrosis in the surrounding tissue. This creates a larger burden for the remaining, viable myocardium to sustain the energy requirements of pump function.

Reperfusion Causes Elevated ROS Production

Upon the reversal of ischemia, oxygen is rapidly reintroduced to cardiac tissue diffusing through the endothelium and quickly into the myocardium. A prominent observation at reperfusion is a burst of ROS produced by cellular oxidases and uncoupled nitric oxide synthase (NOS) within minutes of re-established flow. The burst of ROS production initiated at reperfusion is a primary contributor to the activation of cell death pathways. Normal oxidative phosphorylation by the mitochondrial ETC results in a small amount of ROS in the form of superoxide by complex I and complex III that are rapidly metabolized to hydrogen peroxide by mitochondrial superoxide dismutase. During reperfusion, the heavily reduced state of the ETC combined with rapid infusion of oxygen causes a substantial increase

in superoxide production. While the majority of ROS production originates in mitochondria, NADPH oxidase (Nox) enzymes are activated during reperfusion, including Nox1 and Nox2 [7]. Additionally, in prolonged periods of ischemia, the breakdown of ATP causes a buildup of hypoxanthine that catalyzes the reverse reaction of xanthine oxidase to rapidly produce significant quantities of superoxide. ROS contributes to reperfusion-mediated cardiac damage both through oxidative stress signaling pathways and via direct oxidative modification of proteins, lipids, and DNA [7].

The key ROS produced during reperfusion are superoxide ($O_2^{\bullet-}$) and hydrogen peroxide (H_2O_2); however the highly reactive hydroxyl radical (HO^-) can be formed. $O_2^{\bullet-}$ is formed enzymatically via single electron reduction of oxygen by Nox, electron slippage from the ETC to molecular oxygen, and uncoupled eNOS/iNOS. $O_2^{\bullet-}$ may oxidatively modify mitochondrial components, but its short half-life and membrane impermeance (due to its negative charge) limits its involvement in cell signaling. However, $O_2^{\bullet-}$ production has wide-reaching intracellular consequences due to its rapid, spontaneous, and enzymatic conversion to the more stable (and membrane permeable) species H_2O_2. H_2O_2 is considered the primary ROS involved in redox signaling, reacting reversibly with protein thiols on low pK_a cysteine and methionine residues to modulate protein function. This mechanism explains, in part, how ROS produced during reperfusion influences a multitude of cell signaling processes, including ion homeostasis, contraction, hypertrophy, autophagy, cell death, and energy production (which is also affected by the other processes) [7]. The extent to which large quantities of ROS can cause irreversible modifications to biomolecules is dependent on the balance between the amount of ROS produced during reperfusion and the cells' ability to absorb ROS by endogenous antioxidant systems. Importantly, oxidative modifications to DNA interfere with DNA duplication and repair; this will enhance apoptosis rates once the severity of damage passes a certain threshold.

I/R Results in Mitochondrial Dysfunction

Mitochondria constitute nearly a third of cardiomyocyte volume and house a majority of the pathways to generate the ATP needed to sustain cardiac function. Not surprisingly, these organelles play a significant role in cardiac metabolism and are primary targets of I/R injury [8,9]. During ischemia, increased intracellular calcium leads to mitochondrial and organelle swelling (often noted in electron micrographs). The acidic cellular pH prevents the induction of the MPT during ischemia, however at reperfusion this phenomenon of a large, nonspecific pore formation occurs [10]. The MPT dissipates the chemical gradients of the mitochondrion, including the electrochemical gradient that is responsible for creating the potential across the inner membrane ($\Delta\Psi m$), which drives ATP synthesis. There is also release of cytochrome c, a proapoptotic factor. Reperfusion also instigates a burst of $O_2^{\bullet-}$ as previously discussed from the heavily reduced complexes of the ETC that remain partially inhibited when exposed to a rapid increase in oxygen. The proximity of this $O_2^{\bullet-}$ production can incur significant oxidative damage in the mitochondrion to perpetuate defects in oxidative metabolism such as the reversible oxidization of the iron–sulfur center of the citric acid cycle enzyme aconitase, resulting in its inactivation [11]. This has been detected experimentally during reperfusion, and contributes to cytotoxicity and impaired energy production [12]. The extent and severity of mitochondrial dysfunction depends, in part, on the successful turnover of damaged mitochondria through autophagy [13].

I/R Disrupts Calcium Ion Homeostasis and Cardiac Contractility

Cardiac contractility relies on carefully regulated ion transport to propagate the cardiac action potential and, at the molecular level, the calcium-mediated activation of the actin–myosin cross bridge cycle. In physiologic conditions Ca^{2+} enters cardiomyocytes during systole via the L-type Ca^{2+} channel (LTCC) resulting in Ca^{2+}-induced Ca^{2+} release from the SR via the ryanodine receptor, which leads to muscle contraction. Extracellular Ca^{2+} is primarily removed from the cell by the Na^+–Ca^{2+} exchanger (NCE), with a very small contribution by the plasma membrane Ca^{2+}-ATPase (PMCA). Reuptake of SR calcium is largely regulated by the SR Ca^{2+}-ATPase (SERCA).

Altered intracellular Ca^{2+} homeostasis results from the molecular events during cardiac I/R injury. During ischemia, decreased intracellular pH increases Na^+–H^+ exchange and subsequently Ca^{2+} entering the cell via Na^+–Ca^{2+} exchange. Elevated ROS at reperfusion further promotes Ca^{2+} entry by redox-mediated activation of Ca^{2+}/calmodulin-dependent protein kinase (CaMKII), PKA, and PKC [7]. Reduced ATP during ischemia prevents Ca^{2+} reuptake by active transport, resulting in a sustained spike of elevated Ca^{2+}. During reperfusion, extracellular calcium levels rapidly recover, however intracellular levels remain elevated until energy production returns to normal. This altered calcium homeostasis can result in arrhythmias due to action potential conduction slowing, particularly ventricular fibrillation, which is a major contributor to mortality in AMI.

Vascular Responses to I/R: Endothelial Cell Dysfunction and Leukocyte Adhesion

Early into the reperfusion period, endothelial impairment is observed; vascular injury begins with a triggering phase in endothelial cells, followed by a second phase in which neutrophil accumulation is amplified [14]. As nitric oxide production decreases, endothelial cell dysfunction results and leukocytes start to adhere to the endothelium. Neutrophils then migrate across the endothelial cell barrier and migrate toward damaged tissue. Activated neutrophils then release cytokines, proteases, leukotrienes, and free radicals. Experimentally, endothelial dysfunction occurs 4–12 weeks after I/R injury [15]. The combination of endothelial cell dysfunction, platelet activation, and *de novo* thrombosis with edema and oxidative stress contributes to the microvascular damage following I/R injury and may limit the extent of reperfusion. Neutrophil adherence is an important factor in activating leukocytes, leading to the generation of superoxide radicals.

CARDIOPROTECTIVE SIGNALING PATHWAYS IN I/R INJURY

The molecular signaling associated with I/R is a balance between pathologic and protective pathways. While our understanding of the systems that counteract the pathologic pathways is less developed, multiple signaling pathways that may have therapeutic relevance have emerged. By activating the pathways described later in the chapter during cardiac reperfusion, improved cardiac functional outcomes have been reported. In particular, the activation of two intracellular signaling cascades: the reperfusion injury salvage kinase (RISK) pathway mediated by the activation of phosphatidylinositol 3-kinase (PI3K) and extracellular signaling-regulated kinases (ERK)1/2, and the survivor activating factor enhancement (SAFE) pathway mediated by JAK-STAT signaling, reduce MPT and apoptosis to promote cell survival. The activation of other signaling modalities such as AMPK, which alters energy metabolism, Nrf1a/Nrf2 transcription factors, which activate antioxidant pathways, and hypoxia-inducible factors (HIF), which promote angiogenesis and glycolytic metabolism contribute to adaptive processes aimed at protecting the myocardium from I/R injury.

Reperfusion Injury Salvage Kinase (RISK) Pathway

The RISK pathway consists of prosurvival kinases that provide protection from reperfusion injury (see Fig. 11.3). The main mediators of the cardioprotection provided by the RISK pathway are ERK1/2 and PI3K/Akt [16]. A number of agents that protect against reperfusion injury have been found to upregulate the RISK pathway including bradykinin, cardiotrophin-1, insulin, IGF-1, urocortin, atorvastatin, glucogon-like-peptide 1 (GLP-1), and opioid receptor agonists. The stimulation of cell surface receptors (G-coupled protein receptors, GPCR) stimulate the RISK pathway to activate ERK1/2 and PI3K. ERK1/2 is also phosphorylated during ischemic preconditioning. ERK protects against cell death, in part, by mediating the action of Bcl-2 and Bcl-2-associated death (BAD) promoter protein, mitochondrial

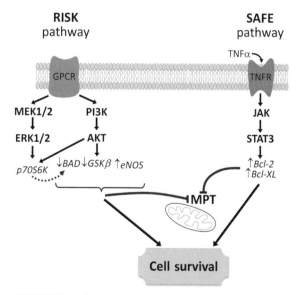

FIGURE 11.3 **The RISK and SAFE cardioprotective signaling pathways invoked by I/R.** The RISK and SAFE pathways are important in cell survival during I/R injury, in addition to pre- and postconditioning cardioprotection. The primary downstream targets include an inhibition of the MPT and upregulation of cell survival pathways.

proteins involved in mediating apoptosis. Bcl-2 prevents MPT and the opening of the mitochondrial apoptosis-induced channel (MAC) to promote cell survival. PI3K catalyzes the activation of phosphatidylinositols (PIPs) via phosphorylation, which is reversed by the phosphatase PTEN (phosphatase and tensin homolog) that is reportedly degraded by ischemic preconditioning. PIPs activate PDK1, which phosphorylates Akt and can activate a number of downstream targets, including mTOR (mammalian target of rapamycin complex), an additional signaling kinase. p70S6K is phosphorylated by the mTORC complex and PDK1 to potentially mediate cardioprotection by inhibiting GSK-3β and Bad to induce autophagy [8]. Akt also activates endothelial and inducible nitric oxide synthase (eNOS/iNOS) by phosphorylation to increase production of the protective gasotransmitter NO. NOS is well-established to be important for cardioprotection; eNOS-knockout mice exhibit reduced cardiac output and larger infarct sizes in response to ischemia [8]. NO is believed to afford cardioprotection by multiple mechanisms [17]. NO is a potent vasodilator, which could provide reperfusion to ischemic tissues more rapidly. NO has antiapoptotic properties by inhibiting caspase-3-like activation and inhibiting the opening of the MPT. NO also inhibits mitochondrial respiration, reducing the contribution of ETC dysfunction to ROS production.

Survivor Activating Factor Enhancement (SAFE) Pathway

The SAFE pathway is a second protective pathway with independent activation to that of the RISK pathway, which consists of the activation of JAK/STAT3 (also summarized in Fig. 11.3). SAFE is reportedly activated by opioids, insulin, and sphingosine-1 as well as by ischemic postconditioning. While high levels of TNFα are known to activate TNFR1 and induce cardiomyocyte death, the SAFE pathway is activated endogenously by low levels of TNFα. TNFα-mediated activation of TNFR2 results in activating phosphorylation of Janus kinase (JAK), a tyrosine kinase that associates with STAT transcription factors. Phosphorylation of STAT3 by JAK induces translocation to the nucleus, activating transcription of prosurvival proteins. Some degree of cross talk exists between the RISK and SAFE pathways; ERK1/2 in the RISK pathway has been demonstrated to activate STAT3. Overexpression of STAT3 in the myocardium results in cardioprotection and STAT3 deficiency results in greater cardiac injury to ischemia. STAT3 promotes protection by inactivating the proapoptotic factor Bad and regulating Bcl2, Bcl-xl. Ultimately, STAT3 activity reduces MPT and cell death.

Antioxidant Pathways

Damage from excessive ROS can be mitigated by biologic antioxidant enzymes. These enzymes are expressed physiologically, but certain signals can induce enhanced expression to protect against oxidative insult. The expression of many antioxidant enzymes including catalase, superoxide dismutase, glutathione peroxidase, peroxiredoxin, and thioredoxin are controlled by the transcription factor Nrf2 via the antioxidant response element (ARE) in their promoter regions. Nrf2 is activated during reperfusion by an oxidative signaling mechanism [18]. Nrf2 is basally inhibited by association with Keap1, an actin-associated scaffold protein. Direct oxidative modification of Keap1 by H_2O_2 releases Nrf2, which translocates to the nucleus to activate transcription of antioxidant genes. Activation of Nrf2 is also enhanced during preconditioning.

AMPK

5′-AMP-activated protein kinase (AMPK) is a serine-threonine kinase that is important in upstream regulation of cellular energy (see Fig. 11.4). AMPK regulates many downstream processes that broadly contribute to cardioprotection such

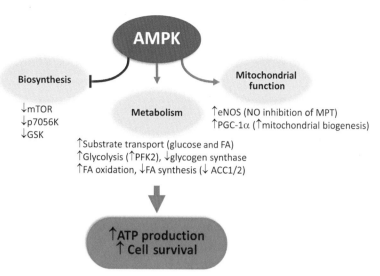

FIGURE 11.4 **The role of AMPK in mediating energy metabolism during cellular stress.** AMPK is a primary sensor for fluctuation in [AMP] signaling energetic stress in the cardiomyocyte. AMPK is activated by I/R to increase ATP production in the cell while downregulating biosynthetic pathways (which require ATP) to conserve energy and promote energy homeostasis.

as protein synthesis (mTOR), energy production (acetyl-CoA carboxylase 1/(ACC)1/ACC2)), mitochondria turnover (regulated by peroxisome proliferator-activated receptor gamma coactivator 1-alpha, PGC-1α) and nitric oxide synthesis (NOS). AMPK is physiologically activated by changes to AMP:ATP ratios, which are seen in nutrient deprivation, and other contexts of cellular stress such as hypoxia/ischemia. While the mechanisms are not yet well understood, activators of AMPK such as metformin, AICAR, and A769662 reduce infarct size in animal models, while mice with inactive AMPK exhibit greater cardiac dysfunction after ischemia.

Hypoxia-Inducible Factor 1

Hypoxia-inducible factors (HIF) are proteins that rapidly accumulate when hypoxia occurs and mediate transcriptional control of a number of targets. The gene programs controlled by HIF1 in the heart participate in the adaptation to reduced oxygen tension in tissue and is involved in the cardioprotective mechanisms invoked during I/R injury [19]. HIF1 is constitutively expressed, but is rapidly ubiquitinated and degraded by the proteasome. Tagging HIF1 to initiate degradation occurs through the activity of prolyl-hydroxylases (PHDs), which requires oxygen as a substrate. During ischemia, HIF1 is stabilized and can act as a transcription factor to upregulate expression of genes involved in the angiogenic/erythropoiesis response (most notably by induction of erythropoietin, EPO) and glucose metabolism. Many of the enzymes of the glycolytic pathway along with the glucose transporters are targeted for transcriptional upregulation by HIF1, providing an adaptive response to lowered oxygen conditions.

THERAPIES AND PHARMACOLOGIC MODULATORS OF I/R INJURY

While clinical trials for therapies to reduce I/R injury have been conducted, there have been few changes to clinical practices. Inapt animal models and poor clinical trial designs likely contribute to the failure of translating promising laboratory therapeutics into the clinic [20]. Developing a therapeutic to limit cardiac damage after I/R is not trivial. Damage is mediated by two different processes, the initial ischemic insult and the subsequent reperfusion injury. Given the necessity of reperfusion in restoring myocardial homeostasis, the emphasis of therapies is reducing the negative consequences of reperfusion without interfering with the restoration of cardiac function.

Pre- and Postconditioning with I/R Injury

A highly efficacious therapy for limiting ischemic damage is the use of preconditioning; several short periods of ischemia and reperfusion induced prior to a prolonged ischemic episode. First described by Murry et al. [21], preconditioning significantly reduced the size of I/R-induced infarcts in canines and has since been corroborated in various animal models and humans. Unfortunately, the unpredictable nature of MI leaves few circumstances where ischemic preconditioning can be employed clinically. However, the observed reduction in cardiac damage suggests that cardioprotective pathways exist that may be activated by other procedures or pharmacologic agents. Postconditioning, initiated by very short bursts of reperfusion and occlusion after prolonged ischemia, also reduces infarct size by activating cardioprotective pathways prior to full restoration of blood flow [22]. Experimental models have shown reductions in infarct size, though clinical trials have yielded conflicting evidence. Postconditioning may not be effective in all types of I/R injury; a randomized trial of MI patients found no improvements in infarct size with postconditioning [23]. As with ischemic preconditioning, postconditioning also has limited clinical use as rapid occlusion and reperfusion in affected arteries would be difficult to implement in a timely fashion. However, recent evidence suggests that cardioprotective benefits are conferred by remote postconditioning using cuff inflation/deflation of distal arteries in the arm or leg. This would not interfere with other interventions during acute critical care, including percutaneous coronary intervention (PCI) or cardiac surgical procedures [24]. How remote postconditioning affords cardioprotection remains unclear, but likely involves a circulating humoral or neurohormonal factor. Regardless, while mechanical methods offer a means of activating protective pathways to reduce infarct size, the limitations have thus far prevented widespread clinical use. Development of pharmaceuticals to enhance cardioprotective signaling prior to MI or reduce reperfusion damage concurrent with reopening of blocked arteries or the dissolution of embolus has become the focus of interventional strategies.

Interventions Minimizing Cellular Damage

Antioxidant therapy has long been a strategy to ameliorate oxidant damage during reperfusion. ROS scavenging or inhibition of the enzymatic release of oxidants during reperfusion would reduce the cellular-oxidative stress burden and inflammation, ultimately diminishing cell death. Many early studies using prophylactic treatment with untargeted antioxidants such as vitamins A, C, and E were ineffective, however, recent trials of targeted antioxidant regimens during acute oxidative stress have proven more efficacious. The antioxidant edaravone has been used in ischemic brain injury; a Japanese trial with the drug found modest improvement with supplementation over conventional treatment [25]. Permeability and subcellular targeting of antioxidants presents the biggest hurdles in development, but mitochondrially targeted antioxidants mitoTEMPO and MitoQ (mitochondria targeted CoQ10) have shown protection in animal models of I/R, though have yet to be used clinically [26,27].

Exogenous activation of cardioprotective signaling pathways has also been investigated.

Activation of adenosine receptors (AR) has been implicated in the cardioprotection via the RISK pathway [28]. While clinical trials with exogenous adenosine have not shown significant improvements [20], targeting agonists to particular AR receptors may offer therapeutic benefit. AR subtypes have unique tissue distributions, and associated signaling as well as differing affinities for adenosine. EPO, an angiogenic factor, has been demonstrated experimentally to reduce I/R injury in rodents when administered during reperfusion through ERK1/2 and PI3K pathway activation [29], however recombinant EPO failed to provide improvement in ST-segment elevation MI (STEMI) patients [20].

In addition to cardioprotective pathway activation, inhibition of ischemia-related cell death pathways has been explored. Inhibition of protein kinase C delta (PKCδ) by intracoronary and intravenous peptide inhibitors prevented damage in rodent and porcine models of I/R by reducing apoptosis via GSK3β, however recent human trials found no discernable difference compared to placebo [20]. Given the role of MPT in I/R myocyte death, there has been interest in whether its inhibition affords cardioprotection. While experimental animal data is quite promising, it is not yet known whether MPT inhibitors are effective in humans. Clinical trials are ongoing with the MPT inhibitor cyclosporine A (clinically used as an immunosuppressant) given prior to PCI in STEMI patients. The Na$^+$–H$^+$ exchange inhibitor cariporide has also been tested and in non-STEMI patients infarct size was reduced, but with no change in long-term cardiac outcomes [20]. The K$^+$ channel activator nicorandil was also proposed to reduce intracellular calcium overload during reperfusion, but no benefit was observed in a subset of MI patients [20].

Metabolic Modulation

I/R injury elicits dynamic changes in energy metabolism that ultimately have a negative impact on cardiac function. Numerous pharmacologic approaches have been tested to re-establish and/or improve the balance between energetic needs and ATP production. β-adrenergic blockers, such as atenolol and carvedilol, lower heart rate, contractility, and/or blood pressure to reduce cardiac workload, thus diminishing oxygen and energetic demand (and may also reduce catecholamine-mediated lipolysis and subsequent elevations in circulating FA). Vasodilatory drugs, such as adenosine and nitrates, promote greater delivery of oxygen and nutrients to the heart to support energy production. Recent investigation into the regulation of cardiac energy metabolism has prompted investigations to improve more-efficient ATP production in the compromised myocardium.

Direct metabolic modulation shifts metabolism away from fatty acid β-oxidation toward improved glucose oxidation. Current approaches include increasing glucose availability and enhancing glucose oxidation, in addition to partially inhibiting FAO. One of the first metabolic manipulations, glucose–insulin–potassium (GIK) infusion, was developed based on clinical observations and has been used for >40 years during cardiac surgery and in acute MI. Variable clinical outcomes with GIK have called into question its efficacy, but this is likely due to differences in treatment application [30]. GIK therapy has positive effects on cardiac function when administered at reperfusion, however experimental data suggests insulin may not be beneficial in the presence of elevated FAs, commonly noted in MI patients [31]. Dichloroacetate (DCA) has been long-established as a metabolic modulator through its action as an inhibitor of PDH kinase 4 that phosphorylates, thus inactivates, PDH preventing pyruvate entry into the TCA cycle. Experimental data has shown DCA to be effective at increasing glucose oxidation, reducing lactate, and improving cardiac function when given at reperfusion [32,33]. Unfortunately, limited clinical data exists on DCA use in I/R injury, but it is effective in alleviating lactate production in inherited mitochondrial disorders and did improve cardiac output in a small human study [34,35].

Agents that directly suppress FAO have also shown promise for use in I/R injury [36]. As FA transport into the cardiomyocyte is not fully understood, altering cardiomyocyte FA uptake has not been pursued, however, a number of partial FAO inhibitors have been developed. Oxfenicine was one of the first agents used to suppress FAO by inhibiting carnitine palmitoyltransferase I (CPT1). Etomoxir and perhexiline are additional CPT1 inhibitors with experimental and clinical data demonstrating improved glucose oxidation with beneficial effects in heart failure, however some potentially adverse effects require investigation into appropriate dosing regimens [37]. Another class of FAO inhibitors are piperazidine derivatives, including trimetazidine and ranolazine, which target the FAO pathway directly. Trimetazidine has been used for many years in Europe as an antianginal drug and improves cardiac function in heart failure [38]. Trimetazidine is a competitive inhibitor of long chain 3-ketoacyl-CoA thiolase that promotes an increase in glucose oxidation. Interestingly, trimetazidine appears to only modestly suppress FAO suggesting secondary metabolic targets or other mechanisms of action. Ranolazine is also a partial FAO inhibitor that increases PDH activity, and subsequently glucose oxidation without increase in glycolysis, suggesting an improved coupling of glucose metabolism. It is approved in the United States in the treatment of chronic stable angina and in experimental models reduce infarct size in addition to long-term reductions in remodeling. Like trimetazidine, ranolizidine has off-target effects that complicate interpretation of whether metabolic modulation is the only benefit of the drug [39].

While several approaches have been identified in experimental models to reduce reperfusion injury, none have exhibited sufficient benefit to warrant regular clinical use. In addition, much of the data for these therapeutics in human studies is inconclusive regarding the long-term effects of treatment. However, given the therapeutic benefit of reducing reperfusion-induced cardiac damage evidenced by significant reductions in infarct size by mechanical and pharmacological therapies, optimal use of these therapies is feasible. Furthermore, the promising results from pharmacologic agents targeted at optimizing energy metabolism in the heart following ischemic damage will undoubtedly play a role in future therapeutic interventions.

References

[1] DeWood MA, Spores J, Hensley GR, Simpson CS, Eugster GS, Sutherland KI, Grunwald RP, Shields JP. Coronary arteriographic findings in acute transmural myocardial infarction. Circulation 1983;68:I39–49.
[2] Jennings RB. Historical perspective on the pathology of myocardial ischemia/reperfusion injury. Circ Res 2013;113:428–38.
[3] Bak MI, Ingwall JS. Contribution of Na+/H+ exchange to Na+ overload in the ischemic hypertrophied hyperthyroid rat heart. Cardiovasc Res 2003;57:1004–14.
[4] Kim M, Tian R. Targeting AMPK for cardiac protection: opportunities and challenges. J Mol Cell Cardiol 2011;51:548–53.
[5] Lopaschuk GD. Amp-activated protein kinase control of energy metabolism in the ischemic heart. Int J Obes 2008;32(Suppl 4):S29–35.
[6] Jaswal JS, Keung W, Wang W, Ussher JR, Lopaschuk GD. Targeting fatty acid and carbohydrate oxidation – a novel therapeutic intervention in the ischemic and failing heart. Biochim Biophys Acta 2011;1813:1333–50.
[7] Brown DI, Griendling KK. Regulation of signal transduction by reactive oxygen species in the cardiovascular system. Circ Res 2015;116:531–49.
[8] Murphy E, Steenbergen C. Mechanisms underlying acute protection from cardiac ischemia-reperfusion injury. Physiol Rev 2008;88:581–609.
[9] Walters AM, Porter GA, Brookes PS. Mitochondria as a drug target in ischemic heart disease and cardiomyopathy. Circ Res 2012;111:1222–36.
[10] Bernardi P, Di Lisa F. The mitochondrial permeability transition pore: molecular nature and role as a target in cardioprotection. J Mol Cell Cardiol 2015;78:100–6.
[11] Gardner PR, Raineri I, Epstein LB, White CW. Superoxide radical and iron modulate aconitase activity in mammalian cells. J Biol Chem 1995;270:13399–405.
[12] Bulteau AL, Lundberg KC, Ikeda-Saito M, Isaya G, Szweda LI. Reversible redox-dependent modulation of

mitochondrial aconitase and proteolytic activity during *in vivo* cardiac ischemia/reperfusion. Proc Natl Acad Sci USA 2005;102:5987–91.

[13] Gustafsson AB, Gottlieb RA. Autophagy in ischemic heart disease. Circ Res 2009;104:150–8.

[14] Tsao PS, Aoki N, Lefer DJ, Johnson G III, Lefer AM. Time course of endothelial dysfunction and myocardial injury during myocardial ischemia and reperfusion in the cat. Circulation 1990;82:1402–12.

[15] Moens AL, Claeys MJ, Timmermans JP, Vrints CJ. Myocardial ischemia/reperfusion-injury: a clinical view on a complex pathophysiological process. Int J Cardiol 2005;100:179–90.

[16] Hausenloy DJ, Yellon DM. Reperfusion injury salvage kinase signalling: taking a risk for cardioprotection. Heart Fail Rev 2007;12:217–34.

[17] Jones SP, Bolli R. The ubiquitous role of nitric oxide in cardioprotection. J Mol Cell Cardiol 2006;40:16–23.

[18] Surh YJ, Kundu JK, Na HK. Nrf2 as a master redox switch in turning on the cellular signaling involved in the induction of cytoprotective genes by some chemopreventive phytochemicals. Planta Med 2008;74:1526–39.

[19] Semenza GL. Hypoxia-inducible factor 1 and cardiovascular disease. Annu Rev Physiol 2014;76:39–56.

[20] Heusch G. Cardioprotection: chances and challenges of its translation to the clinic. Lancet 2013;381:166–75.

[21] Murry CE, Jennings RB, Reimer KA. Preconditioning with ischemia: a delay of lethal cell injury in ischemic myocardium. Circulation 1986;74:1124–36.

[22] Vinten-Johansen J, Yellon DM, Opie LH. Postconditioning: a simple, clinically applicable procedure to improve revascularization in acute myocardial infarction. Circulation 2005;112:2085–8.

[23] Hahn JY, Song YB, Kim EK, Yu CW, Bae JW, Chung WY, Choi SH, Choi JH, Bae JH, An KJ, Park JS, Oh JH, Kim SW, Hwang JY, Ryu JK, Park HS, Lim DS, Gwon HC. Ischemic postconditioning during primary percutaneous coronary intervention: the effects of postconditioning on myocardial reperfusion in patients with st-segment elevation myocardial infarction (post) randomized trial. Circulation 2013;128:1889–96.

[24] Meybohm P, Zacharowski K, Cremer J, Roesner J, Kletzin F, Schaelte G, Felzen M, Strouhal U, Reyher C, Heringlake M, Schon J, Brandes I, Bauer M, Knuefermann P, Wittmann M, Hachenberg T, Schilling T, Smul T, Maisch S, Sander M, Moormann T, Boening A, Weigand MA, Laufenberg R, Werner C, Winterhalter M, Treschan T, Stehr SN, Reinhart K, Hasenclever D, Brosteanu O, Bein B. Remote ischaemic preconditioning for heart surgery. The study design for a multicenter randomized double-blinded controlled clinical trial – the ripheart-study. Eur Heart J 2012;33:1423–6.

[25] Ishibashi A, Yoshitake Y, Adachi H. Investigation of effect of edaravone on ischemic stroke. Kurume Med J 2013;60:53–7.

[26] Adlam VJ, Harrison JC, Porteous CM, James AM, Smith RA, Murphy MP, Sammut IA. Targeting an antioxidant to mitochondria decreases cardiac ischemia-reperfusion injury. FASEB J 2005;19:1088–95.

[27] Liang HL, Sedlic F, Bosnjak Z, Nilakantan V. Sod1 and mitotempo partially prevent mitochondrial permeability transition pore opening, necrosis, and mitochondrial apoptosis after ATP depletion recovery. Free Radic Biol Med 2010;49:1550–60.

[28] Laubach VE, French BA, Okusa MD. Targeting of adenosine receptors in ischemia-reperfusion injury. Expert Opin Ther Targets 2011;15:103–18.

[29] Bullard AJ, Govewalla P, Yellon DM. Erythropoietin protects the myocardium against reperfusion injury *in vitro* and *in vivo*. Basic Res Cardiol 2005;100:397–403.

[30] Apstein CS, Opie LH. A challenge to the metabolic approach to myocardial ischaemia. Eur Heart J 2005;26:956–9.

[31] Folmes CD, Clanachan AS, Lopaschuk GD. Fatty acids attenuate insulin regulation of 5'-amp-activated protein kinase and insulin cardioprotection after ischemia. Circ Res 2006;99:61–8.

[32] Bersin RM, Stacpoole PW. Dichloroacetate as metabolic therapy for myocardial ischemia and failure. Am Heart J 1997;134:841–55.

[33] Ussher JR, Wang W, Gandhi M, Keung W, Samokhvalov V, Oka T, Wagg CS, Jaswal JS, Harris RA, Clanachan AS, Dyck JR, Lopaschuk GD. Stimulation of glucose oxidation protects against acute myocardial infarction and reperfusion injury. Cardiovasc Res 2012;94:359–69.

[34] Stacpoole PW, Kurtz TL, Han Z, Langaee T. Role of dichloroacetate in the treatment of genetic mitochondrial diseases. Adv Drug Deliv Rev 2008;60:1478–87.

[35] Wargovich TJ, MacDonald RG, Hill JA, Feldman RL, Stacpoole PW, Pepine CJ. Myocardial metabolic and hemodynamic effects of dichloroacetate in coronary artery disease. Am J Cardiol 1988;61:65–70.

[36] Jaswal JS, Keung W, Wang W, Ussher JR, Lopaschuk GD. Targeting fatty acid and carbohydrate oxidation - a novel therapeutic intervention in the ischemic and failing heart. Biochim Biophys Acta 2011;1813:1333–50.

[37] Killalea SM, Krum H. Systematic review of the efficacy and safety of perhexiline in the treatment of ischemic heart disease. Am J Cardiovasc Drugs 2001;1:193–204.

[38] Kalra BS, Roy V. Efficacy of metabolic modulators in ischemic heart disease: an overview. J Clin Pharmacol 2012;52:292–305.

[39] Parang P, Singh B, Arora R. Metabolic modulators for chronic cardiac ischemia. J Cardiovasc Pharmacol Ther 2005;10:217–23.

Metabolic Remodeling in the Development of Heart Failure

Tien Dung Nguyen

Department of Cardiothoracic Surgery, Jena University Hospital,
Friedrich Schiller University of Jena, Jena, Germany

Heart failure (HF) is a clinical syndrome characterized by a mismatch between cardiac pump function and an individual's physiological requirements. The clinical presentation includes pulmonary congestion, dyspnea, and fatigue. HF may result from various disorders with hypertension and myocardial infarction being the two most common causes [1]. Historically, the term HF mostly refers to impaired systolic function. In HF preceded by cardiac hypertrophy (commonly seen in hypertensive heart disease), stages of cardiac hypertrophy without systolic dysfunction have usually been termed "compensated hypertrophy." However, it has become clear that HF symptoms can also exist despite preserved systolic function. This condition is mainly caused by diastolic dysfunction that may even be present in compensated hypertrophy. For the sake of simplicity and because most studies on cardiac metabolism in HF did not investigate diastolic function, we use the term "compensated hypertrophy" for hypertrophied hearts with normal systolic function, and

the term HF to indicate the presence of systolic dysfunction.

There are two major points that should be considered while interpreting study results on metabolic alterations in HF. First, metabolic phenotypes and their mechanisms differ between HF of different etiologies. Second, because the progression to HF is often long and complex, the time point of assessment (i.e., compensated hypertrophy with or without diastolic dysfunction vs. manifest systolic dysfunction) will influence metabolic changes that are observed.

Figure 12.1 illustrates major processes of metabolic remodeling in the development of HF with a focus on their interactions with each other and with structural remodeling.

SUBSTRATE UTILIZATION

Table 12.1 summarizes reported changes in substrate utilization in various models of HF.

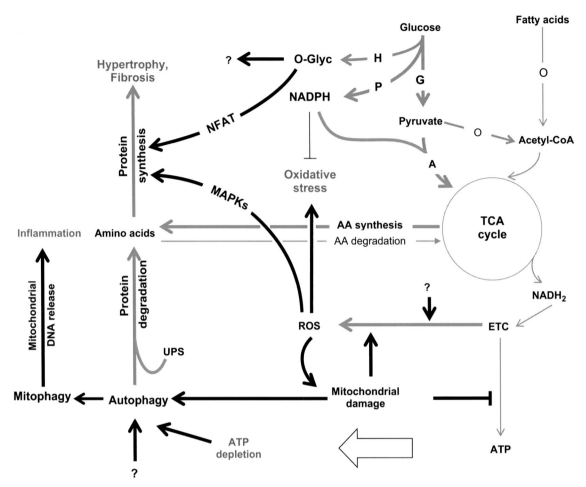

FIGURE 12.1 **Overview of metabolic remodeling and proposed mechanisms linking it to other processes in the development of HF.** H, Hexosamine biosynthetic pathway (HBP); P, Pentose phosphate pathway (PPP); G, Glycolysis; A, Anaplerosis; O, Oxidation; ETC, Electron transport chain; ROS, reactive oxygen species; UPS, ubiquitin-proteasome system; AA, amino acids; TCA cycle, tricarboxylic acid cycle. Metabolic pathways are blue. Bold lines indicate pathways/processes that are increased or dominant. Thin lines represent pathways/processes that are decreased. The question marks imply unknown causes/effects. In general, metabolic remodeling in cardiac hypertrophy and failure is characterized by a shift away from energy production to activation of biosynthetic pathways required for structural remodeling processes such as ventricular hypertrophy and fibrosis. Particularly, fatty acid oxidation is decreased and may not be sufficiently compensated for given the lack of increase in glucose oxidation. These alterations and further mitochondrial defects result in ATP depletion. Because intermediates of the TCA cycle are increasingly channeled into protein synthesis partly via inhibition of amino acid degradation, pyruvate is preferentially used for anaplerosis to maintain the pool of the TCA cycle instead of being oxidized. Hypertrophic mediators such as MAPKs and NFAT are activated as a result of increased mitochondrial ROS and flux through the HBP, respectively. Overproduction of mitochondrial ROS causes oxidative damage. Although the flux through the PPP is increased, antioxidative defense is inadequate due to the consumption of NADPH by the anaplerotic malic enzyme. Mitochondrial damage and ATP depletion stimulate autophagy. Increased activity of autophagy and the UPS may contribute to hypertrophy by providing amino acids and other metabolites. Increase in mitophagy may trigger myocardial inflammation by releasing mitochondrial DNA.

TABLE 12.1 Changes in Substrate Utilization in Various Models of Cardiac Hypertrophy and Failure

Model of heart disease	Morphological and functional characteristics	FA utilization	Glucose utilization	References
Aortic constriction in rats	Compensated hypertrophy without/with diastolic dysfunction; systolic dysfunction	FAO and mRNA expression of genes regulating FAO decreased in compensated hypertrophy and systolic dysfunction	GO decreased in systolic dysfunction	[2,3]
Aortic constriction in rats	Hypertrophy with diastolic and systolic dysfunction	FAO, mRNA, and protein expression of FAO enzymes decreased	GO decreased, protein expression of PDH complex increased	[4,5]
Aortic constriction in mice	Hypertrophy with systolic dysfunction	FAO unchanged	Glycolysis and GO increased	[6]
Aortic constriction in mice	Hypertrophy with diastolic and systolic dysfunction	FAO unchanged	GU, glucose, and lactate oxidation decreased	[7]
Abdominal aortic constriction in rats	Mild, compensated hypertrophy	FAO and mRNA expression of genes regulating FAO unchanged	Glycolysis increased modestly	[8]
Abdominal aortic constriction in rats	Mild, compensated hypertrophy	FAO and mRNA expression of genes regulating FAO decreased	GO unchanged	[9]
Abdominal aortic constriction in rats	Hypertrophy, heart function ex vivo unchanged	FAO decreased	Glycolysis increased, GO unchanged	[10]
Dahl salt-sensitive rats	Compensated hypertrophy; systolic dysfunction	FA uptake and mRNA expression of genes regulating FA utilization decreased in systolic dysfunction	GU increased, mRNA expression of genes regulating glycolysis, and GO decreased; pentose phosphate pathway flux increased in systolic dysfunction	[11]
Spontaneously hypertensive rats (Sprague Dawley rats as controls)	Hypertrophy, heart function ex vivo unchanged	FAO lower. (Note that some strains of SHRs have a mutation in CD36, which could independently modulate FAO).	GO higher	[12]
Spontaneously hypertensive rats (Wistar rats as controls)	Hypertrophy, minor differences in diastolic and systolic function		Flux through PDH higher	[13]
Myocardial infarction in rats	Compensated hypertrophy; systolic dysfunction	mRNA expression of genes regulating FA utilization decreased in systolic dysfunction	mRNA and protein expression of GLUT-1 increased in systolic dysfunction	[14]
Myocardial infarction in rats	No hypertrophy, left ventricular dilatation, systolic dysfunction	FAO and mRNA expression of genes regulating FAO decreased	GU and GO unchanged	[15]
Myocardial infarction in rats	Mild hypertrophy, systolic dysfunction	FA utilization decreased, protein expression of fatty acid transporters decreased		[16]
Pacing-induced HF in dogs	No hypertrophy, left ventricular dilatation, systolic dysfunction	FA uptake and oxidation, activity of CPT-1 and MCAD decreased	GO increased	[17]
HF patients with IDCM or ischemic heart disease	Systolic dysfunction	mRNA expression of MCAD, LCAD lower		[18]
HF patients with IDCM	Hypertrophy, systolic dysfunction	FA utilization lower	Glucose utilization higher	[19]

IDCM, idiopathic dilated cardiomyopathy; FA, fatty acid; FAO, fatty acid oxidation; GO, glucose oxidation; GU, glucose uptake; MCAD, medium-chain acyl-CoA dehydrogenase; LCAD, long-chain acyl-CoA dehydrogenase; SHR, spontaneously hypertensive rat

Changes in FAs Use

Most studies reveal a reduction in cardiac FA utilization. FA uptake is decreased in HF induced by high-salt diet [11], or by rapid pacing [17]. In line with these findings, studies in other models of HF also observed a reduction in mRNA and protein expression of FA transporters associated with systolic dysfunction [2,16,14]. In early (compensated) stages of left ventricular hypertrophy, we found that FA oxidation rate and the expression of FA oxidation enzymes were already decreased [2]. These results are also consistent with findings in spontaneously hypertensive rats [18,12], and in rats with abdominal constriction [10,9], but not with those in Dahl salt-sensitive rats [11] and in rats with myocardial infarction [14]. However, by the time that systolic dysfunction manifests, there is clear evidence from studies in animal models and human subjects that myocardial FA oxidation is decreased.

Changes in Glucose Use

Compared to FA utilization, data on cardiac glucose utilization are less consistent. In the presence of systolic dysfunction, cardiac glucose uptake was found to be decreased following aortic constriction in mice [7] but unchanged in rats with myocardial infarction [15], and increased in Dahl salt-sensitive rats [11]. In compensated hypertrophy, induced by abdominal aortic constriction, glycolysis was modestly increased without changes in glucose oxidation [9,8]. By assessing substrate oxidation at various time points following aortic constriction in rats, we observed that cardiac glucose oxidation tended to increase initially, but was unchanged in the stage of compensated hypertrophy and ultimately decreased when systolic dysfunction occurred [2]. However, in infarcted rat hearts, we observed no change in glucose oxidation rate despite manifest systolic dysfunction [15]. Interestingly, spontaneously hypertensive rats, at the hypertrophy stage and prior to HF were found

to have higher glucose oxidation rates [12] or increased flux through the PDH complex [13] relative to control rats. In addition, Osorio et al. showed increased glucose oxidation rates in failing dog hearts induced by rapid pacing [17], and Dávila-Román et al. demonstrated higher total rates of glucose utilization in patients with idiopathic dilated cardiomyopathy [19]. It is unlikely that these differences can be explained by methodological differences. We therefore suggest that in contrast to changes in FA oxidation, glucose oxidation does not correlate with contractile function in HF, but that changes in glucose oxidation may depend on both the stage and the etiology of HF.

Modulating Substrate Utilization

Due to the strong decrease in FA oxidation in the hypertrophied or failing heart, there is a switch in cardiac substrate selection toward more reliance of ATP production on glucose utilization. This substrate selection pattern resembles that of the fetal heart and is often considered part of the reactivation of the fetal gene program. Oxidation of FA has been associated with ROS production and the induction of uncoupling proteins [20,21]. Furthermore, glucose oxidation is more oxygen-efficient than FA oxidation. Therefore, the phenomenon of substrate switch has been suggested as an adaptive response that shifts cardiac energy metabolism to a more favorable phenotype [22,23].

Many attempts have been made to evaluate the role of the cardiac substrate switch. In the normal heart, inhibition of FA oxidation may cause a substrate switch by inducing a compensatory increase in glucose oxidation. This interaction is also referred to as the Randle cycle and has been exploited to modulate cardiac substrate selection [24]. A large number of drugs targeting inhibition of FA utilization, for example, etomoxir, perhexiline, trimetazidine have been developed (see review by Lopaschuk et al. [25] for an overview). While most studies observed beneficial effects of these drugs in models of HF and in patients, it is still unclear if the putative inhibition of FA utilization was indeed the underlying mechanism. Recently, the therapeutic concept of inhibiting FA utilization in the hypertrophied or failing heart has been challenged by evidence showing improved cardiac function following increased myocardial FA supply or oxidation. For example, hypertensive Dahl salt-sensitive rats subjected to a high-fat diet showed decreased cardiac hypertrophy and improved contractile function [26]. Furthermore, increasing cardiac FA oxidation by deleting acetyl-CoA carboxylase in the heart attenuated ventricular hypertrophy and fibrosis and preserved cardiac energetics and contractile function in mice subjected to pressure overload [27]. Therefore, while modulating FA utilization may potentially affect cardiac function, available findings are contradictory as to whether this metabolic pathway should be enhanced or inhibited.

Cardiac substrate preference may also be modulated by stimulating glucose utilization directly. In mice subjected to pressure overload, cardiac-specific overexpression of GLUT-1 increased glucose uptake and glycolysis and improved cardiac energetics and function [28]. Pharmacological strategies to stimulate glucose utilization directly are limited with dichloroacetate (DCA) being the most intensively studied compound. DCA indirectly activates the PDH complex and thus glucose oxidation. In patients with advanced HF, Bersin et al. showed that a 30-min-infusion of DCA stimulated myocardial lactate consumption, decreased oxygen consumption and increased cardiac work [29]. In rats with hyperthyroidism, long-term DCA treatment also increased flux through PDH and reduced cardiac hypertrophy [30]. Of note, although these approaches improved cardiac glucose oxidation, possible changes in accessory pathways of glucose metabolism such as the pentose phosphate pathway and the hexosamine biosynthetic pathway could also play a role but have not yet been assessed.

Taken together, there is a strong evidence suggesting that approaches targeting utilization of FA or glucose in the hypertrophied and failing heart may have great therapeutic potential. The benefits may involve enhanced cardiac ATP production, improved oxygen-efficiency and reduced cardiac hypertrophy. However, while studies consistently suggest improving cardiac glucose utilization, it is still a matter of debate how FA utilization in the diseased heart should be modulated. Finally, the role of cardiac substrate preference remains controversial and warrants further investigations.

AMINO ACID METABOLISM

While the majority of studies on cardiac metabolisms has focused on changes in the utilization of glucose and fatty acids, little is known about alterations in amino acid metabolism in cardiac hypertrophy and HF. Recently, the use of metabolomics has revealed profound changes in amino acid metabolism suggesting its role in metabolic remodeling. In mice subjected to pressure overload or myocardial infarction, Sansbury et al. reported significant increases in levels of branched-chain amino acids. Interestingly, their levels were negatively correlated with left ventricular ejection fraction [31]. By combining transcriptomic and quantitative targeted metabolomic profiling, Lai et al. found that expression of genes involved in the degradation of proline, alanine, tryptophan, and branched-chain amino acids were decreased in mice with compensated hypertrophy or HF [32]. These changes are of particular interest for at least two reasons. First, because the heart undergoes hypertrophy in the progression to HF, the inhibition of amino acid degradation may support protein synthesis required for hypertrophic growth. Second, considering that these amino acids can serve as sources of anaplerotic input into the tricarboxylic acid (TCA) cycle, a decrease in their degradation may reduce the pool of TCA cycle and thereby affect energy production. Thus, although specific changes in amino acid metabolism are still poorly understood, first data suggest an important role of amino acid metabolism in the regulation of cardiac hypertrophy and contractile function.

ANAPLEROSIS

Characteristic changes were also identified at the level of anaplerosis, a crucial process comprising metabolic pathways that replenish the pool of the TCA cycle. In hypertrophied rat hearts induced by aortic constriction, Sorokina et al. showed that glucose oxidation by PDH is unchanged despite increased glycolysis. More importantly, they identified an increase in anaplerotic flux into the TCA cycle probably via pyruvate carboxylation by malic enzyme [33]. The increase in this alternative route of pyruvate may account for the mismatch between glycolysis and glucose oxidation that is commonly seen in models of pressure-overload induced cardiac hypertrophy. Of note, this mismatch may also result from increased lactate formation, although this possibility remains to be evaluated in the hypertrophied heart.

Hypertrophy-associated anaplerotic changes have also been confirmed in mice with aortic constriction [27] and in hyperthyroid rats [26]. Thus, induction of anaplerotic pathways appears to be a hallmark of metabolic remodeling in cardiac hypertrophy. Considering that hypertrophic growth requires increased supply of amino acids and nucleic acids derived from precursors in the TCA cycle, then activation of anaplerotic pathways may be required to maintain TCA cycle function. However, this compensatory mechanism might also have unfavorable consequences (see Fig. 12.1 for illustration). For example, increased flux of pyruvate through anaplerotic pathways reduces its availability for oxidation by the PDH complex. As a result, pyruvate oxidation might not sufficiently compensate for the impaired FA oxidation,

leading to energetic inefficiency of the TCA cycle [33]. Furthermore, increased flux through malic enzyme may also affect cardiac function by consuming NADPH [34], which is required for triglyceride synthesis and for defending against oxidative stress. Therefore, the role of anaplerotic changes particularly in the regulation of cardiac hypertrophy and energy metaboslim remains incompletely understood. This knowledge gap is due in part to the technical challenges inherent in measuring anaplerosis.

THE PENTOSE PHOSPHATE PATHWAY

The regulation of the cellular redox environment is tightly linked to substrate metabolism via the PPP as an important source of NADPH. NADPH is required for the generation of cytosolic ROS, which at low levels, are involved in proliferation and survival signaling. Conversely, NADPH also maintains the pool of reduced antioxidants such as glutathione, which are crucial defenses against oxidative stress [35]. Hence, the PPP may play a dual role in the regulation of redox homeostasis. While much effort has been made to characterize glycolysis and glucose oxidation, little is known about the regulation of the PPP in HF. In dogs subjected to pacing-induced HF, increased superoxide levels were attributed to increased activity of G6PD, the key oxidative enzyme of the PPP [36]. If the effects of superoxide are indeed damaging, this activation of the PPP in the failing heart could be considered detrimental. However, G6PD deficient mice developed higher oxidative stress and worsened contractile function following myocardial infarction or aortic constriction [37]. In Dahl salt-sensitive rats, Kato et al. also found that the flux through the PPP progressively increased during the development of HF. Importantly, they demonstrated that treatment with DCA further increased this flux, which was associated with improved cardiac function [11]. Taken together, the PPP is activated in HF models. Although superoxide production may consequently increase, currently available data support the notion that higher flux through the PPP in HF may represent a compensatory mechanism whose further activation could be of therapeutic relevance.

THE HEXOSAMINE BIOSYNTHETIC PATHWAY

The hexosamine biosynthetic pathway (HBP), which is linked to metabolism of glucose, FAs, and amino acids, has been implicated in various models of heart disease. Recently, the role of the HBP in HF has also received special interest. By employing various models of HF including aortic constriction, myocardial infarction, and hypertensive rats, Lunde et al. demonstrated that global O-GlcNAcylation was increased by at least 40% in hypertrophied and failing hearts. They also confirmed these findings in patients with aortic stenosis [38]. In endothelial cells, increased flux through the HBP has been attributed to overproduction of mitochondrial superoxide [39]. Considering that mitochondrial ROS formation also increases in cardiac hypertrophy [40], mitochondrial ROS might represent a potential mechanism for activation of the HBP in the heart.

Whereas the induction of O-GlcNAcylation in the progression of HF is well described, less is known about the functional relevance of these changes. In a cardiomyocyte model of hypertrophy, Facundo et al. found that increased flux through the HBP and the resulting increase in O-GlcNAcylation were responsible for hypertrophic growth via activation of the nuclear factor of activated T-cells (NFAT) [41]. This finding is highly relevant because it indicates that induction of the HBP is an early event that triggers myocardial remodeling. Because NFAT signaling has been associated with pathological hypertrophy [42], one might consider this activation of the HBP to be detrimental. However, reducing O-GlcNAcylation

in mice with myocardial infarction exacerbated ventricular dysfunction [43]. Thus, although the HBP activates signaling pathways that initiate cardiac remodeling, it is possible that additional signaling pathways linked to the HBP also promote favorable effects. Further studies are therefore needed that focus on specific targets of O-GlcNAcylation in the context of clinical and experimental models of HF.

OTHER CELLULAR PROCESSES POTENTIALLY LINKED TO METABOLIC REMODELING

Mitochondrial Biogenesis and Function

Mitochondria are the center of energy metabolism. Perturbations in mitochondrial structure and function may therefore compromise cardiac energetics significantly. In rat hearts with systolic dysfunction induced by aortic constriction, we identified abnormal mitochondrial morphology, reduced mitochondrial volume density, and altered levels of most ETC proteins with the majority being decreased [4]. Mitochondrial proteomic remodeling in HF has also been demonstrated in mice with pressure overload, which is primarily characterized by altered abundance of proteins involved in the ETC and substrate metabolism [44]. Under normal conditions, mitochondrial material is continuously replenished by a process termed mitochondrial biogenesis. These results therefore suggest that perturbed mitochondrial biogenesis is an important feature of pressure overload-induced HF.

Consistent with this notion, the expression of PGC-1α, the master regulator of mitochondrial biogenesis, is decreased in systolic HF [4,45,46,6]. There is limited evidence for mitochondrial alterations in other models of HF. In failing rat hearts caused by myocardial infarction, we found normal mRNA expression of PGC-1α. However, the expression of p38 MAPK, which could modulate PGC-1α activity was reduced [15]. In HF patients of diverse etiologies, Karamanlidis et al. observed decreased mitochondrial DNA (mtDNA) content accompanied by reductions in mtDNA-encoded proteins. Of note, these changes were associated with increased abundance of PGC-1α protein but decreased expression of ERRα and Tfam, both at mRNA and protein levels [47]. These findings suggest that mitochondrial biogenesis signaling is depressed in advanced HF and that posttranslational modulation of PGC-1α may play a role.

Little is known about the regulation of mitochondrial biogenesis in compensated hypertrophy. We found a significant decline in mRNA expression of PGC-1α, ERRα, and Tfam at times when left ventricular ejection fraction was still preserved (unpublished observations), suggesting repressed mitochondrial biogenesis pathways in compensated hypertrophy. By contrast, in hypertrophied mouse hearts with normal function induced by chronic angiotensin II infusion, Dai et al. reported increased mRNA expression of PGC-1α, ERRα, and Tfam. However, these hearts exhibited increased mitochondrial damage and mitophagy [40]. Therefore, the induction of mitochondrial biogenic signaling could represent a compensatory mechanism triggered by mitochondrial injury and loss. Collectively, although the regulation of mitochondrial biogenesis and its signaling in cardiac hypertrophy and failure is still incompletely understood, available evidence supports the concept that cardiac mitochondria are affected early and undergo progressive remodeling during the development of cardiac hypertrophy and failure.

Mitochondrial dysfunction as evidenced by decreased mitochondrial respiration and reduced ATP production is consistently observed when systolic dysfunction occurs in various models of HF [2,48,49]. Interestingly, others and we found that mitochondrial respiration in pressure-overloaded hearts initially increased and did not fall until systolic dysfunction developed [2].

Considering that mitochondrial damage is probably an early event in the progression to HF, this biphasic response of mitochondrial respiratory capacity might reflect unknown compensatory mechanisms, which can sustain oxidative phosphorylation during the stage of compensated hypertrophy but are lost as cardiac dysfunction develops.

Due to the potential role of mitochondrial dysfunction in HF, many approaches targeting stimulation of mitochondrial biogenesis in the heart, with the focus on activation of PGC-1α, have been tested. Interestingly, mice with constitutive or inducible cardiac overexpression of PGC-1α exhibited mitochondrial abnormalities and dilated cardiomyopathy [50,51]. Despite these results, recent studies have evaluated the effects of modest overexpression of PGC-1α in the heart in the setting of pressure overload. Karamanlidis et al. found that a threefold overexpression of PGC-1α in pressure-overloaded mice even exacerbated contractile dysfunction and reduced survival [3]. Furthermore, Pereira et al. demonstrated that maintaining PGC-1α levels by a 40%-overexpression in mice did not prevent mitochondrial and contractile dysfunction [52]. Thus, forced activation of PGC-1α alone and via overexpression is unlikely a promising therapeutic approach for HF. Because the regulation of PGC-1α is highly complex, future studies should aim at its posttranslational modification as well as its interactions with other cellular processes. It is possible that a combined modulation of PGC-1α and additional targets is necessary to improve function outcome.

Mitochondrial ROS and its Link to Cardiac Metabolism

Increased cardiac ROS levels have been implicated in HF, but the role of ROS in the pathophysiology of HF was unclear as a result of disappointing outcomes of antioxidant interventions in human studies [53,54]. However, in a recent animal study, Dai et al. demonstrated that angiotensin II increased mitochondrial ROS, leading to mitochondrial damage, activation of mitogen-activated protein kinases and finally cardiac hypertrophy and fibrosis [40]. Another study by the same group also showed that mitochondria-targeted antioxidant treatment but not nontargeted ROS scavenging with N-acetyl cysteine ameliorated cardiomyopathy induced by chronic angiotensin II infusion [55]. Of note, hypertrophied hearts in this study presented mild diastolic dysfunction without changes in left ventricular ejection fraction [40], indicating an early stage of compensated hypertrophy. These results suggest that increased mitochondrial ROS might be an early event triggering structural remodeling and mitochondrial defects in hypertensive heart disease.

The regulation of cellular ROS homeostasis is complex and only partially understood. In the mitochondria, ROS are produced mainly from the electron transport chain (ETC) particularly from complex I and III. Therefore, changes in the ETC may favor "electron leakage" and consequently ROS formation [35]. The elevation in ROS production may further affect the function of the ETC leading to a vicious cycle (Fig. 12.1). Studies in various tissues have linked FA utilization to increased ROS levels [56,57]. In diabetic cardiomyopathy, ROS have also been associated with myocardial lipid accumulation, lipotoxicity, and decreased cardiac efficiency on the basis of mitochondrial uncoupling [58]. The failing heart is believed to be subjected to higher levels of free fatty acids, probably as a result of increased lipolysis [20,21]. In addition, models of HF have been associated with myocardial lipid accumulation [59,60]. As a result, FA utilization has been considered unfavorable for the stressed heart because of increasing ROS production [21]. However, in animal models with cardiac specific loss of ACC-2, increased myocardial FA utilization did not adversely impact left ventricular remodeling following pressure overload. Furthermore, cardiac FA oxidation is depressed during HF progression and there is

little direct evidence in HF models that FA utilization may increase ROS production. Thus, the causes of increased mitochondrial ROS in the progression to HF remain unclear and warrants further investigations.

Autophagy and its Link to Cardiac Metabolism

Metabolic remodeling in HF is also associated with changes in autophagy. Autophagy is a highly conserved process by which organelles and large cellular components are degraded. The products of autophagy (amino acids, fatty acids, sugars, and nucleosides) may then be channeled into both energy generating and biosynthesis pathways. Under normal conditions, basal autophagy is crucial by eliminating damaged organelles and misfolded proteins. In states of nutrient deprivation such as starvation or ischemia, autophagic activity is increased, which may support cell function by mobilizing endogenous nutritional sources [61]. In models of HF including pressure overload [62] and myocardial infarction [63], autophagy has also been shown to be induced. Mechanisms for the activation of autophagy in HF are less clear. A number of changes observed in the hypertrophied and failing heart such as energy depletion, AMPK activation [64], mitochondrial ROS overproduction, and damaged mitochondria are strong mediators of autophagy in various settings [61,65]. However, the potential contribution of these events to autophagic activation in HF remains to be verified.

Autophagy has also been implicated in ventricular remodeling. In an elegant study, Cao et al. provided evidence indicating that the activation of autophagy, although degradative in nature, is essential for cardiac hypertrophy [66]. Since cell growth requires stimulation of biosynthesis pathways, these data may appear paradoxical at first glance. However, they are supported by earlier studies showing that blunting of protein degradation mediated by the

UPS may attenuate cardiac hypertrophy [5,67]. Together, these data indicate that cardiac hypertrophy and remodeling are dynamic reconstructive processes, in which degradation of certain cellular components by autophagy and the UPS is required for the biosynthesis of new structures.

Due to the essential role of autophagy in ventricular remodeling, one might expect that inhibiting autophagy may prevent cardiac hypertrophy and failure. However, studies in knockout models with suppressed autophagy have delivered inconsistent results [68,69]. Given the highly complex regulation of autophagy, it is reasonable to assume that specific changes in the induction or targeting of autophagy but not the autophagic flux itself may affect functional outcome. For example, Oka et al. demonstrated in mice with aortic constriction that mtDNA that escapes from autophagy causes cardiac inflammation and failure [70]. Therefore, the regulation and the role of autophagy in HF remain to be fully elucidated. Because both the induction of autophagy and the processing of autophagic products are linked to metabolism, the role of metabolic remodeling in the regulation of autophagy is an attractive target for future studies.

CONCLUSIONS

HF is associated with profound changes in cardiac metabolism. Metabolic remodeling in HF is characterized by defects at the level of substrate utilization and mitochondrial biogenesis. Furthermore, it involves changes in metabolic pathways that regulate numerous essential cellular processes such as growth, redox homeostasis, and autophagy. Therefore, modulating cardiac metabolism may also affect these critical processes and improve cardiac function by mechanisms beyond the energetic aspect. However, the exact mechanisms linking metabolic changes to other pathological processes in HF

are still poorly understood. Future mechanistic investigations are therefore needed to decipher this complex network and improve the effectiveness of metabolic therapies for HF.

References

[1] Ho KK, Pinsky JL, Kannel WB, Levy D. The epidemiology of heart failure: the Framingham study. J Am Coll Cardiol 1993;22(4 Suppl A):6A–13A.

[2] Doenst T, Pytel G, Schrepper A, Amorim P, Farber G, Shingu Y, et al. Decreased rates of substrate oxidation *ex vivo* predict the onset of heart failure and contractile dysfunction in rats with pressure overload. Cardiovasc Res 2010;86(3):461–70.

[3] Karamanlidis G, Garcia-Menendez L, Kolwicz SC Jr, Lee CF, Tian R. Promoting PGC-1alpha-driven mitochondrial biogenesis is detrimental in pressure-overloaded mouse hearts. Am J Physiol Heart Circ Physiol 2014; 307(9).

[4] Bugger H, Schwarzer M, Chen D, Schrepper A, Amorim PA, Schoepe M, et al. Proteomic remodelling of mitochondrial oxidative pathways in pressure overload-induced heart failure. Cardiovasc Res 2010;85(2):376–84.

[5] Depre C, Wang Q, Yan L, Hedhli N, Peter P, Chen L, et al. Activation of the cardiac proteasome during pressure overload promotes ventricular hypertrophy. Circulation 2006;114(17):1821–8.

[6] Riehle C, Wende AR, Zaha VG, Pires KM, Wayment B, Olsen C, et al. PGC-1beta deficiency accelerates the transition to heart failure in pressure overload hypertrophy. Circ Res 2011;109(7):788–93.

[7] Zhabyeyev P, Gandhi M, Mori J, Basu R, Kassiri Z, Clanachan A, et al. Pressure-overload-induced heart failure induces a selective reduction in glucose oxidation at physiological afterload. Cardiovasc Res 2013; 97(4):676–85.

[8] Degens H, de Brouwer KF, Gilde AJ, Lindhout M, Willemsen PH, Janssen BJ, et al. Cardiac fatty acid metabolism is preserved in the compensated hypertrophic rat heart. Basic Res Cardiol 2006;101(1):17–26.

[9] Akki A, Smith K, Seymour AM. Compensated cardiac hypertrophy is characterised by a decline in palmitate oxidation. Mol Cell Biochem 2008;311(1–2):215–24.

[10] Allard MF, Schonekess BO, Henning SL, English DR, Lopaschuk GD. Contribution of oxidative metabolism and glycolysis to ATP production in hypertrophied hearts. Am J Physiol 1994;267(2 Pt 2):H742–50.

[11] Kato T, Niizuma S, Inuzuka Y, Kawashima T, Okuda J, Tamaki Y, et al. Analysis of metabolic remodeling in compensated left ventricular hypertrophy and heart failure. Circ Heart Fail 2010;3(3):420–30.

[12] Christe ME, Rodgers RL. Altered glucose and fatty acid oxidation in hearts of the spontaneously hypertensive rat. J Mol Cell Cardiol 1994;26(10):1371–5.

[13] Dodd MS, Ball DR, Schroeder MA, Le Page LM, Atherton HJ, Heather LC, et al. *In vivo* alterations in cardiac metabolism and function in the spontaneously hypertensive rat heart. Cardiovasc Res 2012;95(1): 69–76.

[14] Rosenblatt-Velin N, Montessuit C, Papageorgiou I, Terrand J, Lerch R. Postinfarction heart failure in rats is associated with upregulation of GLUT-1 and downregulation of genes of fatty acid metabolism. Cardiovasc Res 2001;52(3):407–16.

[15] Amorim PA, Nguyen TD, Shingu Y, Schwarzer M, Mohr FW, Schrepper A, et al. Myocardial infarction in rats causes partial impairment in insulin response associated with reduced fatty acid oxidation and mitochondrial gene expression. J Thorac Cardiovasc Surg 2010;140(5):1160–7.

[16] Heather LC, Cole MA, Lygate CA, Evans RD, Stuckey DJ, Murray AJ, et al. Fatty acid transporter levels and palmitate oxidation rate correlate with ejection fraction in the infarcted rat heart. Cardiovasc Res 2006;72(3):430–7.

[17] Osorio JC, Stanley WC, Linke A, Castellari M, Diep QN, Panchal AR, et al. Impaired myocardial fatty acid oxidation and reduced protein expression of retinoid X receptor-alpha in pacing-induced heart failure. Circulation 2002;106(5):606–12.

[18] Sack MN, Rader TA, Park S, Bastin J, McCune SA, Kelly DP. Fatty acid oxidation enzyme gene expression is downregulated in the failing heart. Circulation 1996;94(11):2837–42.

[19] Davila-Roman VG, Vedala G, Herrero P, de las Fuentes L, Rogers JG, Kelly DP, et al. Altered myocardial fatty acid and glucose metabolism in idiopathic dilated cardiomyopathy. J Am Coll Cardiol 2002;40(2):271–7.

[20] Ashrafian H, Frenneaux MP, Opie LH. Metabolic mechanisms in heart failure. Circulation 2007;116(4):434–48.

[21] Opie LH, Knuuti J. The adrenergic-fatty acid load in heart failure. J Am Coll Cardiol 2009;54(18):1637–46.

[22] Rajabi M, Kassiotis C, Razeghi P, Taegtmeyer H. Return to the fetal gene program protects the stressed heart: a strong hypothesis. Heart Fail Rev 2007;12(3–4):331–43.

[23] Taegtmeyer H, Sen S, Vela D. Return to the fetal gene program: a suggested metabolic link to gene expression in the heart. Ann N Y Acad Sci 2010;1188:191–8.

[24] Hue L, Taegtmeyer H. The Randle cycle revisited: a new head for an old hat. Am J Physiol Endocrinol Metab 2009;297(3):E578–91.

[25] Lopaschuk GD, Ussher JR, Folmes CD, Jaswal JS, Stanley WC. Myocardial fatty acid metabolism in health and disease. Physiol Rev 2010;90(1):207–58.

[26] Okere IC, Chess DJ, McElfresh TA, Johnson J, Rennison J, Ernsberger P, et al. High-fat diet prevents cardiac hypertrophy and improves contractile function in the hypertensive dahl salt-sensitive rat. Clin Exp Pharmacol Physiol 2005;32(10):825–31.

[27] Kolwicz SC Jr, Olson DP, Marney LC, Garcia-Menendez L, Synovec RE, Tian R. Cardiac-specific deletion of acetyl CoA carboxylase 2 prevents metabolic remodeling during pressure-overload hypertrophy. Circ Res 2012;111(6):728–38.

[28] Liao R, Jain M, Cui L, D'Agostino J, Aiello F, Luptak I, et al. Cardiac-specific overexpression of GLUT1 prevents the development of heart failure attributable to pressure overload in mice. Circulation 2002;106(16):2125–31.

[29] Bersin RM, Wolfe C, Kwasman M, Lau D, Klinski C, Tanaka K, et al. Improved hemodynamic function and mechanical efficiency in congestive heart failure with sodium dichloroacetate. J Am Coll Cardiol 1994;23(7):1617–1624.

[30] Atherton HJ, Dodd MS, Heather LC, Schroeder MA, Griffin JL, Radda GK, et al. Role of pyruvate dehydrogenase inhibition in the development of hypertrophy in the hyperthyroid rat heart: a combined magnetic resonance imaging and hyperpolarized magnetic resonance spectroscopy study. Circulation 2011;123(22):2552–61.

[31] Sansbury BE, DeMartino AM, Xie Z, Brooks AC, Brainard RE, Watson LJ, et al. Metabolomic analysis of pressure-overloaded and infarcted mouse hearts. Circ Heart Fail 2014;7(4):634–42.

[32] Lai L, Leone TC, Keller MP, Martin OJ, Broman AT, Nigro J, et al. Energy metabolic reprogramming in the hypertrophied and early stage failing heart: a multisystems approach. Circ Heart Fail. 2014;7(6):1022–31.

[33] Sorokina N, O'Donnell JM, McKinney RD, Pound KM, Woldegiorgis G, LaNoue KF, et al. Recruitment of compensatory pathways to sustain oxidative flux with reduced carnitine palmitoyltransferase I activity characterizes inefficiency in energy metabolism in hypertrophied hearts. Circulation 2007;115(15):2033–41.

[34] Pound KM, Sorokina N, Ballal K, Berkich DA, Fasano M, Lanoue KF, et al. Substrate-enzyme competition attenuates upregulated anaplerotic flux through malic enzyme in hypertrophied rat heart and restores triacylglyceride content: attenuating upregulated anaplerosis in hypertrophy. Circ Res 2009;104(6):805–12.

[35] Burgoyne JR, Mongue-Din H, Eaton P, Shah AM. Redox signaling in cardiac physiology and pathology. Circ Res 2012;111(8):1091–106.

[36] Gupte SA, Levine RJ, Gupte RS, Young ME, Lionetti V, Labinskyy V, et al. Glucose-6-phosphate dehydrogenase-derived NADPH fuels superoxide production in the failing heart. J Mol Cell Cardiol 2006;41(2):340–9.

[37] Hecker PA, Lionetti V, Ribeiro RF Jr, Rastogi S, Brown BH, O'Connell KA, et al. Glucose 6-phosphate dehydrogenase deficiency increases redox stress and moderately accelerates the development of heart failure. Circ Heart Fail 2013;6(1):118–26.

[38] Lunde IG, Aronsen JM, Kvaloy H, Qvigstad E, Sjaastad I, Tonnessen T, et al. Cardiac O-GlcNAc signaling is increased in hypertrophy and heart failure. Physiol Genomics 2012;44(2):162–72.

[39] Du XL, Edelstein D, Rossetti L, Fantus IG, Goldberg H, Ziyadeh F, et al. Hyperglycemia-induced mitochondrial superoxide overproduction activates the hexosamine pathway and induces plasminogen activator inhibitor-1 expression by increasing Sp1 glycosylation. Proc Natl Acad Sci USA 2000;97(22):12222–6.

[40] Dai DF, Johnson SC, Villarin JJ, Chin MT, Nieves-Cintron M, Chen T, et al. Mitochondrial oxidative stress mediates angiotensin II-induced cardiac hypertrophy and Galphaq overexpression-induced heart failure. Circ Res 2011;108(7):837–46.

[41] Facundo HT, Brainard RE, Watson LJ, Ngoh GA, Hamid T, Prabhu SD, et al. O-GlcNAc signaling is essential for NFAT-mediated transcriptional reprogramming during cardiomyocyte hypertrophy. Am J Physiol Heart Circ Physiol 2012;302(10).

[42] Wilkins BJ, Dai YS, Bueno OF, Parsons SA, Xu J, Plank DM, et al. Calcineurin/NFAT coupling participates in pathological, but not physiological, cardiac hypertrophy. Circ Res 2004;94(1):110–8.

[43] Watson LJ, Facundo HT, Ngoh GA, Ameen M, Brainard RE, Lemma KM, et al. O-linked beta-N-acetylglucosamine transferase is indispensable in the failing heart. Proc Natl Acad Sci USA 2010;107(41):17797–802.

[44] Dai DF, Hsieh EJ, Liu Y, Chen T, Beyer RP, Chin MT, et al. Mitochondrial proteome remodelling in pressure overload-induced heart failure: the role of mitochondrial oxidative stress. Cardiovasc Res 2012;93(1):79–88.

[45] Garnier A, Fortin D, Delomenie C, Momken I, Veksler V, Ventura-Clapier R. Depressed mitochondrial transcription factors and oxidative capacity in rat failing cardiac and skeletal muscles. J Physiol 2003;551(Pt 2):491–501.

[46] Arany Z, Novikov M, Chin S, Ma Y, Rosenzweig A, Spiegelman BM. Transverse aortic constriction leads to accelerated heart failure in mice lacking PPAR-gamma coactivator 1alpha. Proc Natl Acad Sci USA 2006;103(26):10086–91.

[47] Karamanlidis G, Nascimben L, Couper GS, Shekar PS, del Monte F, Tian R. Defective DNA replication impairs mitochondrial biogenesis in human failing hearts. Circ Res 2010;106(9):1541–8.

[48] Faerber G, Barreto-Perreia F, Schoepe M, Gilsbach R, Schrepper A, Schwarzer M, et al. Induction of heart failure by minimally invasive aortic constriction in mice: reduced peroxisome proliferator-activated receptor gamma coactivator levels and mitochondrial dysfunction. J Thorac Cardiovasc Surg 2011;141(2):492–500.

[49] Gundewar S, Calvert JW, Jha S, Toedt-Pingel I, Ji SY, Nunez D, et al. Activation of AMP-activated protein kinase by metformin improves left ventricular function and survival in heart failure. Circ Res 2009;104(3):403–11.

[50] Lehman JJ, Barger PM, Kovacs A, Saffitz JE, Medeiros DM, Kelly DP. Peroxisome proliferator-activated receptor gamma coactivator-1 promotes cardiac mitochondrial biogenesis. J Clin Invest 2000;106(7):847–56.

[51] Russell LK, Mansfield CM, Lehman JJ, Kovacs A, Courtois M, Saffitz JE, et al. Cardiac-specific induction of the transcriptional coactivator peroxisome proliferator-activated receptor gamma coactivator-1alpha promotes mitochondrial biogenesis and reversible cardiomyopathy in a developmental stage-dependent manner. Circ Res 2004;94(4):525–33.

[52] Pereira RO, Wende AR, Crum A, Hunter D, Olsen CD, Rawlings T, et al. Maintaining PGC-1alpha expression following pressure overload-induced cardiac hypertrophy preserves angiogenesis but not contractile or mitochondrial function. FASEB J 2014;28(8):3691–702.

[53] Yusuf S, Dagenais G, Pogue J, Bosch J, Sleight P. Vitamin E supplementation and cardiovascular events in high-risk patients. The heart outcomes prevention evaluation study investigators. N Engl J Med 2000;342(3):154–60.

[54] Lonn E, Bosch J, Yusuf S, Sheridan P, Pogue J, Arnold JM, et al. Effects of long-term vitamin E supplementation on cardiovascular events and cancer: a randomized controlled trial. JAMA 2005;293(11):1338–47.

[55] Dai DF, Chen T, Szeto H, Nieves-Cintron M, Kutyavin V, Santana LF, et al. Mitochondrial targeted antioxidant peptide ameliorates hypertensive cardiomyopathy. J Am Coll Cardiol 2011;58(1):73–82.

[56] Inoguchi T, Li P, Umeda F, Yu HY, Kakimoto M, Imamura M, et al. High glucose level and free fatty acid stimulate reactive oxygen species production through protein kinase C-dependent activation of NAD(P)H oxidase in cultured vascular cells. Diabetes 2000;49(11):1939–45.

[57] Schonfeld P, Wojtczak L. Fatty acids decrease mitochondrial generation of reactive oxygen species at the reverse electron transport but increase it at the forward transport. Biochim Biophys Acta 2007;1767(8):1032–40.

[58] Boudina S, Sena S, Theobald H, Sheng X, Wright JJ, Hu XX, et al. Mitochondrial energetics in the heart in obesity-related diabetes: direct evidence for increased uncoupled respiration and activation of uncoupling proteins. Diabetes 2007;56(10):2457–66.

[59] Sharma S, Adrogue JV, Golfman L, Uray I, Lemm J, Youker K, et al. Intramyocardial lipid accumulation in the failing human heart resembles the lipotoxic rat heart. FASEB J 2004;18(14):1692–700.

[60] Krishnan J, Suter M, Windak R, Krebs T, Felley A, Montessuit C, et al. Activation of a HIF1alpha-PPARgamma axis underlies the integration of glycolytic and lipid anabolic pathways in pathologic cardiac hypertrophy. Cell Metab 2009;9(6):512–24.

[61] Rabinowitz JD, White E. Autophagy and metabolism. Science 2010;330(6009):1344–8.

[62] Rothermel BA, Hill JA. Autophagy in load-induced heart disease. Circ Res 2008;103(12):1363–9.

[63] Kanamori H, Takemura G, Goto K, Maruyama R, Tsujimoto A, Ogino A, et al. The role of autophagy emerging in postinfarction cardiac remodelling. Cardiovasc Res 2011;91(2):330–9.

[64] Tian R, Musi N, D'Agostino J, Hirshman MF, Goodyear LJ. Increased adenosine monophosphate-activated protein kinase activity in rat hearts with pressure-overload hypertrophy. Circulation 2001;104(14):1664–9.

[65] Huang J, Lam GY, Brumell JH. Autophagy signaling through reactive oxygen species. Antioxid Redox Signal 2011;14(11):2215–31.

[66] Cao DJ, Wang ZV, Battiprolu PK, Jiang N, Morales CR, Kong Y, et al. Histone deacetylase (HDAC) inhibitors attenuate cardiac hypertrophy by suppressing autophagy. Proc Natl Acad Sci USA 2011;108(10): 4123–8.

[67] Hedhli N, Depre C. Proteasome inhibitors and cardiac cell growth. Cardiovasc Res 2010;85(2):321–9.

[68] Nakai A, Yamaguchi O, Takeda T, Higuchi Y, Hikoso S, Taniike M, et al. The role of autophagy in cardiomyocytes in the basal state and in response to hemodynamic stress. Nat Med 2007;13(5):619–24.

[69] Zhu H, Tannous P, Johnstone JL, Kong Y, Shelton JM, Richardson JA, et al. Cardiac autophagy is a maladaptive response to hemodynamic stress. J Clin Invest 2007;117(7):1782–93.

[70] Oka T, Hikoso S, Yamaguchi O, Taneike M, Takeda T, Tamai T, et al. Mitochondrial DNA that escapes from autophagy causes inflammation and heart failure. Nature 2012;485(7397):251–5.

13

Energetics in the Hypertrophied and Failing Heart

Craig A. Lygate

Radcliffe Department of Medicine, Division of Cardiovascular Medicine,
University of Oxford, Oxford, UK

THE CREATINE KINASE PHOSPHAGEN SYSTEM – THE BASICS

Cells with high-energy requirements, such as cardiomyocytes, possess a phosphagen system for the purposes of optimizing energy transport and providing a buffer at times of high demand. In all vertebrates, this role is taken by the creatine kinase (CK) system, with CK catalyzing the reversible transfer of a high-energy phosphoryl group from ATP to creatine to form phosphocreatine (PCr) and ADP [1]: Cr + ATP ↔ PCr + ADP + H$^+$.

Phosphocreatine accumulates to high levels and is more readily diffusible than ATP and therefore represents a mobile form of energy storage, which is preferentially used to maintain normal ATP levels when energy demand outstrips supply. This negates the need for ATP and ADP diffusion and helps to maintain favorable localized metabolite levels, for example, low (ATP/ADP) ratio at the mitochondria to stimulate energy production and high (ATP/ADP)

at sites of energy utilization in order to maximize the energy available from ATP hydrolysis (ΔG_{ATP}) [2–4].

The forward reaction is catalyzed by sarcomeric mitochondrial-creatine kinase (Mt-CK), located within the mitochondrial membrane and the reverse reaction by cytosolic CK dimers with the muscle-isoform predominant (MM-CK; 55–67%) along with low abundance (1–3%) MB- and BB-isoenzymes [5]. Cardiomyocytes do not normally express the enzymes for creatine biosynthesis, so uptake via the creatine transporter (CrT) is the sole mechanism for creatine to enter the cell.

Approximately 10% of myocardial phosphotransfer is via the activity of adenylate kinase (AK) [6] which catalyses the reversible reaction 2ADP ↔ ATP + AMP [7]. This allows for interchange between adenine nucleotides (i.e., AMP, ADP, and ATP, the sum of which is termed total adenine nucleotides, TAN). AK serves at least two additional important functions during metabolic stress: as a mechanism to utilize both the γ- and β-phosphoryl groups of ATP, thus doubling the energy available from ATP, [8] and

The Scientist's Guide to Cardiac Metabolism
http://dx.doi.org/10.1016/B978-0-12-802394-5.00013-3

as an important metabolic signaling and sensing molecule via changes in cytosolic AMP availability, in turn stimulating the AMPK-PGC-1α signaling cascade [9].

IMPAIRED ENERGETICS IN THE FAILING HEART

There is abundant observational data to indicate that a reduced total CK activity with commensurate reduction in total creatine levels (creatine + PCr) are hallmarks of the failing heart [3]. Observations in humans date back to 1939 [10] and have since been replicated in a wide variety of species and disease models (summarized in Table 13.1). Typically, similar reductions in both Mt-CK and MM-CK are observed, and in some models a small (in absolute terms) compensatory increase in BB-CK activity [11,12]. AK activity is less prone to change even in severe congestive heart failure, [13] which means the relative importance to phosphotransfer is likely increased [6].

The signaling mechanisms leading to downregulation remain to be fully elucidated. However, loss of CK activity has been linked to reversible posttranslational modification, [12] while reduced creatine levels are a consequence of CrT downregulation, with insufficient creatine-uptake to counter normal creatine loss [11,20]. It is important to note that ATP levels are protected by the CK reaction favoring ATP production at the expense of PCr, such that ATP only falls in end-stage heart failure when there is progressive loss of adenine nucleotides [2,21]. This makes the PCr/ATP ratio a convenient and sensitive indicator of early energetic dysfunction, which can be quantified by Phosphorus-31-nuclear magnetic resonance (NMR).

There is strong correlative evidence linking energetic changes to heart function. For example, PCr/ATP ratio is low in patients with dilated cardiomyopathy, correlating with ejection fraction and adverse remodeling, and improving when symptoms improve [22,23]. Such changes also have prognostic value; with both low PCr/ATP and ATP transfer rate predicting increased mortality in human heart failure [24,25]. These correlations suggest gradations in energetic compromise between normal, mild dysfunction, and heart failure, and it seems likely that a deficit does exist in the hypertrophied heart, although detecting it will depend

TABLE 13.1   Impaired Cardiac Energetics are a Consistent Finding in the Failing Heart

Species and etiology of heart failure	Total Cr	Total CK activity	Mito-CK	MM-CK	MB-CK	BB-CK
Mouse – aortic banding [11]	↓11%	↓30%	↓32%	↓29%	↓29%	↔
Mouse – myocardial infarction [11]	↓16%	↓26%	↓27%	↓29%	↔	↑95%
Rat – myocardial infarction [14,15]	↓35%	↓19%	↓35%	↓21%	↔	↔
Syrian cardiomyopathic hamster [16]	↓56%	↓34%	↓18%	↓42%	↔	NR
Turkey – induced cardiomyopathy [17]	↓25%	↓34%	NR	NR	NR	NR
Pig – aortic banding [18]	↓26%	↓48%	↓28%	NR	NR	NR
Dog – pacing [12]	↓39%	↓25%	↓30%	↓33%	↑76%	↑109%
Human – dilated cardiomyopathy [19]	↓33%	↓18%	↔	↓28%	↔	Trace

This table is illustrative of the literature and is not meant to be exhaustive. Percentage changes in CK system in heart failure shown relative to controls. Only statistically significant changes are shown. NR denotes "not reported." CK is total creatine kinase; Cr is total creatine levels. *Adapted from Ref. [11].*

on the sensitivity of the parameters measured. For example, mice with LV hypertrophy due to transverse aortic constriction only developed significant changes in CK activity and Cr with the development of congestive heart failure, [11] but low PCr/ATP is already detectable in this model at 3 weeks, when there is LV hypertrophy but before overt dysfunction [26]. Similarly, patients with hypertensive LV hypertrophy were shown to have low PCr/ATP and diastolic dysfunction at a time when systolic function was normal [27]. Early PCr/ATP impairment has also been observed in obese patients [28] and patients with type 2 diabetes [29]. Measuring ATP synthesis rates (CK flux) may be even more sensitive than measuring metabolite levels, since it was able to detect distinct energetic differences between normal, hypertrophic, and failing hearts [30].

Such studies suggest that energetic impairment is an early phenomenon that could have a causal role in driving subsequent adverse cardiac remodeling and dysfunction. However, evidence from loss-of-function experiments have been equivocal and suggests a more nuanced relationship.

DOES IMPAIRED ENERGETICS CAUSE CARDIAC DYSFUNCTION *PER SE?*

A large number of studies have sought to gain insight into the relative importance of the CK system by performing loss-of-function experiments, either with pharmacological approaches or using genetic knockout models. A detailed discussion is beyond the scope of this review, however, taken collectively, a general pattern emerges that inhibition of the CK system has very little effect on resting baseline function, but affects the ability of the heart to attain maximal workloads, that is, that CK is a major determinant of contractile reserve [2,3]. It would seem that under the right conditions

energetically compromised mice may even develop heart failure (eventually), but this is not a universal finding and depends on a permissive genetic background. Some consideration of the methodological limitations and key results are outlined in the following section.

Creatine Depletion by β-Guanidinopropionic Acid

β-GPA is a competitive inhibitor of creatine uptake, but a poor substrate for CK, with only 1% of the reaction velocity [31]. An important advantage is the ability to give in food or water to live animals, but it requires chronic dosing over weeks since creatine loss is slow (~2% per day), [32] potentially allowing sufficient time for compensatory adaptations (e.g., increased mitochondrial density) [33]. Furthermore, there is always some residual creatine (10–50%), which may be sufficient to support near normal function and may explain some of the experimental variability [34]. As with any pharmacological approach, accumulation of β-GPA may have toxic or off-target effects and whole body creatine-deficiency leads to a loss of 10–15% in body weight, which is likely to have confounding metabolic consequences. Nevertheless, *ex vivo* systolic dysfunction in creatine-depleted rat hearts was partially rescued by introducing creatine to the perfusate and this was accompanied by a simultaneous increase in intracellular PCr [35].

CK Inhibition by Iodoacetamide

IA is a rapidly acting irreversible inhibitor of all the CK isoenzymes [36]. As an alkylating agent, IA is inherently toxic with multiple potential off-target effects, and its use has therefore been mostly confined to acute studies in isolated perfused hearts. For example, it can also inhibit glyceraldehyde 3-phosphate dehydrogenase (GAPDH), although apparently without altering glycolysis rates, suggesting that this is not a confounding factor at the workloads achievable *ex vivo* [37]. Graded inhibition by IA has been used to demonstrate that the energy available from ATP hydrolysis (ΔG_{ATP}) is dependent on CK activity and limits maximal cardiac performance [38]. This infers that the reduced contractile reserve observed in the failing heart may be a direct consequence of impaired energetics.

CK Knockout Mice

The cardiac phenotype has been studied in M-CK, Mt-CK, and double M/Mt-CK knockout mice. These are constitutive whole-body knockouts so there may be systemic metabolic effects and multiple potentially compensatory adaptations have been described [39,40]. This may explain why phenotypes have been relatively mild, particularly in the M-CK$^{-/-}$, which appears to have no energetic or functional deficit even at high workloads [41–45]. *Ex vivo* studies in M/Mt-CK$^{-/-}$ mice have demonstrated impaired energetics, including ATP synthesis rate, but without a functional correlate, [43] and again this may reflect the limitations of low workloads *ex vivo*.

In vivo findings have been inconsistent, probably reflecting sex and age differences between studies, but most fundamentally, as a result of a mixed and diverging genetic background. For example, extensive LV hypertrophy was noted in some studies of M/Mt-CK$^{-/-}$ mice, [44] which was absent in earlier studies arising from the same colony [46]. These strains have since been back-crossed onto a pure C57BL/6 background and compared directly with the mixed background in the same laboratory. At one year of age, male M/Mt-CK$^{-/-}$ mice on the mixed background developed adverse LV remodeling and congestive heart failure, while the phenotype in females was much less severe. In stark contrast, mice on a pure genetic background exhibited no structural remodeling and only very mild hemodynamic compromise, while still maintaining contractile reserve [45]. These findings are open to interpretation. On one level, they can be taken as evidence that a primary defect in the CK system is sufficient, in itself, to cause heart failure. On the other, there are important caveats, it requires a permissive genetic background, takes a very long time to develop, and occurs at a level of CK knockdown much greater than occurs in the failing heart.

Creatine Deficiency by Genetic Modification

Mice with knockout of guanidinoacetate methyltransferase (GAMT$^{-/-}$) have a whole-body absolute creatine-deficiency. This results in a very low body weight due to reduced lean mass and body fat, [47] which may have metabolic consequences for the heart. Again there remains potential for long-term adaptations, although if these exist, they are not obvious [48,49]. A major potential confounder is the accumulation of the creatine precursor guanidinoacetate, which can be phosphorylated and utilized under extreme conditions such as ischemia; however, phosphotransfer was below the limit of detection under normal conditions [49]. This does not rule out potential toxic effects from guanidinoacetate accumulation. Notably, ATP levels are maintained despite zero PCr [50] and there is no LV structural remodeling even up to 1 year [51]. *In vivo* cardiac function is normal at baseline with the exception of lower LV systolic pressure, while contractile reserve is impaired [49].

A new model of creatine-deficiency has recently been described by genetic ablation of arginine: glycine amidinotransferase (AGAT), the first enzyme in creatine biosynthesis [52]. It was expected that this would provide a cleaner model without the accumulation of guanidinoacetate that occurs in GAMT$^{-/-}$; however the AGAT enzyme has since been shown to also synthesise homoarginine, [53,54] so these mice have chronic low levels of plasma homoarginine, which may be a confounding factor.

DOES IMPAIRED CARDIAC ENERGETICS CONTRIBUTE TO PROGRESSION OF HEART FAILURE?

Initial studies using β-GPA feeding suggested that creatine-deficient rats are unable to survive acute myocardial infarction, [55,56] however GAMT$^{-/-}$ mice have normal survival in the same MI model [49]. The most likely explanation is enhanced arrhythmia, [55] which is not evident in the mouse model due to the intrinsically low arrhythmic susceptibility in this species. When β-GPA feeding was initiated immediately after myocardial infarction with follow-up 8 weeks later, treated rats had 87% lower PCr and 13% lower ATP levels compared to control infarct animals and yet cardiac dysfunction was not exacerbated [56]. This is in agreement with a study of post-MI heart failure in the creatine-free GAMT$^{-/-}$ mouse, which demonstrated that survival, LV remodeling and function were no worse than controls [49]. Similar results have been obtained for M/Mt-CK$^{-/-}$ mice in the same post-MI model using MRI follow-up at 4 weeks [44,57]. These four independent studies seriously question whether the extent of CK system downregulation observed in the failing heart makes any meaningful contribution to disease progression. It has been suggested that creatine loss is itself a compensatory mechanism to maintain ΔG_{ATP} when ATP levels are reduced, [58] but this does not explain why creatine levels decline well in advance of ATP in the failing heart.

AUGMENTATION OF CARDIAC ENERGETICS AS A THERAPEUTIC STRATEGY IN THE FAILING HEART

Increasing myocardial Cr has been achieved by constitutively over-expressing the creatine transporter in the heart (CrT-OE mice). Very high levels of Cr >twofold increase caused cardiac hypertrophy and dysfunction due to impaired glycolysis and an inability to keep the augmented creatine pool adequately phosphorylated [59,60]. However, creatine levels up to twofold are safe even over a prolonged 18 month period [61]. Mice within this range were subjected to myocardial infarction, but despite augmented creatine levels throughout, there was no measureable benefit on LV remodeling or function at 6 weeks [61]. In contrast, mice over-expressing M-CK in the heart had increased CK flux and exhibited improved survival and contractile function in the transverse aortic constriction model of pressure-overload heart failure. These favorable effects were lost when the transgene was switched off part way through the protocol [62]. There is a clear rationale for testing a combined approach of creatine elevation with enhanced CK activity. An effective method of preserving the total adenine nucleotide pool would also be theoretically favorable.

CARDIAC ENERGETICS IN ISCHEMIA/REPERFUSION (I/R)

It should be noted that an intact CK system is fundamentally important in protecting the heart against ischemia and the damaging effects of reperfusion. Both CK$^{-/-}$ and creatine-deficient mice have impaired functional recovery in *ex vivo* I/R experiments [50,63]. Furthermore, both creatine elevation and M-CK overexpression have independently been shown to be cardioprotective in this setting [61,64].

SUMMARY AND CONCLUSIONS

There is a large body of evidence that impairment of cardiac energetics is a hallmark of the failing heart and that the CK system is important in attaining high workloads and in determining contractile reserve. There is also evidence that altered cardiac energetics can directly result in cardiac dysfunction, although this is highly dependent on other factors. Despite this, neither creatine nor CK deficiency exacerbates the development of heart failure in animal models, thereby questioning the significance of the much smaller changes that occur in the failing heart. Nevertheless, the finding that increasing CK activity is beneficial in heart failure marks a new era promising therapeutics that specifically target cardiac energetics.

Acknowledgment

Work in the author's laboratory is funded by the British Heart Foundation.

References

[1] Wyss M, Kaddurah-Daouk R. Creatine and creatinine metabolism. Physiol Rev 2000;80:1107–213.

[2] Ingwall JS, Weiss RG. Is the failing heart energy starved? On using chemical energy to support cardiac function. Circ Res 2004;95:135–45.

[3] Neubauer S. The failing heart—an engine out of fuel. New Engl J Med 2007;356:1140–51.

[4] Wallimann T, Wyss M, Brdiczka D, Nicolay K, Eppenberger HM. Intracellular compartmentation, structure and function of creatine kinase isoenzymes in tissues with high and fluctuating energy demands: the 'phosphocreatine circuit' for cellular energy homeostasis. Biochem J 1992;281(Pt 1):21–40.

[5] Schlattner U, Tokarska-Schlattner M, Wallimann T. Mitochondrial creatine kinase in human health and disease. Biochim Biophys Acta 2006;1762:164–80.

[6] Dzeja PP, Vitkevicius KT, Redfield MM, Burnett JC, Terzic A. Adenylate kinase-catalyzed phosphotransfer in the myocardium: increased contribution in heart failure. Circ Res 1999;84:1137–43.

[7] Dzeja PP, Terzic A. Phosphotransfer networks and cellular energetics. J Exp Biol 2003;206:2039–47.

[8] Dzeja P, Kalvenas A, Toleikis A, Praskevicius A. The effect of adenylate kinase activity on the rate and efficiency of energy transport from mitochondria to hexokinase. Biochem Int 1985;10:259–65.

[9] Frederich M, Balschi JA. The relationship between AMP-activated protein kinase activity and AMP concentration in the isolated perfused rat heart. J Biol Chem 2002;277:1928–32.

[10] Herrmann G, Decherd GM. The chemical nature of heart failure. Ann Intern Med 1939;12:1233–44.

[11] Lygate CA, Fischer A, Sebag-Montefiore L, Wallis J, Ten Hove M, Neubauer S. The creatine kinase energy transport system in the failing mouse heart. J Mol Cell Cardiol 2007;42:1129–36.

[12] Shen W, Spindler M, Higgins MA, Jin N, Gill RM, Bloem LJ, Ryan TP, Ingwall JS. The fall in creatine levels and creatine kinase isozyme changes in the failing heart are reversible: complex post-transcriptional regulation of the components of the CK system. J Mol Cell Cardiol 2005;39:537–44.

[13] Aksentijevic D, Lygate CA, Makinen K, Zervou S, Sebag-Montefiore L, Medway D, Barnes H, Schneider JE, Neubauer S. High-energy phosphotransfer in the failing mouse heart: role of adenylate kinase and glycolytic enzymes. Eur J Heart Fail 2010;12:1282–9.

[14] Laser A, Ingwall JS, Tian R, Reis I, Hu K, Gaudron P, Ertl G, Neubauer S. Regional biochemical remodeling in non-infarcted tissue of rat heart post-myocardial infarction. J Mol Cell Cardiol 1996;28:1531–8.

[15] Neubauer S, Frank M, Hu K, Remkes H, Laser A, Horn M, Ertl G, Lohse MJ. Changes of creatine kinase gene expression in rat heart post-myocardial infarction. J Mol Cell Cardiol 1998;30:803–10.

[16] Nascimben L, Friedrich J, Liao R, Pauletto P, Pessina AC, Ingwall JS. Enalapril treatment increases cardiac performance and energy reserve via the creatine kinase reaction in myocardium of Syrian myopathic hamsters with advanced heart failure. Circulation 1995;91:1824–33.

[17] Liao R, Nascimben L, Friedrich J, Gwathmey JK, Ingwall JS. Decreased energy reserve in an animal model of dilated cardiomyopathy. Relationship to contractile performance. Circ Res 1996;78:893–902.

[18] Ye Y, Gong G, Ochiai K, Liu J, Zhang J. High-energy phosphate metabolism and creatine kinase in failing hearts: a new porcine model. Circulation 2001;103:1570–6.

[19] Nascimben L, Ingwall JS, Pauletto P, Friedrich J, Gwathmey JK, Saks V, Pessina AC, Allen PD. Creatine kinase system in failing and nonfailing human myocardium. Circulation 1996;94:1894–901.

[20] Neubauer S, Remkes H, Spindler M, Horn M, Wiesmann F, Prestle J, Walzel B, Ertl G, Hasenfuss G, Wallimann T. Downregulation of the Na(+)-creatine cotransporter in failing human myocardium and in experimental heart failure. Circulation 1999;100:1847–50.

[21] Shen W, Asai K, Uechi M, Mathier MA, Shannon RP, Vatner SF, Ingwall JS. Progressive loss of myocardial ATP due to a loss of total purines during the development of heart failure in dogs: a compensatory role for the parallel loss of creatine. Circulation 1999;100:2113–8.

[22] Neubauer S, Krahe T, Schindler R, Horn M, Hillenbrand H, Entzeroth C, Mader H, Kromer EP, Riegger GA, Lackner K, et al. 31P magnetic resonance spectroscopy in dilated cardiomyopathy and coronary artery disease. Altered cardiac high-energy phosphate metabolism in heart failure. Circulation 1992;86:1810–8.

[23] Neubauer S, Horn M, Pabst T, Gödde M, Lübke D, Jilling B, Hahn D, Ertl G. Contributions of 31P-magnetic resonance spectroscopy to the understanding of dilated heart muscle disease. Eur Heart J 1995;16:115–8.

[24] Neubauer S, Horn M, Cramer M, Harre K, Newell JB, Peters W, Pabst T, Ertl G, Hahn D, Ingwall JS, Kochsiek K. Myocardial phosphocreatine-to-ATP ratio is a predictor of mortality in patients with dilated cardiomyopathy. Circulation 1997;96:2190–6.

[25] Bottomley PA, Panjrath GS, Lai S, Hirsch GA, Wu K, Najjar SS, Steinberg A, Gerstenblith G, Weiss RG. Metabolic rates of ATP transfer through creatine kinase (CK flux) predict clinical heart failure events and death. Sci Transl Med 2013;5:215re3.

[26] Maslov MY, Chacko VP, Stuber M, Moens AL, Kass DA, Champion HC, Weiss RG. Altered high-energy phosphate metabolism predicts contractile dysfunction and subsequent ventricular remodeling in pressure-overload hypertrophy mice. Am J Physiol Heart Circ Physiol 2007;292:H387–91.

[27] Lamb HJ, Beyerbacht HP, van der Laarse A, Stoel BC, Doornbos J, van der Wall EE, de Roos A. Diastolic dysfunction in hypertensive heart disease is associated with altered myocardial metabolism. Circulation 1999;99:2261–7.

[28] Rider OJ, Francis JM, Ali MK, Holloway C, Pegg T, Robson MD, Tyler D, Byrne J, Clarke K, Neubauer S. Effects of catecholamine stress on diastolic function and myocardial energetics in obesity/clinical perspective. Circulation 2012;125:1511–9.

[29] Scheuermann-Freestone M, Madsen PL, Manners D, Blamire AM, Buckingham RE, Styles P, Radda GK, Neubauer S, Clarke K. Abnormal cardiac and skeletal muscle energy metabolism in patients with type 2 diabetes. Circulation 2003;107:3040–6.

[30] Smith CS, Bottomley PA, Schulman SP, Gerstenblith G, Weiss RG. Altered creatine kinase adenosine triphosphate kinetics in failing hypertrophied human myocardium. Circulation 2006;114:1151–8.

[31] Oudman I, Clark JF, Brewster LM. The effect of the creatine analogue beta-guanidinopropionic acid on energy metabolism: a systematic review. PLoS ONE 2013;8: e52879.

[32] Bloch K, Schoenheimer R, Rittenberg D. Rate of formation and disappearance of body creatine in normal animals. J Biol Chem 1941;138:155–66.

[33] Wiesner RJ, Hornung TV, Garman JD, Clayton DA, O'Gorman E, Wallimann T. Stimulation of mitochondrial gene expression and proliferation of mitochondria following impairment of cellular energy transfer by inhibition of the phosphocreatine circuit in rat hearts. J Bioenerg Biomembr 1999;31:559–67.

[34] Kapelko VI, Kupriyanov VV, Novikova NA, Lakomkin VL, Steinschneider AY, Severina MY, Veksler VI, Saks VA. The cardiac contractile failure induced by chronic creatine and phosphocreatine deficiency. J Mol Cell Cardiol 1988;20:465–79.

[35] Zweier JL, Jacobus WE, Korecky B, Brandejs-Barry Y. Bioenergetic consequences of cardiac phosphocreatine depletion induced by creatine analogue feeding. J Biol Chem 1991;266:20296–304.

[36] Hamman BL, Bittl JA, Jacobus WE, Allen PD, Spencer RS, Tian R, Ingwall JS. Inhibition of the creatine kinase reaction decreases the contractile reserve of isolated rat hearts. Am J Physiol Heart Circ Physiol 1995;269: H1030–6.

[37] Tian R, Christe ME, Spindler M, Hopkins JC, Halow JM, Camacho SA, Ingwall JS. Role of MgADP in the development of diastolic dysfunction in the intact beating rat heart. J Clin Invest 1997;99:745–51.

[38] Tian R, Ingwall JS. Energetic basis for reduced contractile reserve in isolated rat hearts. Am J Physiol 1996; 270:H1207–16.

[39] Boehm E, Ventura-Clapier R, Mateo P, Lechene P, Veksler V. Glycolysis supports calcium uptake by the sarcoplasmic reticulum in skinned ventricular fibres of mice deficient in mitochondrial and cytosolic creatine kinase. J Mol Cell Cardiol 2000;32:891–902.

[40] Kaasik A, Veksler V, Boehm E, Novotova M, Minajeva A, Ventura-Clapier R. Energetic crosstalk between organelles: architectural integration of energy production and utilization. Circ Res 2001;89:153–9.

[41] Ventura-Clapier R, Kuznetsov AV, d'Albis A, van Deursen J, Wieringa B, Veksler VI. Muscle creatine kinase-deficient mice. I. Alterations in myofibrillar function. J Biol Chem 1995;270:19914–20.

[42] Van Dorsten FA, Nederhoff MG, Nicolay K, Van Echteld CJ. 31P NMR studies of creatine kinase flux in M-creatine kinase-deficient mouse heart. Am J Physiol Heart Circ Physiol 1998;275:H1191–9.

[43] Saupe KW, Spindler M, Tian R, Ingwall JS. Impaired cardiac energetics in mice lacking muscle-specific isoenzymes of creatine kinase. Circ Res 1998;82:898–907.

[44] Nahrendorf M, Spindler M, Hu K, Bauer L, Ritter O, Nordbeck P, Quaschning T, Hiller KH, Wallis J, Ertl G, Bauer WR, Neubauer S. Creatine kinase knockout mice show left ventricular hypertrophy and dilatation, but unaltered remodeling post-myocardial infarction. Cardiovasc Res 2005;65:419–27.

[45] Lygate CA, Medway DJ, Ostrowski PJ, Aksentijevic D, Sebag-Montefiore L, Hunyor I, Zervou S, Schneider JE, Neubauer S. Chronic creatine kinase deficiency eventually leads to congestive heart failure, but severity is dependent on genetic background, gender and age. Basic Res Cardiol 2012;107:276.

[46] Spindler M, Niebler R, Remkes H, Horn M, Lanz T, Neubauer S. Mitochondrial creatine kinase is critically necessary for normal myocardial high-energy phosphate metabolism. Am J Physiol Heart Circ Physiol 2002;283:H680–7.

[47] Schmidt A, Marescau B, Boehm EA, Renema WK, Peco R, Das A, Steinfeld R, Chan S, Wallis J, Davidoff M, Ullrich K, Waldschutz R, Heerschap A, De Deyn PP, Neubauer S, Isbrandt D. Severely altered guanidino

compound levels, disturbed body weight homeostasis and impaired fertility in a mouse model of guanidino-acetate N-methyltransferase (GAMT) deficiency. Hum Mol Genet 2004;13:905–21.

[48] Branovets J, Sepp M, Kotlyarova S, Jepihhina N, Sokolova N, Aksentijevic D, Lygate CA, Neubauer S, Vendelin M, Birkedal R. Unchanged mitochondrial organization and compartmentation of high-energy phosphates in creatine-deficient GAMT-/- mouse hearts. Am J Physiol Heart Circ Physiol 2013;305:H506–20.

[49] Lygate CA, Aksentijevic D, Dawson D, Ten Hove M, Phillips D, de Bono JP, Medway DJ, Sebag-Montefiore L, Hunyor I, Channon KM, Clarke K, Zervou S, Watkins H, Balaban RS, Neubauer S. Living without creatine: unchanged exercise capacity and response to chronic myocardial infarction in creatine-deficient mice. Circ Res 2013;112:945–55.

[50] ten Hove M, Lygate CA, Fischer A, Schneider JE, Sang AE, Hulbert K, Sebag-Montefiore L, Watkins H, Clarke K, Isbrandt D, Wallis J, Neubauer S. Reduced inotropic reserve and increased susceptibility to cardiac ischemia/reperfusion injury in phosphocreatine-deficient guanidinoacetate-N-methyltransferase-knockout mice. Circulation 2005;111:2477–85.

[51] Schneider JE, Stork LA, Bell JT, Hove MT, Isbrandt D, Clarke K, Watkins H, Lygate CA, Neubauer S. Cardiac structure and function during ageing in energetically compromised Guanidinoacetate N-methyltransferase (GAMT)-knockout mice – a one year longitudinal MRI study. J Cardiovasc Magn Reson 2008;10:9.

[52] Choe CU, Nabuurs C, Stockebrand MC, Neu A, Nunes P, Morellini F, Sauter K, Schillemeit S, Hermans-Borgmeyer I, Marescau B, Heerschap A, Isbrandt D. l-arginine:glycine amidinotransferase deficiency protects from metabolic syndrome. Hum Mol Genet 2013;22:110–23.

[53] Davids M, Ndika JDT, Salomons GS, Blom HJ, Teerlink T. Promiscuous activity of arginine:glycine amidino-transferase is responsible for the synthesis of the novel cardiovascular risk factor homoarginine. FEBS Lett 2012;586:3653–7.

[54] Choe CU, Atzler D, Wild PS, Carter AM, Böger RH, Ojeda F, Simova O, Stockebrand M, Lackner K, Nabuurs C, Marescau B, Streichert T, Müller C, Lüneburg N, De Deyn PP, Benndorf RA, Baldus S, Gerloff C, Blankenberg S, Heerschap A, Grant PJ, Magnus T, Zeller T, Isbrandt D, Schwedhelm E. Homoarginine levels are regulated by l-arginine:glycine amidinotransferase and affect stroke outcome: results from human and murine studies. Circulation 2013;128:1451–61.

[55] Lorentzon M, Ramunddal T, Bollano E, Soussi B, Waagstein F, Omerovic E. In vivo effects of myocardial creatine depletion on left ventricular function, morphology, and energy metabolism–consequences in acute myocardial infarction. J Card Fail 2007;13:230–7.

[56] Horn M, Remkes H, Stromer H, Dienesch C, Neubauer S. Chronic phosphocreatine depletion by the creatine analogue beta-guanidinopropionate is associated with increased mortality and loss of ATP in rats after myocardial infarction. Circulation 2001;104:1844–9.

[57] Nahrendorf M, Streif JU, Hiller KH, Hu K, Nordbeck P, Ritter O, Sosnovik DE, Bauer L, Neubauer S, Jakob PM, Ertl G, Spindler M, Bauer WR. Multimodal functional cardiac MR imaging in creatine kinase deficient mice reveals subtle abnormalities in myocardial perfusion and mechanics. Am J Physiol Heart Circ Physiol 2006;290:H2516–21.

[58] Shen W, Vatner DE, Vatner SF, Ingwall JS. Progressive loss of creatine maintains a near normal [Delta]G~ATP in transgenic mouse hearts with cardiomyopathy caused by overexpressing Gs[alpha]. J Mol Cell Cardiol 2010; 48:591–9.

[59] Wallis J, Lygate CA, Fischer A, Ten Hove M, Schneider JE, Sebag-Montefiore L, Dawson D, Hulbert K, Zhang W, Zhang MH, Watkins H, Clarke K, Neubauer S. Supranormal myocardial creatine and phosphocreatine concentrations lead to cardiac hypertrophy and heart failure: insights from creatine transporter-overexpressing transgenic mice. Circulation 2005;112:3131–9.

[60] Phillips D, Ten Hove M, Schneider JE, Wu CO, Sebag-Montefiore L, Aponte AM, Lygate CA, Wallis J, Clarke K, Watkins H, Balaban RS, Neubauer S. Mice over-expressing the myocardial creatine transporter develop progressive heart failure and show decreased glycolytic capacity. J Mol Cell Cardiol 2010;48:582–90.

[61] Lygate CA, Bohl S, Ten Hove M, Faller KM, Ostrowski PJ, Zervou S, Medway DJ, Aksentijevic D, Sebag-Montefiore L, Wallis J, Clarke K, Watkins H, Schneider JE, Neubauer S. Moderate elevation of intracellular creatine by targeting the creatine transporter protects mice from acute myocardial infarction. Cardiovasc Res 2012;96:466–75.

[62] Gupta A, Akki A, Wang Y, Leppo MK, Chacko VP, Foster DB, Caceres V, Shi S, Kirk JA, Su J, Lai S, Paolocci N, Steenbergen C, Gerstenblith G, Weiss RG. Creatine kinase-mediated improvement of function in failing mouse hearts provides causal evidence the failing heart is energy starved. J Clin Invest 2012;122: 291–302.

[63] Spindler M, Meyer K, Stromer H, Leupold A, Boehm E, Wagner H, Neubauer S. Creatine kinase-deficient hearts exhibit increased susceptibility to ischemia-reperfusion injury and impaired calcium homeostasis. Am J Physiol Heart Circ Physiol 2004;287:H1039–45.

[64] Akki A, Su J, Yano T, Gupta A, Wang Y, Leppo MK, Chacko VP, Steenbergen C, Weiss RG. Creatine kinase over-expression improves atp kinetics and contractile function in post-ischemic myocardium. Am J Physiol Heart Circ Physiol 2012;303:H844–52.

Cardiac Metabolism – The Link to Clinical Practice

Paul Christian Schulze*, Peter J. Kennel*, Torsten Doenst†, Linda R. Peterson**

*Department of Medicine, Division of Cardiology, Columbia University Medical Center, New York, New York; †Department of Cardiothoracic Surgery, Jena University Hospital, Friedrich Schiller University Jena, Jena, Germany; **Department of Medicine, Cardiovascular Division, Washington University School of Medicine, St. Louis, Missouri, USA

INTRODUCTION

Metabolic processes are of outmost importance for both the normal and pathologic structure and function of the cardiovascular system. Yet, most findings are based on tissue analysis and experimental animal models since noninvasive quantification of myocardial metabolism using nuclear and magnetic resonance techniques has not been available or fully established in many centers for the assessment of clinical changes in humans. Characterization of metabolic processes in the cardiovascular system is still not a standard procedure and the majority of metabolic imaging is semiquantitative. However, many findings related to pathophysiologic changes do not require imaging or quantification in humans in order to be tested for their clinical relevance. Besides limitations in the availability and standardization of noninvasive imaging of cardiac metabolism, interventions for the

therapeutic modulation of cardiac metabolism are also limited. There is, however, a growing portfolio of drugs and experimental compounds that utilize metabolic modulation for their clinical effects. This chapter will illustrate the links between the basic principles of cardiac metabolism and clinical relevance and current practice. We will review noninvasive cardiac imaging for the assessment of metabolic processes and provide an overview of the current spectrum of medications with effects on human cardiac metabolism as well as metabolic compounds investigated for the treatment of human disease.

NONINVASIVE ANALYSIS OF CARDIAC METABOLISM – IMAGING

Investigators have measured human myocardial metabolism since 1949 [1]. However, in these early studies, measurements were made using invasive coronary catheterization and coronary sinus cannulation. Now, myocardial metabolic flux rates in human tissue can be measured noninvasively using radionuclear imaging and magnetic resonance spectroscopy techniques. Nuclear imaging of myocardial metabolism requires radionuclear tracers. These are generally divided into two groups: those that are made by a generator and those that are cyclotron-made. Tracers made by a generator tend to have longer half-lives and to be cheaper in general than cyclotron-made tracers. In contrast, cyclotron-developed radionuclear tracers can give more detailed information about intracellular myocardial metabolic processes such as oxidation and esterification. A comprehensive list of all tracers is beyond the scope of this chapter; however, a list of many available tracers is available in a recent review [2].

SPECT Imaging

Clinically, single photon emission computed tomography (SPECT) is most commonly used to determine blood flow and myocardial ischemia/infarction. However, it can also be used to evaluate myocardial fatty acid and/or glucose metabolic rates. SPECT imaging has certain advantages: it is a widely available technique, and it can be used with ECG-gating to generate functional data at the same time [3]. Clinically, SPECT may be used for detection of the metabolic signatures of stunned or hibernating myocardium [4].

Glucose Metabolism

There are no specific SPECT tracers for glucose metabolism, but with specific detection schemes and/or collimators, glucose uptake can be assessed using SPECT imaging after [^{18}F]2-deoxy-2-fluoro-d-glucose (FDG) injection (more detail on FDG later in the chapter) [3]. SPECT imaging after FDG injection (in conjunction with blood flow imaging) can be used for the detection of viable but ischemic myocardium.

A perfusion-FDG metabolism "mismatch" (see PET section later in the chapter) is indicative of viable myocardium as opposed to scar tissue. However, SPECT has a relatively low temporal and spatial resolution in comparison with positron emission tomography (PET). In addition, SPECT imaging does not lend itself to the measurement of intracellular glucose metabolism.

Fatty Acid Metabolism

Similarly, SPECT can be used with 15-(*p*-iodophenyl)-pentadecanoic acid (IPPA) or ^{123}I-beta-methyl-p-iodophenylpentadecanoic acid (BMIPP) for measurement of myocardial fatty acid metabolism. IPPA can rapidly accumulate in the myocardium and IPPA clearance rates correlate with β-oxidation of fatty acids [3]. In patients with coronary artery disease, tracer uptake/washout are reduced in ischemic segments. However, because of the poor temporal resolution of SPECT, image quality is reduced [3]. BMIPP, on the other hand, allows for better image quality owing to better radiotracer retention [3]. (Like FDG, BMIPP is taken up by the myocardium but is then trapped in the tissue rather than metabolized) [3].

One of the most common uses of BMIPP imaging clinically is in the evaluation of ischemia. BMIPP uptake of an ischemic segment may remain abnormally low even after resolution of chest pain and normalization of myocardial blood flow. BMIPP uptake is low because there is a relatively slow recovery of fatty acid metabolism after ischemia, a condition in which the myocardium preferentially refers to glucose metabolism. This phenomenon is termed "ischemic memory" and can help clinicians make the diagnosis of coronary artery disease [5]. For example, in one multicenter study of patients in the emergency room with recent chest pain, the addition of rest BMIPP SPECT imaging increased the sensitivity of the diagnosis of ischemia from 43% (based on EKG and clinical assessment) to 81%. The negative predictive value changed from 62% to 83%. Recently, one small SPECT imaging study of myocardial metabolism has shown SPECT myocardial fatty acid imaging to be useful in the diagnosis of Takotsubo cardiomyopathy [6]. In this study, SPECT imaging using BMIPP showed much greater apical defect than in patients with acute myocardial infarction [6]. Decreased fatty acid metabolism, as measured using BMIPP with SPECT imaging, is also a poor prognostic indicator in patients with end-stage renal disease, presumably due to silent ischemia [7]. However, there are limitations of SPECT for metabolic imaging. First, measurements are only semiquantitative. Second, SPECT has relatively poor spatial and temporal resolution and correction for photon attenuation [3]. Thus, the use of SPECT for research in myocardial metabolic imaging is somewhat limited.

Positron Emission Tomography Imaging

The primary imaging technique for *quantification* of myocardial metabolism is PET. PET also has better spatial resolution than SPECT imaging. Unlike SPECT, PET imaging also allows for correction of the attenuation caused by tissue overlying the heart. PET uses positron-emitting radionuclides such as ^{15}O, ^{11}C, ^{13}N, and ^{18}F, which can be incorporated into molecules the heart uses in metabolism, such as 1-^{11}C-palmitate. Thus, the molecule used as a tracer is extremely similar to the nonradiolabeled, endogenous compound being traced (e.g., palmitate). This means that the body should treat the radiolabeled tracer molecule in the same way as the compound traced. Moreover, these radionuclides can be incorporated into radiopharmaceuticals, so that specific receptors can be evaluated [8]. Also, now there are new combination scanners – PET/CT and PET/MRI, which allow for structural imaging in addition to the metabolic information provided by PET alone. The CT can also be used to determine the attenuation correction needed for the PET. Specific recommendations regarding PET instrumentation, quality control, image acquisition, image processing, patient preparation, image display, protocols, and clinical reporting are beyond the scope of this chapter but are outlined in the guidelines issued by the American Society of Nuclear Cardiology (ASNC) [9].

Myocardial Glucose Uptake and Metabolism Measurement – Assessment of Hibernating Myocardium and Beyond

[^{18}F]2-deoxy-2-fluoro-d-glucose (FDG), a generator-made radiolabeled tracer, is the most common PET tracer used for measuring myocardial glucose uptake, especially for clinical purposes. Clinically, FDG–PET is the gold standard (used in conjunction with a tracer of blood flow such as $^{13}NH_3$) for determining the presence of viable "hibernating myocardium." This is myocardium that is viable and metabolically active (takes up FDG) but has reduced blood flow and does not contract significantly ("meat that eats but doesn't beat"). In these PET viability studies, one looks for a "flow-metabolism mismatch [10]." That is to say, where FDG uptake is preserved relative to decreased flow there is viable myocardium [10]. However, several factors can affect FDG uptake by the heart in these clinical viability studies. Dietary state, cardiac work, sympathetic stimulation, and insulin sensitivity will all affect myocardial FDG uptake [10]. One way of overcoming the effects of these factors on FDG uptake is to perform the PET scan during a hyperinsulinemic-euglycemic clamp (see Fig. 14.1). The increased plasma insulin level helps to increase the uptake of FDG even in the setting of insulin resistance. The use of FDG/$^{13}NH_3$ PET is a Class I indication for the prediction of improvement of left-ventricular function after revascularization [10]. In general, if FDG uptake rate is ≥ 0.25 micromol/g/min, it suggests that a hibernating myocardial segment will likely improve at least one functional grade after revascularization [10].

There are, however, some limitations to using FDG as a tracer of glucose uptake, especially for research purposes. Using FDG for calculating myocardial glucose uptake requires the use of a "lumped constant." This ratio is derived from the use of 2-deoxyglucose relative

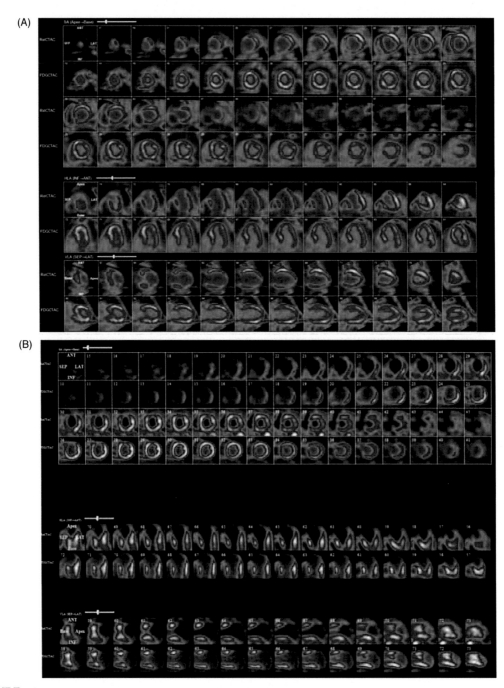

FIGURE 14.1 **Panel A. An example of hibernating myocardium.** Resting myocardial blood flow after attenuation correction by computed tomography injection with 15N-ammonia, labeled RestTAC. Resting myocardial glucose metabolism after attenuation is shown in the rows labeled FDGTAC. The top 4 rows are from the short-axis view (SAX). The next 2 rows are in the horizontal long-axis view and the bottom 2 rows of Panel A are in the vertical long axis view. The blood flow (RestTAC) images show a large defect in the apical, apical and mid-septal, apical and mid-inferior, and apical and mid-posterior walls. The images of myocardial metabolism, however, do not match those of the blood flow. Instead, almost all of the aforementioned segments have FDG uptake, indicative of viable myocardium. **Panel B. An example of irreversible myocardial injury/scar.** These RestTAC images and images after FDG injection were obtained from a different patient with a history of myocardial infarction. In these sets of images the defect seen in blood flow (RestTAC) matches with that seen in the FDG images. Thus, there appears to be no viable myocardium, only scar.

to glucose. The uptake rate of FDG over time using the lumped constant is linear. Unfortunately, the lumped constant can vary based on metabolic condition [11,12]. Also, because FDG is trapped in the cytosol after uptake, FDG can only be used to estimate glucose uptake, but not oxidation [3].

[11]C-glucose, on the other hand, is a cyclotron-derived tracer that can also be used to quantify myocardial glucose uptake *and oxidation*. The time activity curves of the myocardium and the blood generated from PET after ^{11}C-glucose injection are used in conjunction with compartmental models to determine the intracellular fate of glucose that is taken up by the heart. The paper by Herrero et al. lists the details of this multicompartmental model [13]. This model was validated against direct arterial coronary sinus sampling in canine hearts under multiple conditions such as hyperinsulinemic euglycemic clamp, clamp + intralipid infusion, and fasting [13]. From this model, one can calculate myocardial glucose uptake, oxidation, as well as glycolysis and glycogen deposition rates, while taking into account the contribution of ^{11}C-lactate generation [13].

Myocardial Lactate Metabolism Measurement

The heart is an avid consumer of plasma lactate. Lactate is a significant source of fuel for the heart especially during exercise and other conditions of increased plasma lactate. Until 2007, measurement of human myocardial lactate uptake and utilization could only be performed by the invasive, coronary artery–coronary sinus catheterization method. However, with the development of the physiologic tracer, L-3-^{11}C-lactate and the corresponding mathematical models, noninvasive measurement of myocardial metabolism of exogenous lactate is now feasible [14].

Myocardial Oxygen Consumption Measurement

Total oxygen consumption may be measured using PET and either ^{15}O-oxygen or ^{11}C-acetate. Oxygen is the final electron receptor in aerobic metabolism, so oxygen extraction can be measured and then multiplied by myocardial blood flow and arterial oxygenation in order to generate oxygen consumption (MVO_2). The half-life of ^{15}O-oxygen is very short, and its use requires complex kinetic modeling [3]. ^{11}C-acetate is perhaps more commonly used to measure MVO_2. Since acetate is rapidly converted to acetyl-coA and then enters the Krebs cycle, which is closely linked to oxidative phosphorylation, the flux of ^{11}C-acetate reflects total MVO_2. Although MVO_2 alone is not typically used in clinical situations, it is useful for translational research of both MVO_2 and, when combined with a measure of cardiac work, MVO_2 can be used to calculate efficiency. For example, MVO_2 has been shown to be increased and efficiency decreased with increasing obesity [15]. Moreover, the change in MVO_2 that occurs with weight loss predicts the improvement in left-ventricular diastolic function [16].

Myocardial Fatty Acid Metabolism (Flux) Measurement

In general, fatty acids are the preferred substrate for the adult human myocardium. Fatty acids are thought to provide 60–70% of the energy needed by the adult heart in the fasting state. There are both generator- and cyclotron-derived fatty acid tracers. The generator-derived tracers include fatty acid analogs such as 14-(R,S)-^{18}F-fluoro-6-thiaheptadecanoic acid (FHTA) and 16-^{18}F-fluoro-4-thia-palmitate (FTP). The latter is more responsive to hypoxic conditions. Like FDG, the FTP-derived measurement of myocardial fatty acid requires the use of a lumped constant. ^{11}C-palmitate, on the other hand, is a cyclotron-generated tracer of fatty acid metabolism. Quantification of fatty acid uptake, utilization, and oxidation using radiolabeled palmitate requires kinetic modeling, fitting of the blood and myocardial time-activity curves, and correction for blood [11]CO_2 [17]. There has been a renewed interest in quantification of myocardial fatty acid uptake, utilization, and oxidation because of the discovery of the associations between excessive fatty acid metabolism and decreased left-ventricular function in animal models of diseases such as type II diabetes [18]. Congruent with the findings in animal models, studies in humans also show increased myocardial fatty acid uptake, utilization, and oxidation flux rates with obesity and/or type II diabetes [15,19].

Magnetic Resonance Spectroscopy (MRS)

Myocardial Fatty Acid Deposition Measurement

Fat deposition (as opposed to flux measured using PET) in the heart can now be measured noninvasively using ^{1}H-magnetic resonance spectroscopy (^{1}H-MRS). The use of this technique for cardiac lipids in humans was pioneered by Lidia Szczepaniak in 2003 [20]. It takes advantage of the magnetic spin of ^{1}H for generating distinct resonance peaks for molecules containing hydrogen. The largest peak on the spectra of the heart is water. Other peaks such as those from lipid species are much smaller but can be relatively quantified in relation to the water peak. The interventricular septum is the preferred region of interest for interrogation for lipid deposition because it is the farthest from overlying subcutaneous and pericardial fat. This noninvasive fat-measuring technique correlates well with biochemical measurements of fat in myocardial biopsies [21].

MR with Hyperpolarization and Dynamic Nuclear Polarization (DNP)

This is a relatively new technique. Currently, this is used for cardiovascular research purposes only. However, in 2010, this technique was first applied to cancer patients, so in the future this technique may be applied clinically [22]. MR with hyperpolarization and DNP can be used with ^{13}C-labeled tracers (in supraphysiologic levels) for the noninvasive evaluation of pathways such as the Krebs cycle and glycolysis. DNP requires a very high magnetic field and very low temperatures to force most nuclear spins to point in the same direction, thus increasing the MR signal dramatically. A full description of this technique was reviewed by Schroeder et al. [23].

Myocardial Oxygenation with MRI

There are now MR-based methods for measuring tissue oxygenation. However, these are generally reserved for research purposes. The most common MR-based method is called "BOLD" – the blood oxygen level-dependent technique. This method takes advantage of the differing MR signals from oxygenated and deoxygenated blood [2]. Another MR-based method is performed after administration of ^{17}O-enriched agent, so that the ^{17}O-water made from metabolic processes may be measured. However, both techniques are technologically challenging and are generally reserved for research purposes.

The role of prandial states and plasma substrate levels on measures of myocardial metabolism. When interpreting myocardial metabolic flux or deposition data, it is important to consider the different fasting/fed states of the subject because the heart is so responsive to its metabolic milieu and the substrates being presented to it. For example, myocardial *deposition* increases markedly with prolonged fasting, presumably because plasma-free fatty acid levels are elevated due to peripheral lipolysis [24]. (On the other hand, eating a high-fat diet and increasing plasma triglycerides does not appear to alter myocardial fat deposition) [24]. Plasma substrate levels also affect myocardial fatty acid *flux*, as measured by PET. For example, when rosiglitazone or acipimox decrease plasma-free fatty acid levels, myocardial fatty acid uptake and utilization decrease [25,26].

Clinically, when evaluating for myocardial viability using FDG–PET imaging, patients are often given an oral glucose (25–100 g) load after fasting for ~ 6 h, to increase insulin response. This will enhance FDG uptake in the myocardium [9]. Diabetic patients often require exogenous intravenous glucose and insulin in the form of a hyperinsulinemic/euglycemic clamp to enhance FDG uptake. Several standard protocols for enhancement of FDG uptake are outlined by the 2009 American Society of Nuclear Cardiology guidelines [9].

Reading the previous text, it becomes clear, that the noninvasive imaging and/or quantification is complex. However, it is of extreme importance to eventually be able to assess the impact of metabolic therapy in humans. As illustrated later in the chapter, there are many drugs with metabolic effect, yet, the relative contribution of this metabolic effect to the treatment effect is not always clear.

THERAPEUTIC INTERVENTIONS – MODULATORS OF FATTY ACID UPTAKE AND METABOLISM

An extensive body of research is available characterizing metabolic derangements of glucose and fatty acid metabolism in the failing heart. Fundamental changes in fatty acid metabolism have been observed, whereas the impact on glucose metabolism remains controversial. The healthy heart has the highest oxygen consumption rate per weight of all organs [27] and its high-ATP turnover relies mainly on fatty acid oxidation, which contributes up to 90% of the substrate for oxidative phosphorylation, and, to a lesser extent, glucose oxidation [28]. For example, the ATP yield from substrate turnover of 1 mol glucose or 1 mol palmitate, respectively, is 38 mol ATP/mol glucose and 129 mol ATP/mol palmitate. However, in this calculation, it has to be taken into account, that 2 mol ATP are needed per mol for a palmitate-to-palmitate-CoA activation for further metabolization of the fatty acid [29]. In the healthy myocardium, with unrestricted coronary artery flow, oxygen delivery is not a limiting factor and the myocardial oxygen extraction is highly efficient matching its high oxygen demand. However, in myocardial ischemia, differences in the substrates' ATP/O ratio, which reflects the ratio of energy produced per oxygen consumed, have to be taken into account. This ratio is higher for glucose than for palmitate: complete glucose oxidation requires 6 mol O_2 per mol glucose, whereas oxidation of 1 mol of palmitate demands for 23 mol O_2 [30]. The high demand of oxygen for fatty acid substrate utilization raised the question whether pharmacologic inhibition of fatty acid oxidation and a compensatory increase of glucose metabolization has protective effects on the heart under ischemic conditions. For this purpose, a variety of pharmacologic agents affecting the fatty acid oxidation pathway were developed, with several of them having been studied in clinical trials (Table 14.1) [31]. Fatty acid oxidation is an elaborately orchestrated process with several enzymatic key reactions on the level of fatty acid transmembrane translocation via conjugation with carnitine, acyl-CoA-esterification, and along the β-oxidation pathway: after acyl-CoA esterification, the fatty acids are transported into the mitochondria via carnitine-shuttling, with carnitine palmitoyl-transferase 1 and 2 (CPT1 and 2) as pacemaker-enzymes for this step. In the mitochondria, the acyl-CoA-carbon backbone is oxidized with 2 carbons being cleaved with every cycle.

The investigated fatty acid oxidation inhibitors can be classified by their mechanism of action in CPT inhibitors, with the substances perhexiline, etomoxir, amiodarone, and metoprolol being the most important agents, and, with the exception of etomoxir, being approved and marketed, inhibitors of steps in the β-oxidation cycle such as ranolazine, trimetazidine, and carnitine synthesis inhibitors such as Met-88.

CPT Inhibitors

This is the largest group of fatty acid oxidation modulators. Perhexiline is a long-known antianginal agent with potent CPT-1 as well as CPT-2 inhibitory function [31,52]. The largest

TABLE 14.1   Pharmacologic Agents Modulating Myocardial Fatty Acid Oxidation

Substance	Mechanism of action on FA metabolism	Adverse effects	Trials	Marketed
Perhexiline	CPT1/2 inhibition	CYP-metabolized/ GIT adverse effects/ neurotoxicity/ hepatotoxicity	Beadle (50 patients) [32]/Cole (17 patients) [33]/Abozguia (46 patients) [34]/Phan (151 patients) [35]	Yes
Ranolazine	PDH activation/ 3-Ketoacyl-CoA thiolase inhibition [36]	CYP-metabolized/ GIT adverse effects/QT prolongation?	Ling (150 patients) [37]/ Shammas (28 patients) [38]	Yes
Trimetazidine	3-Ketoacyl-CoA thiolase inhibition [39]	CYP-metabolized/ GIT adverse effects/QT prolongation? Parkinsonoid symptoms?	Fragasso (669 patients) [40]/ Fragasso (55 patients) [41]/ Di Napoli (51 patients) [42]/Winter (60 patients) [43]/Tuunanen (19 patients) [44]	Yes
Amiodarone	CPT1 inhibition	CYP-metabolized. ILD, Iodine deposits	Kennedy (animal study) [45]	Yes (as Class III AR)
Metoprolol	CPT1 inhibition, inhibition of peripheral adipose tissue lipolysis decrease of myocardial demand	Related to β-blockade	Wallhaus (9 patients) [46]/ Podbregar (26 patients) [47]	Yes (as β-blocker)
MET-88 – Meldonium – Mildronate	Gamma-butyrobetaine dioxygenase inhibition	Allergic skin reactions	Dzerve (512 patients) [48]	Yes
Etomoxir	CPT1-inhibition/ carnitine-acyltransferase inhibition	Hepatotoxicity	Holubarsch (121 patients) [49]/Schmidt-Schweda (10 patients) [50]	No (terminated in Phase II)
Oxfenicine	CPT1 inhibition		Drake-Holland (animal study) [51]	No (preclinical)
Emeriamine	CPT2 inhibition			No (preclinical)

ILD, interstitial lung disease; GIT, gastrointestinal; AR, antiarrhythmic drug.

study to date, retrospectively analyzing a cohort of 151 CHF patients, reports a symptomatic improvement without changes in mortality outcome [35] for this drug. In another trial, Abozguia et al. [34] report an improved myocardial phosphocreatine to adenosine triphosphate ratio, improved diastolic function, and increased exercise capacity in their cohort of 46 patients with hypertrophic cardiomyopathy treated with perhexiline compared to placebo. In a recent study, Beadle et al. report a significant improvement in cardiac energetics and NYHA class in a double-blind study in 50 patients with nonischemic systolic HF compared to placebo.

Etomoxir is an irreversible CPT1-specific inhibitor preventing the transport from fatty acids into the mitochondrial matrix for further metabolization. The compound has been developed in the 1980s and reached Phase II clinical trials such as the ERGO-trial [49], which however had to be terminated prematurely because of elevated transaminase levels in a portion of patients. In the small cohort who completed the study, the investigators reported a trend toward an improved exercise capacity of the CHF patients.

Amiodarone and Metoprolol are well-established drugs in the CHF therapy regimen. While these medications are used for their primary mechanism of action as Class III antiarrhythmic drug and β-blocker, respectively, it has been shown that Amiodarone [45] and β-blockers such as Metoprolol [47] also have an inhibitory effect on CPT. The β-blocking effect of both drugs also likely decreases myocardial fatty acid uptake due to inhibition of peripheral lipolysis thereby decreasing plasma fatty acid availability for the heart. The pleiotropic effect of these drugs may contribute to their proven beneficial effect in CHF therapy.

β-Oxidation Inhibitors

To inhibit pivotal steps of β-oxidation, 3-ketoacyl-CoA thiolase inhibitiors have been developed. These inhibitors act on the final step of the β-oxidation cycle in which the acyl-CoA is cleaved by another CoA yielding acetyl-CoA and an acyl-CoA shortened by 2 carbons. Ranolazine as the main agent of this group has been approved by the FDA in 2006 for the treatment of angina alone or with other drugs. In a study including 150 patients with refractory angina pectoris, ranolazine treatment achieved a significant improvement in CCS angina class and the number of hospitalizations were significantly lower during ranolazine therapy [37].

Trimetazidine, another drug of the thiolase-inhibitor class, has been studied in several large clinical trials. A recent multicenter retrospective analysis of 669 patients concluded that trimetazidine reduces mortality and event-free survival in patients with CHF and that the addition of the drug to optimal medical therapy improves long-term survival in CHF patients [40]. Trimetazidine is approved by the EMA and available for the treatment of patients with angina pectoris as a second-line, add-on therapy.

Carnitine Synthesis Inhibitors

Fatty acids are transported into the mitochondrial matrix to undergo β-oxidation via the carnitine–acylcarnitine translocase as carnitine fatty acid complexes. Mildronate is a Gamma-butyrobetaine dioxygenase (BBOX)-inhibitor blocking the last step of L-carnitine biosynthesis. This drug is marketed in several eastern European countries, but not approved by the EMA or FDA. Despite its clinical use, few robust clinical trials are available. One randomized, double-blind, placebo controlled study of 512 patients with stable CHF reports an improvement of total exercise time in the treatment group compared to placebo [48].

MODULATORS OF GLUCOSE UPTAKE AND METABOLISM

Originating from the observations as outlined previously, another approach to alter cardiac energy substrate metabolism is to increase cardiac glucose utilization. To achieve this, several concepts have been tested in experimental and clinical studies (Table 14.2). Approaches, therefore, are to increase cellular glucose uptake via Glucose–Insulin–Potassium (GIK-) Infusion or increase endogenous insulin release, activity, or sensitivity using antidiabetic agents, also taking advantage of their pleiotropic and complex effects on several metabolic pathways.

Glucose–Insulin–Potassium Infusion

The concept of increasing glucose supply to the ischemic or failing myocardium has been investigated since the 1970s but attracts interest to date. The underlying concept of GIK therapy

TABLE 14.2   Pharmacologic Agents Modulating Myocardial Glucose Metabolism

Substance/ intervention	Mechanism of action	Adverse effects	Selected trials	Marketed
GIK infusion	Pleiotropic insulin effects (PI3K-NOS, antiinflammatory, antiapoptotic, FFA metabolization modulation)/ substrate shift?	Fluid overload hyperkalemia	IMMEDIATE trial 2012 (871 patients) [53–55]/Howell [56]/Rasoul (889 patients) [57] GIPS-II study [58]	In-patient therapy
Biguanides – Metformin	Pleiotropic effects on glucose and FA metabolism (ETC complex I, AMPK, and others)	GIT adverse effects, lactic acidosis	Ladeiras-Lopes MET-DIME [59]/Lexis [60]/Wong [61]/ Aguilar [62]	Yes as OAD
Incretin mimetics (GLP analogs and DPP4-inhibitors)	Increased contractiliy via cAMP increase; insulin release	GIT adverse effects questionable increased risk for endocrine neoplasias?	Monami [63]/Jorsal [64]/Marguiles [65]/ Longborg [66]/Read [67]/Halbrik [68]/ Sokos [69]	Yes as antidiabetic
Thiazolidinediones	PPARγ-agonism	Contraindicated in diabetic patients with HF, fluid retention, neoplasias?	Hernandez/Singh [70,71]	Yes as OAD

GIT, gastrointestinal; OAD, oral antidiabetic; PPAR, peroxisome proliferator-activated receptor.

are the pleiotropic effects of insulin on glucose utilization, its inhibitory effect on fatty acid metabolism, and its regulatory capacity on various other metabolic pathways such as the PI3-Kinase–AMPK pathway. In recent years, several large clinical trials have explored possible beneficial effects of this intervention in acute myocardial infarction and in heart failure decompensation. For the HINGE trial, Howell et al. compared the effectiveness of perioperative-GIK infusion to prevent low-cardiac-output syndrome in patients with left-ventricular hypertrophy undergoing aortic valve replacement [56]. Interestingly, the authors found not only a significant beneficial effect with treatment compared to placebo, but could also demonstrate an increase of AKT and AMPK phosphorylation with GIK intervention. In another large clinical trial ("IMMEDIATE"), Selker et al. analyzed the effects of in-field GIK administration in patients

with suspected acute coronary syndrome. In their study, randomizing a total of 911 patients to GIK infusion or placebo who were likely to have an ACS event, the investigators found a significantly reduced infarct size and an improvement in a composite of cardiac arrest or in-hospital mortality at 30 days postintervention [53]. However, a 1 year postevent analysis failed to show significant differences in outcome with the exception of the STEMI subgroup, in which the composite endpoints of cardiac arrest, mortality, or HF hospitalization within 1 year were significantly reduced [54]. Interestingly, in an auxiliary study analyzing correlations between GIK-effectiveness and genetic variants on glucose response, Ellis et al. identified several alleles, which, they concluded, may modify glucose response to GIK infusion, which contributes to GIK-therapy response [55]. In another trial with 889 STEMI patients receiving

GIK infusion or placebo, no differences in clinical endpoints could be observed [57]. In summary, to date, results of studies investigating the potential benefit of GIK infusion for myocardial protection are still ambiguous.

Biguanides – Metformin

Metformin is a cornerstone of oral antidiabetic therapy and has manifold effects on metabolic pathways. A strong correlation between heart failure and insulin resistance is well established, leading to a vicious circle of deteriorating cardiac function and metabolic homeostasis. Several large clinical trials investigated the effect of metformin therapy in heart failure patients with insulin resistance, manifested diabetes mellitus, metabolic syndrome, or without those comorbidities. In a 2-year prospective propensity score study enrolling 6185 patients with diabetes and heart failure, Aguilar et al. found a significantly decreased mortality in the metformin-treated cohort compared to no metformin medication [62]. In another small, double-blind, placebo controlled study in HF patients with insulin resistance, the primary endpoint of an improvement in peak oxygen uptake was not met, however, the investigators report a significantly different VE/VCO2 slope in the treatment arm, implying a metformin effect on a metabolic level in the HF patients [61]. In 2014, the GIPS-III study, a double-blinded, placebo-controlled study about the effect of metformin on left-ventricular function in patients after STEMI without diabetes, found no difference between the treatment groups [60]. Currently, a new study is conceived investigating the effect of metformin in diastolic dysfunction in patients with metabolic syndrome ("MET-DIME") [59].

Incretin Mimetics – GLP Analogs and DPP4-Inhibitors

The underlying concept of using the relatively new class of incretin mimetics follows the same reasoning as using antidiabetics with different mechanisms of action: the failing heart, as an energy-starved organ, is supported by facilitating glucose metabolism and countering insulin resistance. However, it has also been shown that GLP-receptors are present in coronary endothelium and cardiomyocytes and that GLP1- receptor activation increases myocardial cAMP levels [72]. A direct effect of incretins on the heart has further been elucidated by using a murine GLP1-R knockout model: these animals presented LV dysfunction. Several initial small clinical trials, mainly infusing GLP1 in diabetic patients after AMI, showed promising results with improvements in EF [68,69,72,73].

Several clinical trials are currently conceived investigating the use of incretin mimetics in the heart failure population. The FIGHT trial [65], enrolling 300 patients with and without diabetes ("Functional Impact of GLP-1 for Heart Failure Treatment") is a randomized, double-blind, placebo-controlled clinical trial in patients with reduced ejection fraction who will be treated for 180 days with placebo or the GLP1 agonist liraglutide. The primary endpoint of FIGHT is conceptualized to be a global rank endpoint including time to death, time to HF hospitalization, and changes in BNP.

A similar study, the "liraglutide on left-ventricular function in chronic heart failure patients – LIVE" trial [64], enrolling 240 patients with heart failure with and without diabetes, will further elucidate the effect of GLP-agonism. Encouraging results of an initial study, administering exenatide to patients with STEMI after reperfusion therapy have been published by Lonborg et al. [66]. Periprocedual exenatide infusion resulted in an improved myocardial salvage index in the exenatide group, compared to placebo.

DPP4-inhibitors act by decreasing the breakdown of endogenous GLP1 and can be administered orally. An initial study by Read et al. [67] appears to confirm the concept of myocardial energy salvation and improvement in contractility by GLP augmentation: in 14 patients with CAD, the inhibition of DPP4 improved global and regional LV contractility in a dobutamine stress test and mitigated postischemic stunning.

However, a comprehensive meta-analysis of clinical trials assessing the risk of heart failure decompensation in patients with diabetes, treated with DPP4-inhibitors, stated that the use of this class of agents could be associated with an increased risk of HF decompensation [63]. In summary, further studies are needed to improve our understanding of benefits and risks of GLP-agonism in heart failure.

Thiazolidinediones

In the context of substrate preference of the failing heart and a potential energetic advantage of glucose utilization over fatty acid oxidation, PPARγ-agonists have a promising mechanism of action. This class of drugs affects substrate metabolism on the transcriptional level, decreasing triglyceride metabolism with an increase of glucose utilization. Despite this mechanism of action being well in line with the underlying concepts as outlined earlier, clinical evidence suggests that PPARγ-agonism has detrimental effects in heart failure. Over the years since its introduction in the 1990s, TZD use in diabetic patients has been thoroughly studied and several large clinical trials and meta-analyses suggest an increased risk of heart failure incidence with the use of this class of drugs [70,71]. The proposed underlying mechanism is an increase in renal sodium reabsorption and vascular permeability caused by thiazolidinediones.

CONCLUSIONS

In summary, the concept of modulating myocardial energy generation by shifting substrate utilization toward glucose and, with this, potentially decreasing oxygen demand through decreased fatty acid metabolism, has led to the development of a variety of pharmacologic approaches. Many of those agents have shown promising results in mostly small clinical trials, but to date, their validation through large clinical trials are still lacking. Consequently, several large multicenter trials are currently under way, trying to elucidate the role of modulators of fatty acid and glucose metabolism in heart failure therapy. However, well-established drugs, such as β-blockers and biguanides, with their pleiotropic effect on cardiac metabolism also have shown an advantageous effect on cardiac substrate utilization.

The growing understanding of cardiac metabolism and its clinical implications for the therapy of heart failure goes in line with the development of sophisticated noninvasive imaging techniques of cardiac metabolic processes. The visualization of cardiac metabolism is a well-established principle for the management of cardiac ischemia, such as myocardial viability studies using radiolabeled substrates. New approaches are the noninvasive evaluation of metabolic pathways such as the Krebs cycle and glycolysis, which will further improve our understanding of the derangements of cardiac metabolic pathways in the failing heart. New concepts and an expansion of established principles of noninvasive cardiac imaging are of outmost importance for the utilization of current and introduction of novel compounds for the therapeutic modulation of cardiac metabolism in human disease states.

References

[1] Bing RJ. The metabolism of the heart. Trans Am Coll Cardiol 1955;5:8–14.
[2] Gropler RJ. Recent advances in metabolic imaging. J Nucl Cardiol 2013;20(6):1147–72.
[3] Peterson LR, Gropler RJ. Radionuclide imaging of myocardial metabolism. Circ Cardiovasc Imaging 2010;3(2):211–22.
[4] Taki JM, Wakabayashi I, Inaki H, Kinuya AS. In: Gaze DC, editor. Role of fatty acid imaging with 123I-β-methyl-p-123I- iodophenyl-pentadecanoic acid (123I-BMIPP) in ischemic heart diseases. Ischemic Heart Disease; 2013.
[5] Kontos MC, Dilsizian V, Weiland F, et al. Iodofiltic acid I 123 (BMIPP) fatty acid imaging improves initial diagnosis in emergency department patients with suspected acute coronary syndromes: a multicenter trial. J Am Coll Cardiol 2010;56(4):290–9.

[6] Matsuo S, Nakajima K, Kinuya S, Yamagishi M. Diagnostic utility of 123I-BMIPP imaging in patients with Takotsubo cardiomyopathy. J Cardiol 2014;64(1): 49–56.

[7] Nishimura M, Tsukamoto K, Hasebe N, Tamaki N, Kikuchi K, Ono T. Prediction of cardiac death in hemodialysis patients by myocardial fatty acid imaging. J Am Coll Cardiol 2008;51(2):139–45.

[8] Solingapuram Sai KK, Kil KE, Tu Z, et al. Synthesis, radiolabeling and initial in vivo evaluation of [(11)C] KSM-01 for imaging PPAR-alpha receptors. Bioorganic Med Chem Lett 2012;22(19):6233–6.

[9] Dilsizian MV, Bacharach PSL, Beanlands MRS, et al. PET myocardial perfusion and metabolism clinical imaging. J Nucl Cardiol 2009;16(4):651.

[10] Camici PG, Prasad SK, Rimoldi OE. Stunning, hibernation, and assessment of myocardial viability. Circulation 2008;117(1):103–14.

[11] Botker HE, Bottcher M, Schmitz O, et al. Glucose uptake and lumped constant variability in normal human hearts determined with [18F]fluorodeoxyglucose. J Nucl Cardiol 1997;4(2 Pt 1):125–32.

[12] Rhodes CG, Camici PG, Taegtmeyer H, Doenst T. Variability of the lumped constant for [18F]2-deoxy-2-fluoroglucose and the experimental isolated rat heart model: clinical perspectives for the measurement of myocardial tissue viability in humans. Circulation 1999 Mar 9;99(9):1275–6.

[13] Herrero P, Kisrieva-Ware Z, Dence CS, et al. PET measurements of myocardial glucose metabolism with 1-11C-glucose and kinetic modeling. J Nucl Med 2007;48(6):955–64.

[14] Herrero P, Dence CS, Coggan AR, Kisrieva-Ware Z, Eisenbeis P, Gropler RJ. L-3-11C-lactate as a PET tracer of myocardial lactate metabolism: a feasibility study. J Nucl Med 2007 Dec;48(12):2046–55.

[15] Peterson LR, Herrero P, Schechtman KB, et al. Effect of obesity and insulin resistance on myocardial substrate metabolism and efficiency in young women. Circulation 2004;109(18):2191–6.

[16] Lin CH, Kurup S, Herrero P, et al. Myocardial oxygen consumption change predicts left ventricular relaxation improvement in obese humans after weight loss. Obesity 2011;19(9):1804–12.

[17] Bergmann SR, Weinheimer CJ, Markham J, Herrero P. Quantitation of myocardial fatty acid metabolism using PET. J Nucl Med 1996;37(10):1723–30.

[18] Zhou YT, Grayburn P, Karim A, et al. Lipotoxic heart disease in obese rats: implications for human obesity. Proc Natl Acad Sci USA 2000;97(4):1784–9.

[19] McGill JB, Peterson LR, Herrero P, et al. Potentiation of abnormalities in myocardial metabolism with the development of diabetes in women with obesity and insulin resistance. J Nucl Cardiol 2011;18(3):421–9. quiz 432–3.

[20] Szczepaniak LS, Dobbins RL, Metzger GJ, et al. Myocardial triglycerides and systolic function in humans: in vivo evaluation by localized proton spectroscopy and cardiac imaging. Magn Reson Med 2003;49(3):417–23.

[21] O'Connor RD, Gropler RJ, Peterson L, Schaffer J, Ackerman JJ. Limits of a localized magnetic resonance spectroscopy assay for ex vivo myocardial triacylglycerol. J Pharm Biomed Anal 2007;45(3):382–9.

[22] Peterson LR, Schilling J, Taegtmeyer H. Alterations in cardiac metabolism. Heart Failure: A Companion to Braunwald's Heart Disease. 3rd ed. Elsevier, Philadelphia.

[23] Schroeder MA, Clarke K, Neubauer S, Tyler DJ. Hyperpolarized magnetic resonance: a novel technique for the in vivo assessment of cardiovascular disease. Circulation 2011;124(14):1580–94.

[24] Reingold JS, McGavock JM, Kaka S, Tillery T, Victor RG, Szczepaniak LS. Determination of triglyceride in the human myocardium by magnetic resonance spectroscopy: reproducibility and sensitivity of the method. Am J Physiol Endocrinol Metab 2005;289(5): E935–9.

[25] Lyons MR, Peterson LR, McGill JB, et al. Impact of sex on the heart's metabolic and functional responses to diabetic therapies. Am J Physiol Heart Circ Physiol 2013;305(11):H1584–91.

[26] Tuunanen H, Engblom E, Naum A, et al. Free fatty acid depletion acutely decreases cardiac work and efficiency in cardiomyopathic heart failure. Circulation 2006;114(20):2130–7.

[27] Kolwicz SC Jr, Purohit S, Tian R. Cardiac metabolism and its interactions with contraction, growth, and survival of cardiomyocytes. Circ Res 2013;113(5):603–16.

[28] Doenst T, Nguyen TD, Abel ED. Cardiac metabolism in heart failure: implications beyond ATP production. Circ Res 2013;113(6):709–24.

[29] Lodish HB, Berk A, Zipursky SL, et al. Oxidation of glucose and fatty acids to CO_2. Molecular Cell Biology. 4th ed. New York: W.H. Freeman; 2000.

[30] McDonagh T, Gardner RS, Clark AL, Dargie H. Oxford Textbook of heart failure. Oxford University Press, Oxford; UK 2011.

[31] Rupp H, Zarain-Herzberg A, Maisch B. The use of partial fatty acid oxidation inhibitors for metabolic therapy of angina pectoris and heart failure. Herz 2002;27(7):621–36.

[32] Beadle RM, Williams LK, Kuehl M, et al. Improvement in cardiac energetics by perhexiline in heart failure due to dilated cardiomyopathy. JACC Heart Fail 2015;3(3):202–11.

[33] Cole PL, Beamer AD, McGowan N, et al. Efficacy and safety of perhexiline maleate in refractory angina. A double-blind placebo-controlled clinical trial of a novel antianginal agent. Circulation 1990;81(4):1260–70.

[34] Abozguia K, Elliott P, McKenna W, et al. Metabolic modulator perhexiline corrects energy deficiency and improves exercise capacity in symptomatic hypertrophic cardiomyopathy. Circulation 2010;122(16):1562–9.

[35] Phan TT, Shivu GN, Choudhury A, et al. Multi-centre experience on the use of perhexiline in chronic heart failure and refractory angina: old drug, new hope. Eur J Heart Fail 2009;11(9):881–6.

[36] Bhandari B, Subramanian L. Ranolazine, a partial fatty acid oxidation inhibitor, its potential benefit in angina and other cardiovascular disorders. Recent Pat Cardiovasc Drug Discov 2007;2(1):35–9.

[37] Ling H, Packard KA, Burns TL, Hilleman DE. Impact of ranolazine on clinical outcomes and healthcare resource utilization in patients with refractory angina pectoris. Am J Cardiovasc Drugs 2013;13(6):407–12.

[38] Shammas NW, Shammas GA, Keyes K, Duske S, Kelly R, Jerin M. Ranolazine versus placebo in patients with ischemic cardiomyopathy and persistent chest pain or dyspnea despite optimal medical and revascularization therapy: randomized, double-blind crossover pilot study. Ther Clin Risk Manag 2015;11:469–74.

[39] Lionetti V, Stanley WC, Recchia FA. Modulating fatty acid oxidation in heart failure. Cardiovasc Res 2011;90(2):202–9.

[40] Fragasso G, Rosano G, Baek SH, et al. Effect of partial fatty acid oxidation inhibition with trimetazidine on mortality and morbidity in heart failure: results from an international multicentre retrospective cohort study. Int J Cardiol 2013;163(3):320–5.

[41] Fragasso G, Montano C, Perseghin G, et al. The anti-ischemic effect of trimetazidine in patients with postprandial myocardial ischemia is unrelated to meal composition. Am Heart J 2006;151. 1238.e1–8.

[42] Di Napoli P, Taccardi AA, Barsotti A. Long term cardioprotective action of trimetazidine and potential effect on the inflammatory process in patients with ischaemic dilated cardiomyopathy. Heart 2005;91(2):161–5.

[43] Winter JL, Castro PF, Quintana JC, et al. Effects of trimetazidine in nonischemic heart failure: a randomized study. J Card Fail 2014;20(3):149–54.

[44] Tuunanen H, Engblom E, Naum A, et al. Trimetazidine, a metabolic modulator, has cardiac and extra-cardiac benefits in idiopathic dilated cardiomyopathy. Circulation 2008;118(12):1250–8.

[45] Kennedy JA, Unger SA, Horowitz JD. Inhibition of carnitine palmitoyltransferase-1 in rat heart and liver by perhexiline and amiodarone. Biochem Pharmacol 1996;52(2):273–80.

[46] Wallhaus TR, Taylor M, DeGrado TR, et al. Myocardial free fatty acid and glucose use after carvedilol treatment in patients with congestive heart failure. Circulation 2001;103(20):2441–6.

[47] Podbregar M, Voga G. Effect of selective and nonselective beta-blockers on resting energy production rate and total body substrate utilization in chronic heart failure. J Card Fail 2002;8(6):369–78.

[48] Dzerve V. A dose-dependent improvement in exercise tolerance in patients with stable angina treated with mildronate: a clinical trial "MILSS I". Medicina (Kaunas) 2011;47(10):544–51.

[49] Holubarsch CJ, Rohrbach M, Karrasch M, et al. A double-blind randomized multicentre clinical trial to evaluate the efficacy and safety of two doses of etomoxir in comparison with placebo in patients with moderate congestive heart failure: the ERGO (etomoxir for the recovery of glucose oxidation) study. Clin Sci (Lond) 2007;113(4):205–12.

[50] Schmidt-Schweda S, Holubarsch C. First clinical trial with etomoxir in patients with chronic congestive heart failure. Clin Sci 2000;99:27–35.

[51] Drake-Holland AJ, Passingham JE. The effect of Oxfenicine on cardiac carbohydrate metabolism in intact dogs. Basic Res Cardiol 1983;78(1):19–27.

[52] Kennedy JA, Kiosoglous AJ, Murphy GA, Pelle MA, Horowitz JD. Effect of perhexiline and oxfenicine on myocardial function and metabolism during low-flow ischemia/reperfusion in the isolated rat heart. J Cardiovasc Pharmacol 2000;36(6):794–801.

[53] Selker HP, Beshansky JR, Sheehan PR, et al. Out-of-hospital administration of intravenous glucose-insulin-potassium in patients with suspected acute coronary syndromes: the IMMEDIATE randomized controlled trial. JAMA 2012;307(18):1925–33.

[54] Selker HP, Udelson JE, Massaro JM, et al. One-year outcomes of out-of-hospital administration of intravenous glucose, insulin, and potassium (GIK) in patients with suspected acute coronary syndromes (from the IMMEDIATE [Immediate Myocardial Metabolic Enhancement During Initial Assessment and Treatment in Emergency Care] Trial). Am J Cardiol 2014;113(10):1599–605.

[55] Ellis KL, Zhou Y, Beshansky JR, Ainehsazan E, Yang Y, Selker HP, Huggins GS, Cupples LA, Peter I, et al. Genetic variation at glucose and insulin trait loci and response to glucose-insulin-potassium (GIK) therapy: the IMMEDIATE trial. Pharmacogenomics J 2015;15(1):55–62.

[56] Howell NJ, Ashrafian H, Drury NE, et al. Glucose-insulin-potassium reduces the incidence of low cardiac output episodes after aortic valve replacement for aortic stenosis in patients with left ventricular hypertrophy: results from the Hypertrophy, Insulin, Glucose, and Electrolytes (HINGE) trial. Circulation 2011;123(2):170–7.

[57] Rasoul S, Ottervanger JP, Timmer JR, et al. One year outcomes after glucose-insulin-potassium in ST

elevation myocardial infarction. The Glucose-insulin-potassium study II. Int J Cardiol 2007;122(1):52–5.

[58] Timmer JR, Svilaas T, Ottervanger JP, et al. Glucose-insulin-potassium infusion in patients with acute myocardial infarction without signs of heart failure: the Glucose-Insulin-Potassium Study (GIPS)-II. J Am Coll Cardiol 2006;47(8):1730–1.

[59] Ladeiras-Lopes R, Fontes-Carvalho R, Bettencourt N, Sampaio F, Gama V, Leite-Moreira AF. METformin in DIastolic Dysfunction of MEtabolic syndrome (MET-DIME) trial: rationale and study design : MET-DIME trial. Cardiovasc Drugs Ther 2014;28(2):191–6.

[60] Lexis CP, van der Horst IC, Lipsic E, et al. Effect of metformin on left ventricular function after acute myocardial infarction in patients without diabetes: the GIPS-III randomized clinical trial. JAMA 2014;311(15):1526–35.

[61] Wong AK, Symon R, AlZadjali MA, et al. The effect of metformin on insulin resistance and exercise parameters in patients with heart failure. Eur J Heart Fail 2012;14(11):1303–10.

[62] Aguilar D, Chan W, Bozkurt B, Ramasubbu K, Deswal A. Metformin use and mortality in ambulatory patients with diabetes and heart failure. Circ Heart Fail 2011;4(1):53–8.

[63] Monami M, Dicembrini I, Mannucci E. Dipeptidyl peptidase-4 inhibitors and heart failure: a meta-analysis of randomized clinical trials. Nutr Metab Cardiovasc Dis 2014;24(7):689–97.

[64] Jorsal A, Wiggers H, Holmager P, et al. A protocol for a randomised, double-blind, placebo-controlled study of the effect of LIraglutide on left VEntricular function in chronic heart failure patients with and without type 2 diabetes (The LIVE Study). BMJ Open 2014;4(5):e004885.

[65] Margulies KB, Anstrom KJ, Hernandez AF, et al. GLP-1 agonist therapy for advanced heart failure with reduced ejection fraction: design and rationale for the functional impact of GLP-1 for heart failure treatment study. Circ Heart Fail 2014;7(4):673–9.

[66] Lonborg J, Vejlstrup N, Kelbaek H, et al. Exenatide reduces reperfusion injury in patients with ST-segment elevation myocardial infarction. Eur Heart J 2012;33(12):1491–9.

[67] Read PA, Khan FZ, Heck PM, Hoole SP, Dutka DP. DPP-4 inhibition by sitagliptin improves the myocardial response to dobutamine stress and mitigates stunning in a pilot study of patients with coronary artery disease. Circ Cardiovasc Imaging 2010;3(2):195–201.

[68] Sokos GG, Nikolaidis LA, Mankad S, Elahi D, Shannon RP. Glucagon-like peptide-1 infusion improves left ventricular ejection fraction and functional status in patients with chronic heart failure. J Card Fail 2006;12(9):694–9.

[69] Halbirk M, Norrelund H, Moller N, et al. Cardiovascular and metabolic effects of 48-h glucagon-like peptide-1 infusion in compensated chronic patients with heart failure. Am J Physiol Heart Circ Physiol 2010;298(3):H1096–102.

[70] Hernandez AV, Usmani A, Rajamanickam A, Moheet A. Thiazolidinediones and risk of heart failure in patients with or at high risk of type 2 diabetes mellitus: a meta-analysis and meta-regression analysis of placebo-controlled randomized clinical trials. Am J Cardiovasc Drugs 2011;11(2):115–28.

[71] Singh S, Loke YK, Furberg CD. Thiazolidinediones and heart failure: a teleo-analysis. Diab Care 2007;30(8):2148–53.

[72] Oyama J, Node K. Incretin therapy and heart failure. Circ J 2014;78(4):819–24.

[73] Vest AR. Incretin-related drug therapy in heart failure. Curr Heart Fail Rep 2015;12(1):24–32.

Historical Perspectives

Terje S. Larsen

Cardiovascular Research Group, Department of Medical Biology,
UiT the Arctic University of Norway, Tromsø, Norway

INTRODUCTION

The human mind has probably been engaged on the fundamental processes governing the heartbeat ever since the acknowledgment of the heart's key role for individual life. Already in ancient Greek culture Aristotle hypothesized that cardiac function was somehow associated with heat, and that the principle of heat and nutrition were connected [1,2]. The views and concepts of the Greek cultures survived for centuries, and it was not until the anatomical studies by Andreas Vesalius (1514–1564) and William Harvey (1578–1657) that we got a better fundamental understanding of the function of the heart [3,4]. Vesalius resurrected the use of human dissection, regardless of the strict ban by the Catholic Church. His influential book about human anatomy "De Humani Commis Fabrica" (*The Structure of The Human Body*) which was published in 1543 contained over 200 anatomical illustrations. The work was the earliest known precise presentation of human anatomy. William Harvey was the first person to correctly describe that arteries and veins form a complete circuit, and that the circuit starts at the heart and leads back to the heart. In 1628 Harvey published his masterpiece – usually referred to as "De Moto Cordis" (*The Motion of the Heart*).

Today we know that the heart needs a constant and plentiful source of fuel to maintain normal pumping function. In the fetus, the heart utilizes lactate and glucose as its main energy substrates, while fatty acids are the main energy substrate in the adults. The transition from carbohydrate to fatty acid metabolism is a complex process which involves maturation of mitochondrial processes [5] and dramatic changes in circulating levels of fatty acids and lactate [6]. However, the source of fuel changes in certain pathologies – fatty acids being the dominating energy source in obesity and diabetes, while carbohydrates seem to dominate in hypertrophy and heart failure [7,8].

The enormous growth of information within metabolic regulation during the last few decades, in particular the appearance and rapid progress of molecular biology and unraveling of signal transduction pathways, have definitely given new insight for understanding the function and metabolism of the heart [9,10]. On the other hand, keeping pace with the rapid development in molecular biology may sometimes be challenging, because it may prevent insight into how integrative physiology and metabolism sustain whole body metabolic homeostasis in healthy as well as in pathological conditions.

The Discovery of Oxygen as the Primary Respiratory Gas

Three persons, namely Carl Wilhelm Scheele (1742–1786), Joseph Priestley (1733–1804), and Antoine Lavoisier (1743–1794) are all connected to the discovery of oxygen. Oxygen was discovered first by Scheele in 1772, while Priestley discovered oxygen independently 2 years later (1774). Priestley published his finding the same year [11], while Scheele's work was not released until 1777 [12]. Today we learn that oxygen consumption is a measure of metabolic rate, but it was experiments performed by Antoine Lavoisier (1743–1794), which first demonstrated the role of oxygen in animal respiration. Lavoisier was the first to recognize oxygen as an element, and in collaboration with Pierre-Simon Laplace (1749–1827) he was able to correlate produced animal heat with consumed oxygen [13]. He also discovered that in animals, oxygen is replaced by another gas, carbon dioxide.

The Discovery of the Three Steps in Cellular Respiration

The fundamental statements by Julius Robert von Mayer (1814–1878) and Hermann von Helmholtz (1821–1894) about energy conservation (an early version of the first law of thermodynamics, namely that energy can be neither created nor destroyed), as well as Helmholtz's founding publication on chemical thermodynamics, played important roles in initiating energy metabolism [14]. However, the discovery of the three steps of energy production was an important breakthrough for the evolution of biochemistry in the nineteenth century. After the initial observation by Hans (1850–1902) and Eduard Buchner (1860–1917) in 1897 that fermentation could be performed by cell-free

yeast extracts [15], important contributions by Gustav Embden and Otto Meyerhof (as well as Carl Neuberg, Jacob Panas, Otto Warburg, and Gerty and Carl Cori) led to the characterization of the enzymatic reactions of the first step in glucose metabolism, namely *anaerobic glycolysis*, in which glucose is broken down to pyruvate [16,17].

The second step, the "Krebs cycle," is attributed to Sir Hans Adolf Krebs (1900–1981) who described the main elements in this process in 1937. He received the Nobel prize in physiology for this discovery in 1953 [18,19]. In this stage of cellular respiration, which occurs in the inner mitochondrial matrix, the oxidation of glucose to CO_2 is completed. This step is also known as the "tricarboxylic acid cycle" or the "citric acid cycle," and it plays a critical role in intermediary metabolism and energy production [20]. Several scientists in addition to Krebs (Albert Szent-Györgyi, Carl Martius, Franz Knoop) contributed to the characterization of the enzymes of this pathway.

The electron transport chain comprises the last step in cellular respiration. Mitochondria were considered as the power stations of the cell, and techniques to isolate these organelles opened up for intensive research on the different components of mitochondrial energy production. Eugene Kennedy and Albert Lehninger were able to demonstrate that the enzymes of fatty acid oxidation of the Krebs cycle and of oxidative phosphorylation are all located within the mitochondria [16,21]. Adenosine triphosphate (ATP) was discovered already in 1929 by Karl Lohmann (1898–1978), and in 1935 Vladimir Engelhardt (1894–1984) observed that muscle contractions require ATP [22,23]. More pieces were added to the puzzle by Herman Kalckar (1908–1991), linking ATP synthase with cell respiration [17], and Fritz Lipmann (1899–1986), who revealed that ATP is the main carrier of chemical energy in the cell [24]. The major breakthrough came in 1961, however, when Peter Mitchell (1920–1992) put forward the so-called chemiosmotic hypothesis [25], which suggested that H[+] ions are actively transported across the inner mitochondrial membrane into the outer mitochondrial compartment. The H[+] ions then flow back through special pores in the membrane, a process that is thought to generate ATP by mitochondrial oxidative phosphorylation. For this work, which explained the function of oxygen in living organisms, Mitchell was awarded the Nobel prize in Chemistry in 1978.

Perfusion of Isolated Organs – A Starting Point for the Study of Cardiac Physiology

Collaborative work between Elias Cyon (1843–1912) and Carl Friedrich Wilhelm Ludwig (1816–1895) at the Carl Ludwig Institute of Physiology in Leipzig (Germany) resulted in the development of the isolated frog heart preparation in 1866 [26]. This work later inspired Sydney Ringer (1836–1909), Henry Martin (1848–1896), and Oskar Langendorff (1853–1908) to develop the technique for perfusing isolated mammalian hearts [27–29], which paved the way for experimental activity that led to important milestones in our knowledge about cardiac function. This included the role of temperature, oxygen, and calcium ions for cardiac contractile function, as well as the origin of cardiac electrical activity in the atrium. Langendorff also demonstrated that the heart receives its nutrients and oxygen from blood via the coronary arteries and that cardiac mechanical function is reflected by changes in the coronary circulation.

In 1914 Ernest Starling (1866–1927), using a dog heart–lung preparation and an apparatus to monitor the volume of the ventricles, showed that ventricular stroke volume was a direct function of the end-diastolic volume. The basis for his discovery was, however, made by Otto Frank [30] (1865–1944), using an improved frog heart preparation; he showed that the maximal isovolumetric pressure showed progressive increase with increased filling of the frog ventricle. The so-called Frank–Starling [31] mechanism of the

heart states that the stroke volume of the heart increases in response to an increase in the volume of blood filling the heart (the end-diastolic volume) when all other factors remain constant.

The Langendorff model is still in use, although it has been modified and further developed. Howard Morgan (1927–2009) and James R. Neely (1935–1988) established an isolated, perfused, working heart preparation from adult rats [32,33], and stimulated by Bing's initial observations on myocardial metabolism (see later in the chapter) they initiated detailed biochemical studies by the use of this heart model [34]. For instance, they were able to demonstrate that insulin accelerates glucose transport into the cell, which in turn increases the rate of glycolysis [35]. In addition, they performed experiments to examine the effect of anoxia on cardiac glucose metabolism [36]. The working heart preparation has been modified and adjusted to specific needs of several other laboratories throughout the world. For instance, it has been used for studies on cardiac metabolism, applying radioisotope techniques [37,38] and nuclear magnetic resonance techniques [39]. In the wake of the genomic era various animal models have been developed (by genetic engineering) to model complex diseases, such as, metabolic syndrome, atherosclerosis, and cancer. Among the various animal models available, mice are the most commonly used for several reasons. The question as to how these conditions affect mechanical and metabolic function of the heart has demanded miniaturized perfusion systems for small mouse hearts, weighing slightly above 100 mg [40,41]. Today, miniaturized conductance catheters are available for assessment of ventricular function (pressure-volume analysis), while fiber-optic oxygen probes allow continuous measurements of myocardial oxygen consumption. In this way, one can study alterations in cardiac efficiency and energetics in response to various pathological conditions, as well as the impact of various metabolic and pharmacological interventions, using the isolated working mouse heart model (see Ref. [42]; Fig. 15.1).

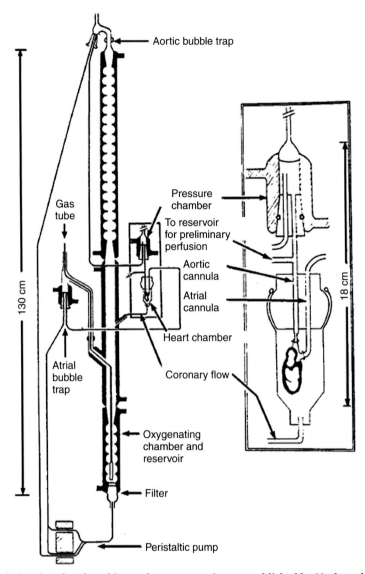

FIGURE 15.1   The isolated perfused working rat heart preparation as established by Neely and Morgan [33].

Richard John Bing – The Beginning of a New Era for Cardiac Metabolism

Richard Bing (1909–2010) is often spoken of as the father of cardiac metabolism [43]. He left his laboratory at the age of 98, having published more than 500 journal articles and medical books.

At the Johns Hopkins School of Medicine, Bing organized the first diagnostic cardiac catheterization laboratory, and it was here that he "accidentally" inserted a catheter into the coronary sinus, which drains deoxygenated blood from the cardiac muscle into the right ventricle. This event was the beginning of a new era in cardiac physiology and metabolism, since it allowed measurements of arterio-venous differences of metabolic substrates. Over the next decade, Bing's research focused on the metabolism of fats, ketones, amino acids, and other substrates in healthy and diseased heart muscles. Bing and his coworkers were able to show that fatty acids were the major substrate for the human heart [44], and their important contribution to cardiac metabolism has been acknowledged by several authors, among others Lionel Opie [45] and Heinrich Taegtmeyer [46] who have also been very influential in putting cardiac metabolism on the map. In addition, Bing's multitalented personality has been emphasized by many; in addition to being productive as a physician and scientist at 100 years, he was also a gifted musician and composer (Fig.15.2).

ENERGY SUBSTRATE SUPPLY OF THE CARDIAC MUSCLE – AND ITS REGULATION

The Randle Glucose–Fatty Acid Cycle

The pioneering work of Bing led to an intense interest in the myocardial metabolism. In 1963 Philip Randle, Peter Garland, Nick Hales. and Eric Newsholme published their landmark paper in the Lancet entitled "The glucose–fatty acid cycle: its role in insulin sensitivity and the metabolic disturbances of diabetes mellitus" [47]. This paper had a great impact on our present understanding of metabolic homeostasis by introducing the fundamental concept of reciprocal substrate competition between glucose and fatty acids. According to the authors "The cycle provides a primitive mechanism, which quite independently of hormonal control, will tend to maintain a constant plasma glucose concentration in animals that feed intermittently." In its simplicity, the glucose–fatty acid cycle predicts that the increased provision of exogenous lipid fuels from increased breakdown of

FIGURE 15.2 **Dr Richard J. Bing, a pioneering cardiologist whose research led to new understandings of blood flow, congenital heart disease, and the mechanics of the heart.**

endogenous triacylglycerol promotes the use of fatty acids as oxidative fuel and, in so doing, glucose utilization is blocked. In addition, the Lancet paper proposed that impaired glucose utilization in diabetes mellitus, instead of being viewed primarily as a disorder of carbohydrate metabolism, could be explained in terms of excessive release of fatty acids from adipose tissue triacylglycerol. Finally, this research article also predicted that lipid derivatives might be the active species in mediating insulin resistance in skeletal muscle (see review by Mary C. Sugden [48]) (Fig.15.3).

FIGURE 15.3   **Sir Philip Randle (1926–2006) who developed the fundamental concept of interplay between carbohydrate and lipid fuels in relation to the requirement for energy utilization and storage.**

Metabolic Signaling in the Heart

The interactions between glucose and fatty acid metabolism, which were described in the glucose–fatty acid cycle, are far more complex than originally proposed, as revealed by new molecular insights. The comprehensive review by Louis Hue and Heinrich Taegtmeyer (The Randle cycle revisited: a new head for an old hat) [49] gives the reader a closer understanding of the mechanisms involved. They include allosteric regulation and reversible phosphorylation controlling glucose uptake, glycolytic flux (phosphofructokinase), and glucose oxidation (pyruvate dehydrogenase) on the one hand, and control of fatty acid uptake (by

malonyl-CoA) and oxidation in the mitochondria on the other. In addition, expression of key enzymes in the metabolic pathway of glucose and lipid metabolism by PPAR, SREBP-1c, and ChREBP play important roles in the long-term regulation of myocardial fuel supply. The elucidation of these interactions as well as of the importance of mitochondrial ROS production has led to a more refined understanding of the mechanisms leading to insulin resistance and type II diabetes.

Myocardial Infarction, Heart Failure, and Diabetes Orchestrating Metabolic Fuel Selection

In the wake of the work by Randle and his coworkers, a number of researchers have made significant contributions to describe the alterations in metabolism taking place in various disease states [49–52]. Whereas the Randle cycle controls fuel selection and adapts the substrate supply and demand in normal tissues, Lionel Opie (University of Cape Town, South Africa) and his coworkers focused on the importance of substrate utilization in the ischemic heart. In collaboration with Eric Newsholme and Hans Krebs (Oxford, England) he worked out ideas about a protective role of glucose already during the late 1950s [53–55]. This work was pursued later to prove the concept that glycolysis is protective of the heart cell membrane [56]. Opie realized very early that myocardial substrate metabolism was not just of theoretical interest, and was eager to translate the findings into clinical understanding and therapy. Thus, he developed the concept that decreased delivery of glucose and glycogen utilization (due to intracellular accumulation of lactate and protons) inhibit glycolytic ATP production during severe myocardial ischemia [57]. In 1970, Opie was the sole author of a Nature paper, which focused on the role of fatty acids on the heart [58], work which was much inspired by collaboration with Michael Oliver (Edinburgh, Scotland).

This work was followed up in collaboration with Joel DeLeiris (France), showing that the adverse effects of fatty acids on experimental myocardial infarction were abolished by glucose and/or insulin [59]. In addition, Demetrio Sodi-Pallares and coworkers reported already in the early 1960s that infusion of glucose–insulin–potassium (GIK) solution reduced electrocardiographic signs of ischemia, limited infarct size, and improved survival in experimental animals, as well as in patients with acute myocardial infarction. Although Sodi-Pallares and his team used GIK to restore potassium depletion in ischemic myocardial cells [60], the rationale for using the solution as a cardio-protective solution was perfect, since it would suppress circulating levels and myocardial uptake of free fatty acids, and improve the efficiency of myocardial energy production through the provision of exogenous glucose. However, the initial enthusiasm was soon met with critique, because one feared that increased lactate production in response to increased glucose and insulin would be harmful instead of beneficial due to inhibition of glycolysis [61]. Opie and coworkers later reported increased tissue concentrations of ATP, phosphocreatine, and glycogen in baboons [62] and in 1999, Carl Apstein and Opie reviewed several clinical trials using GIK and concluded that the data were "not firm nor extensive enough to support the routine use of GIK in patients with AMI." On the other hand, they argued that evidence was strong enough to recommend routine use of GIK for diabetics with acute myocardial infarction [63] (Fig.15.4).

The adult heart is capable of oxidizing a wide range of carbon substrates, and the heart is therefore often referred to as a metabolic omnivore [49]. It is also widely appreciated that shifts in substrate utilization in the heart occur in response to various disease states, such as heart failure and diabetes. Most forms of heart failure are associated with a history of cardiac ischemia, myocardial infarction, hypertension, and/or left ventricular hypertrophy,

FIGURE 15.4 Professor Lionel H. Opie is internationally renowned for his pioneering work on the energy metabolism of the heart, and for his exceptional talent as author and lecturer.

with either normal or decreased ejection fraction. The general concept is that myocardial substrate selection is relatively normal during the early stages of heart failure; however, in the advanced stages activation of a fetal-like gene program causes downregulation in fatty acid oxidation and a concomitant increase in glycolysis and glucose oxidation, reduced respiratory chain activity, and an impaired reserve for mitochondrial oxidative flux. [64]. It has been suggested that this substrate shift, which is associated with reactivation of other fetal-like hallmarks (e.g., myosin heavy chain isoform switching), contributes toward the progression to overt contractile failure [65]. These concepts were thoroughly discussed in a comprehensive review by William Stanley, Fabio Recchia, and Gary Lopaschuk in 2005 [50].

The epidemic increase in obesity, insulin resistance, and diabetes are major risk factors for cardiovascular disease. In the heart, fatty acids enter the cardiomyocytes through specific transporters and are converted to acyl-CoA by acyl-CoA synthetase. Acyl-CoAs might in turn be used for β-oxidation, or they can be diverted to nonoxidative pathways, including esterification and TG synthesis. However, when fatty acid supply is high, myocardial TAG content is increased. In addition, acyl-CoA becomes a source for formation of potential lipotoxic intermediates – ceramides, DAG, and reactive oxygen species (ROS), creating a state of lipotoxicity. It has been argued that insulin resistance may be adaptive when protecting the heart from excess fuel uptake, or maladaptive when associated with ROS formation and activation of signaling pathways of apoptotic cell death. In diabetes and insulin resistance, tissues such as skeletal muscle and heart have lost their capacity to appropriately change between use of lipids in the fasting state and use of carbohydrate in the insulin-stimulated prandial state, a condition which has been termed "metabolic inflexibility" [66,67]. Because the diabetic heart seems to be "starved in the midst of plenty," it has been argued that excess substrate supply may result in impaired transcriptional regulation of proteins constituting the pathways of cardiac energy metabolism and, consequently, in impaired metabolic flexibility [67,68]. Today we know that the heart both adapts and maladapts to various metabolic stresses and that the adaptive responses may be favorably manipulated by the provision of specific fatty acid substrates [69] and ingestion of foods with a low glycemic index. For instance, it has been reported that a low-carbohydrate/high-fat diet can prevent or reduce some of the most serious aspects of heart failure [70,71]. In addition, dietary supplementation with omega-3 polyunsaturated fatty acids prevents development of heart failure [72], while a high-sugar diet further accelerates development of heart failure [73].

How exactly obesity affects the sites of metabolic regulation in the heart is not completely understood. A number of key enzymes important in the regulation of cardiac fatty acid oxidation have been characterized [74] and resulted in metabolic strategies to suppress fatty acid metabolism and protect the heart, for example, following ischemia [75]. Evidence has

also been provided that activation of PPARα drives myocardial FA oxidation in the diabetic heart and leads to impaired glucose uptake and utilization. All these additional discoveries regarding the genetic regulation of metabolism have complemented and reinforced the importance of Randle's glucose–fatty acid cycle by which high concentrations of fatty acids inhibit glucose utilization (for review, see Finck et al. [76]).

Fuel Availability and Cardiac Energetics

In the late 1960s and beginning of the 1970s the importance of free fatty acids as a determinant of myocardial oxygen consumption (MVO_2) was demonstrated in experiments using dog hearts *in situ*. The novel observation was that an acute increase in the myocardial uptake of fatty acids was associated with increased MVO_2, despite unchanged mechanical activity [77]. Furthermore, the fatty acid-induced increase in MVO_2 was considerably higher ($\sim 30\%$) than could be stoichiometrically accounted for based on a switch in substrate oxidation from glucose to the more oxygen-requiring fatty acids.

This finding initiated intensive research on the impact of fatty acids on cardiac function, providing evidence that elevated levels of fatty acids can be harmful for the heart, such as, during ischemic stress [78]. Furthermore, it was also shown that pharmacological strategies, which reduced the supply of fatty acids to the heart attenuated the damage to the heart during subsequent coronary occlusion in dogs [79,80]. The fatty acid-induced increase in MVO_2 has later been confirmed both in pigs and rodents [42,81]. Analysis of the relationships between MVO_2 and total cardiac work (measured by pressure-volume techniques) has established that the increased oxygen consumption can be ascribed to nonmechanical processes, which are still not completely settled [82] and therefore are a challenge to current and new candidates in the field.

FUTURE PROSPECTS

Despite impressive achievements describing many of the molecular mechanisms responsible for the changes in myocardial metabolic phenotype that occur during normal physiology as well as during various disease states, we are probably just beginning to understand the complex regulatory mechanisms governing myocardial energy substrate metabolism. One of the challenges for the immediate future is to identify the mechanistic links between obesity and cardiovascular disease to develop targeted metabolic interventions to ameliorate metabolic inflexibility and glucolipotoxicity in the heart, either in the short term (by changing enzyme activities) or in the long term (by alternating gene expression).

References

[1] Bing R. Coronary circulation and cardiac metabolism. New York: Oxford University Press; 1964.
[2] Vogt AM, Kübler W. Cardiac energy metabolism – a historical perspective. Heart Fail Rev 1999;4:211–9.
[3] Vesalius A, De Humani Corporis Fabrica Libri Septem. Basileae [Basel]: Ex officina Joannis Oporini, 1543.
[4] Harvard Classics. . Part 3. On the motion of the heart and blood in animals: William Harvey's seminal work describing the circulation of blood through the body translated by Robert Willis, 38. New York: P.F. Collier & Son Company; 2001. 1909–1914, New York: Bartleby.Com.
[5] Leone TC, Kelly DP. Transcriptional control of cardiac fuel metabolism and mitochondrial function. Cold Spring Harb Symp Quant Biol 2011;76:175–82.
[6] Lopaschuk GD, Collins-Nakai RL, Itoi T. Developmental changes in energy substrate use by the heart. Cardiovasc Res 1992;26(12):1172–80.
[7] Lopaschuk GD. Metabolic abnormalities in the diabetic heart. Heart Fail Rev 2002;7(2):149–59.
[8] Young ME, Laws FA, Goodwin GW, Taegtmeyer H. Reactivation of peroxisome proliferator-activated receptor alpha is associated with contractile dysfunction in hypertrophied rat heart. J Biol Chem 2001;276(48):44390–5.
[9] Vogt AM, Kübler W. Cardiac energy metabolism – a historical perspective. Heart Fail Rev 1999;4:211–9.
[10] Beloukas AI, Magiorkinis E, Tsoumakas TL, Kosma AG, Diamantis A. Milestones in the history of research on cardiac energy metabolism. Can J Cardiol 2013;29(11):1504–11.
[11] Priestley J. Experiments and observations on different kinds of air. London: J. Johnson; 1974. pp. 1775–1777.
[12] Scheele CW. Chemische abhandlung von der Luft und dem Feuer. Upsala and Leipzig: Leipzig W. Engelmann; 1777.
[13] Lavoisier M, Memoires de Chimie, 6 vols. Paris: Lavoisier; 1864–1893.
[14] Lehninger AL. Bioenergetics: the molecular basis of biological energy transformations. 2nd ed. London: Benjamin/Cummings; 1971.
[15] Kohler RE. The reception of Eduard Buchner's discovery of cell-free fermentation. J Hist Biol 1972;5:327–53.
[16] Fruton J. Molecules and life: historical essays on the interplay of chemistry and biology. New York: Wiley Interscience; 1972.
[17] Kalckar H. Biological phosphorylations: development of concepts. Englewood Cliffs, NJ: Prentice Hall; 1969.
[18] Krebs H. The history of the tricarboxylic acid cycle. Perspect Biol Med 1970;14:154–70.
[19] Krebs H. Control of metabolic process. Endeavour 1957;16:125–32.
[20] Krebs H, Johnson W. The role of citric acid in intermediate metabolism in animal tissues. Enzymologia 1937;4:148–56.
[21] Racker E. From Pasteur to Mitchell: a hundred years of bioenergetics. Fed Proc 1980;39:210–5.
[22] Lohmann K. Über die enzymatische Aufspaltung der Kreatinphosphorsäure; zugleich ein Beitrag zum Mechanismus der Muskelkontraktion. Biochem Z 1934;271:264.
[23] Engelhardt W, Ljubimova F. Myosine and adenosinetriphosphatase. Nature 1939;144:668.
[24] Lipmann F. Metabolic generation and utilization of phosphate bond energy. Adv Enzymol 1941;1:99–162.
[25] Mitchell P. Coupling of phosphorylation to electron and hydrogen transfer by a chemi-osmotic type of mechanism. Nature 1961;191:144–8.
[26] Cyon E. Über den Einfluss der Temperaturänderungen auf Zahl, Dauer und Stärke der Herzschläge. Berichte über die Verhandlungen der Koniglich Sächsischen Gesellschaft der Wissenschaften zu Leipzig. Math Phys Classe 1866;18:256–306.
[27] Ringer S. A further contribution regarding the influence of the different constituents of the blood on the contraction of the heart. J Physiol (Lond) 1883;4:29–42.
[28] Martin HN. The direct influence of gradual variations of temperature upon the rate of beat of the dog's heart. Philos Trans R Soc Lond 1883;174:663–88.
[29] Langendorff O. Untersuchungen am überlebenden Säugetierherzen. Pflügers Arch 1895;61:291–322.
[30] Frank O. Zur Dynamik des Herzmuskels. Z Biol 1895;32:370–437.
[31] Starling E. The linacre lecture on the law of the heart. London: Longman, Green; 1918.

[32] Katz AM, Weisfeldt ML. In memoriam: Howard Morgan (1927–2009). Circ Res 2009;104:1131–2.

[33] Neely JR, Liebermeister H, Battersby EJ, Morgan HE. Effect of pressure development on oxygen consumption by isolated rat heart. Am J Physiol 1967;212:804–14.

[34] Neely JR, Morgan HE. Relationship between carbohydrate and lipid metabolism and the energy balance of heart muscle. Annu Rev Physiol 1974;36:413–59.

[35] Henderson MJ, Morgan HE, Park CR. Regulation of glucose uptake in muscle. V. The effect of growth hormone on glucose transport in the isolated, perfused rat heart. J Biol Chem 1961;236:2157–61.

[36] Morgan HE, Henderson MJ, Regen DM, Park CR. Regulation of glucose uptake in muscle. I. The effects of insulin and anoxia on glucose transport and phosphorylation in the isolated, perfused heart of normal rats. J Biol Chem 1961;236:253–61.

[37] Lopaschuk GD, Barr RL. Measurements of fatty acid and carbohydrate metabolism in the isolated working rat heart. Mol Cell Biochem 1997;172(1–2):137–47.

[38] Belke DD, Larsen TS, Gibbs EM, Severson DL. Glucose metabolism in perfused mouse hearts overexpressing human GLUT-4 glucose transporter. Am J Physiol Endocrinol Metab 2001;280(3):E420–7.

[39] Grainger DJ. Metabolic profiling in heart disease. Heart Metab 2006;32:22–5.

[40] Larsen TS, Belke DD, Sas R, Giles WR, Severson DL, Lopaschuk GD, Tyberg JV. The isolated working mouse heart: methodological considerations. Pflugers Arch 1999;437(6):979–85.

[41] Belke DD, Larsen TS, Lopaschuk GD, Severson DL. Glucose and fatty acid metabolism in the isolated working mouse heart. Am J Physiol 1999;277(4 Pt 2):R1210–7.

[42] Korvald C, Elvenes OP, Myrmel T. Myocardial substrate metabolism influences left ventricular energetics in vivo. Am J Physiol Heart Circ Physiol 2000;278(4):H1345–51.

[43] Bing RJ, Vandam LD, et al. Catheterization of the coronary sinus and the middle cardiac vein in man. Proc Soc Exp Biol Med 1947;66:239.

[44] Bing RJ, Siegel A, Ungar I, Gilbert M. Metabolism of the human heart.II. Studies on fat, ketone and amino acid metabolism. Am J Med 1954;16:504–15.

[45] Opie LH. Myocardial metabolism and the impact of Richard Bing. J Mol Cell Cardiol 1979;11:925–9.

[46] Taegtmeyer H. A Lion at rest. Circ Res 2011;108:9–11.

[47] Randle PJ, Garland PB, Hales CN, Newsholme EA. The glucose fatty-acid cycle. Its role in insulin sensitivity and the metabolic disturbances of diabetes mellitus. Lancet 1963;1:785–9.

[48] Sugden MC. In appreciation of Sir Philip Randle: the glucose-fatty acid cycle. Br J Nutr 2007;97:809–13.

[49] Hue L, Taegtmeyer H. The Randle cycle revisited: a new head for an old hat. Am J Physiol Endocrinol Metab 2009;297:E578–91.

[50] Taegtmeyer H. Metabolism – the lost child of cardiology. J Am Coll Cardiol 2000;36(4):1386–8.

[51] Stanley WC, Recchia FA, Lopaschuk GD. Myocardial substrate metabolism in the normal and failing heart. Physiol Rev 2005;85:1093–129.

[52] Carley AN, Severson DL. What are the biochemical mechanisms responsible for enhanced fatty acid utilization by perfused hearts from type 2 diabetic db/db mice? Cardiovasc Drugs Ther 2008;22:83–9.

[53] Opie LH, Newsholme EA. The inhibition of skeletal-muscle fructose 1,6-diphosphatase by adenosine monophosphate. Biochem J 1967;104(2):353–60.

[54] Opie LH. Metabolism of the heart in health and disease. I. Am Heart J 1968;76(5):685–98.

[55] Opie LH, Thomas M. A new preparation for the study of local metabolic changes in experimental coronary occlusion. J Physiol 1969;202(1):23P–4P.

[56] Cross HR, Opie LH, Radda GK, Clarke K. Is a high glycogen content beneficial or detrimental to the ischemic rat heart? A controversy resolved. Circ Res 1996;78(3):482–91.

[57] Opie LH, Owen P, Thomas M, Young V. Potassium loss, metabolic changes and ventricular arrhythmias following acute experimental coronary occlusion. J Physiol 1969;202(1):44P–5P.

[58] Opie LH. Effect of fatty acids on contractility and rhythm of the heart. Nature 1970;227(5262):1055–6. 5.

[59] De Leiris J, Opie LH, Lubbe WF. Effects of free fatty acid and glucose on enzyme release in experimental myocardial infarction. Nature 1975;253:746–7.

[60] Sodi-Pallares D, Testelli MR, Fishleder BL, et al. Effects of an intravenous infusion of a potassium-glucose-insulin solution on the electrocardiographic signs of myocardial infarction: a preliminary clinical report. Am J Cardiol 1962;9:166–81.

[61] Rovetto MJ, Whitmer JT, Neely JR. Comparison of the effects of anoxia and whole heart ischemia on carbohydrate utilization in isolated working rat hearts. Circ Res 1973;32:699–711.

[62] Opie LH, Bruyneel K, Owen P. Effects of glucose, insulin and potassium infusion on tissue metabolic changes within first hour of myocardial infarction in the baboon. Circulation 1975;52:49–57.

[63] Apstein CS, Opie LH. Glucose-insulin-potassium (GIK) for acute myocardial infarction: a negative study with a positive value. Cardiovasc Drugs Ther 1999;13:185–9.

[64] Allard M, Schonekess B, Henning S, English D, Lopaschuk G. Contribution of oxidative metabolism and glycolysis to ATP production in hypertrophied hearts. Am J Physiol 1994;267:H742–50.

[65] Taegtmeyer H, Sen S, Vela D. Return to the fetal gene program: a suggested metabolic link to gene expression in the heart. Ann N Y Acad Sci 2010;1188:191–8.

[66] Taegtmeyer H, McNulty P, Young ME. Adaptation and maladaptation of the heart in diabetes. Part I: general concepts. Circulation 2002;105(14):1727–33.

[67] Taegtmeyer H, Golfman L, Sharma S, Razeghi P, van Arsdall M. Linking gene expression to function: metabolic flexibility in the normal and diseased heart. Ann N Y Acad Sci 2004;1015:202–13.

[68] Larsen TS, Aasum E. Metabolic (in)flexibility of the diabetic heart. Cardiovasc Drugs Ther 2008;22(2):91–5.

[69] Taegtmeyer H, Stanley WC. Too much or not enough of a good thing? Cardiac glucolipotoxicity versus lipoprotection. J Mol Cell Cardiol 2011;50(1):2–5.

[70] Okere IC, Chess DJ, McElfresh TA, Johnson J, Rennison J, Ernsberger P, Hoit BD, Chandler MP, Stanley WC. High-fat diet prevents cardiac hypertrophy and improves contractile function in the hypertensive dahl salt-sensitive rat. Clin Exp Pharmacol Physiol 2005;32(10):825–31.

[71] Okere IC, Young ME, McElfresh TA, Chess DJ, Sharov VG, Sabbah HN, Hoit BD, Ernsberger P, Chandler MP, Stanley WC. Low carbohydrate/high-fat diet attenuates cardiac hypertrophy, remodeling, and altered gene expression in hypertension. Hypertension 2006;48(6):1116–23.

[72] Stanley WC, Dabkowski ER, Ribeiro RF Jr, O'Connell KA. Dietary fat and heart failure: moving from lipotoxicity to lipoprotection. Circ Res 2012;110(5):764–76.

[73] Sharma N, Okere IC, Duda MK, Johnson J, Yuan CL, Chandler MP, Ernsberger P, Hoit BD, Stanley WC. High fructose diet increases mortality in hypertensive rats compared to a complex carbohydrate or high fat diet. Am J Hypertens 2007;20(4):403–9.

[74] Lopaschuk GD, Stanley WC. Malonyl-CoA decarboxylase inhibition as a novel approach to treat ischemic heart disease. Cardiovasc Drugs Ther 2006;20:433–9.

[75] Lopaschuk GD, Barr R, Thomas PD, Dyck JR. Beneficial effects of trimetazidine in *ex vivo* working ischemic hearts are due to a stimulation of glucose oxidation secondary to inhibition of long-chain 3-ketoacyl coenzyme a thiolase. Circ Res 2003;93(3):e33–7. 8.

[76] Finck BN, Lehman JJ, Leone TC, Welch MJ, Bennett MJ, Kovacs A, Han X, Gross RW, Kozak R, Lopaschuk GD, Kelly DP. The cardiac phenotype induced by PPARalpha overexpression mimics that caused by diabetes mellitus. J Clin Invest 2002;109(1):121–30.

[77] Mjøs OD. Effect of free fatty acids on myocardial function and oxygen consumption in intact dogs. J Clin Invest 1971;50:1386–9.

[78] Kjekshus J, Mjøs OD. Effect of increased afterload on the inotropic state and uptake of free fatty acids in the intact dog heart. Acta Physiol Scand 1972;84:415–27.

[79] Mjøs OD, Miller NE, Riemersma RA, Oliver MF. Effects of p-chlorophenoxy-isobutyrate on myocardial free fatty acid extraction, ventricular blood flow, and epicardial ST-segment elevation during coronary occlusion in dogs. Circulation 1976;54:494–500.

[80] Mjøs OD, Miller NE, Riemersma RA, Oliver MF. Effects of dichloroacetate on myocardial substrate extraction, epicardial ST-segment elevation, and ventricular blood flow following coronary occlusion in dogs. Cardiovasc Res 1976;10:427–36.

[81] How OJ, Aasum E, Kunnathu S, Severson DL, Myhre ES, Larsen TS. Influence of substrate supply on cardiac efficiency, as measured by pressure-volume analysis in *ex vivo* mouse hearts. Am J Physiol Heart Circ Physiol 2005;288(6):H2979–85.

[82] Boardman N, Hafstad AD, Larsen TS, Severson DL, Aasum E. Increased O2 cost of basal metabolism and excitation-contraction coupling in hearts from type 2 diabetic mice. Am J Physiol Heart Circ Physiol 2009;296(5):H1373–9.

Subject Index

Ischemia–reperfusion injury, 25
Ischemic preconditioning, 161
Isolated heart perfusion, 92
Isoleucine, 14

K

3-Ketoacyl-CoA thiolase inhibitors, 199
Ketone(s)
    acetoacetic acid, chemical
        structure, 13
    in heart, 75
    metabolism of, 210
        biochemistry, relevancy, 11–12
Krebs cycle, 23, 25, 48, 93, 195, 208
    acetyl-CoA, 43
    fatty acid oxidation of, 208
    noninvasive evaluation of, 196
    oxidative phosphorylation, 208
Krebs–Henseleit buffer (KHB), 115

L

Lactate
    D/L-isomeric forms, chemical
        structure of, 10
    levels, 74
    metabolically relevant biochemistry,
        10
Lactate dehydrogenase (LDH), 45
Lactate oxidation, 77
Lactose, chemical structure of, 8
Langendorff model, 209
    heart preparation, 115
Laplace, Pierre-Simon, 208
Laplace's law, 139
Lavoisier, Antoine, 208
Lecithin
    chemical structure, 12
    chemical structure of, 12
Leucine, 14
Lipids
    energy sources for cardiac
        contraction, 39
    mitochondrial oxidation of, 73
Lipmann, Fritz, 208
Lipogenic substrates, 75
Lipolysis, 196
Liquid chromatography (LC), 90
Liver ChIPseq information, 132
Long-chain fatty acids, 10
    biological membrane, 15
    saturated and unsaturated, 11
L-type Ca^{2+} channel (LTCC), 160
Lysine, 14
Lysosomal proteins, 31

M

Mammalian target (mTOR), 161
Mass-to-charge ratio, 90
Maternal–fetal gas exchange, 77
Metabolic adaptation, 57
Metabolic control analysis, 57
Metabolic cycles, 39
Metabolic fluxes, 91
Metabolic inflexibility, 213
Metabolic modulator, 165
Metabolic pathways, 2, 39
Metabolic remodeling, in heart
        failure, 169
    amino acid metabolism, 173
    anaplerosis, 174
    autophagy, 177
    cardiac hypertrophy and failure, 171
    cellular processes potentially
        linked, 175
    FA uptake, changes, 172
    glucose, changes, 172
    hexosamine biosynthetic pathway
        (HBP), 175
    increased cardiac ROS levels, 176
    mitochondrial biogenesis/function,
        175
    modulating substrate utilization, 172
    overview of, 170
    pentose phosphate pathway, 174
    substrate utilization, 169
Metabolic substrates, cellular uptake,
        14–16
Metformin, 201
Metformin in diastolic dysfunction
        in patients with metabolic
        syndrome ("MET-DIME"), 201
Metoprolol, 199
Michaelis-Menten kinetics, 16, 88
Mitochondria (mt), 19, 208
Mitochondrial anion carriers, 20
Mitochondrial apoptosis-induced
        channel (MAC), 161
Mitochondrial-creatine kinase
        (Mt-CK), 183
Mitochondrial DNA (mtDNA), 20, 175
Mitochondrial dysfunction, 176
Mitochondrial fission, 27
Mitochondrial oxidative flux, 213
Mitochondrial permeability transition
        (MPT) pore, 158
Mitochondrial respiration, 94
Mitochondrial ribosomes, 20
Mitochondrial superoxide dismutase
        (MnSOD), 26

Mitochondrial transport systems, 20
Mitochondria targeted CoQ10
        (MitoQ), 164
Models to cardiac metabolism
        investigation
    aerobic interval training, 105
    anaerobic exercise training, 105
        strength training, 106
        swimming exercise, 106
    animal models, 103, 104
        advantages/disadvantages, 117
    cell culture models, 116
        adult cardiomyocytes, 117
        HL-1 cells, 117
        neonatal cardiomyocytes, 116
    continuous exercise, 105
    diabetes, 112
    exercise-induced cardiac
        hypertrophy, 104–105
        treadmill training, 105
        voluntary exercise, 105
    ex vivo – isolated heart perfusion,
        114
    genetically determined. *See*
        Genetically determined
        cardiomyopathy
    heart failure (HF). *See* Heart failure
        (HF) models
    in humans, 103
    inflammatory cardiomyopathy, 114
    ischemia in hypoxia, 106
    obesity, 112
Monocarboxylate transporters
        (MCTs), 16
Monosaccharide α-glucose
    chemical structure of, 8
Monosaccharides, 7
MPT inhibitor cyclosporine A, 165
mtDNA mutations, 27
mTOR, protein synthesis, 162
Multisite coordinated control, 66
Myocardial ATP production, 77
Myocardial fatty acid oxidation, 198
    pharmacologic agents modulating,
        198
Myocardial glucose metabolism, 200
Myocardial infarction (MI), 4, 106,
        107, 155
Myocardial ischemia, 158
Myocardial lipid accumulation, 147
Myocardial metabolism, nuclear
        imaging of, 191
Myocardial oxygen consumption
        (MVO$_2$), 77, 214

Printed in the United States
By Bookmasters